T0331129

Core Analysis of Reservoir Rock Systems

Core analysis is an essential source of quantitative data on reservoir properties. These data are used for modeling the distribution and flow of oil and gas reservoirs, CO_2 and hydrogen storage, and flow behavior in geothermal reservoirs. While several books cover core analysis for practitioners, this book fills a gap through introducing laboratory equipment and procedures used in core analysis and the theoretical aspects of the parameters. It offers detailed instructions on experimental execution for those with limited or no experience including risk analysis for high safety.

- Emphasizes the basic characterization of porous materials for multiphase flow, specifically consolidated natural materials
- Features methods commonly used in the special core analysis within the oil and gas industry, extending to the emerging field of gas storage with occasional references to research-oriented equipment
- Offers detailed descriptions of laboratory exercises and instructions on data analysis suitable for student work
- Includes solutions to some exercises to demonstrate the application of measured parameters in reservoir evaluation
- Provides a unique combination of brief discussions on the basic theory of parameters, explanations of experimental principles, detailed experimental procedures according laboratory standards, and data analysis, with multiple laboratory-related example problems

This concise and practical workbook is written for everyone interested in practical measurements of parameters needed for analysis of fluid flow in porous media, specifically students, or for starting staff in the laboratory, preferably with a background in geosciences or fluid mechanics or related topics.

Core Analysis of Reservoir Rock Systems

Hassan Karimaie, Antje van der Net
and Ole Torsæter

CRC Press
Taylor & Francis Group
Boca Raton London New York

CRC Press is an imprint of the
Taylor & Francis Group, an **informa** business

Designed cover image: © 2025 Shutterstock

First edition published 2025
by CRC Press
2385 NW Executive Center Drive, Suite 320, Boca Raton FL 33431

and by CRC Press
4 Park Square, Milton Park, Abingdon, Oxon, OX14 4RN

CRC Press is an imprint of Taylor & Francis Group, LLC

© 2025 by Hassan Karimaie, Antje van der Net and Ole Torsæter

Library of Congress Cataloging-in-Publication Data
Names: Karimaie, Hassan, author. | Net, Antje van der, author. | Torsæter, Ole, author.
Title: Core analysis of reservoir rock systems / by Hassan Karimaie, Antje van der Net and Ole Torsæter.
Description: First edition. | Boca Raton : CRC Press, 2025. | Includes bibliographical references and index.
Identifiers: LCCN 2024022048 (print) | LCCN 2024022049 (ebook) | ISBN 9781032466187 (hbk) | ISBN 9781032466194 (pbk) | ISBN 9781003382584 (ebk)
Subjects: LCSH: Drill core analysis. | Hydrocarbon reservoirs—Mathematical models. | Hydrocarbon reservoirs—Statistical methods.
Classification: LCC TN870.57 .K37 2025 (print) | LCC TN870.57 (ebook) | DDC 622/.338—dc23/eng/20241009
LC record available at https://lccn.loc.gov/2024022048
LC ebook record available at https://lccn.loc.gov/2024022049

ISBN: 978-1-032-46618-7 (hbk)
ISBN: 978-1-032-46619-4 (pbk)
ISBN: 978-1-003-38258-4 (ebk)

DOI: 10.1201/9781003382584

Typeset in Times
by Apex CoVantage, LLC

Contents

About the Authors

Hassan Karimaie is an internationally recognized expert in petroleum reservoir engineering. He holds a PhD from NTNU, Norway, where he also completed post-doctoral research. With extensive hands-on experience in core-related work, Hassan has served in roles ranging from field engineer to lab engineer and worked in several core laboratories worldwide, including RIPI, NTNU, ResLab, Sintef, PLRS, and KNUST. Throughout his academic career, Hassan has made significant contributions by teaching graduate-level reservoir engineering courses at various universities and to industry professionals. He has published over 60 papers in esteemed journals and supervised numerous master's and PhD students. With more than 30 years of industrial experience, Hassan has held staff positions and provided consultancy services in reservoir engineering to multiple oil companies. He was honored with the SPE/EAGE Award in 2009 and was selected as a distinguished lecturer by EAGE in 2019. Currently, Hassan serves as Chair of Petroleum Engineering and Associate Professor at Kwame Nkrumah University of Science and Technology (KNUST).

Antje van der Net has been since 2019 Associate Professor in Reservoir Engineering at the Department of Geoscience of the Norwegian University of Science and Technology (NTNU), Trondheim, Norway. She is also a research associate at PoreLab, a Norwegian Centre of Excellence. She received a master's degree in petroleum engineering from Delft Technical University in 2005. She graduated from Trinity College Dublin, Ireland in 2008 with a thesis on the experimental physics on generation, characterization, and solidification of microcrystalline foams. She worked for 11 years in industry, in BASF SE, as laboratory team leader in the physics research department of Wintershall Holding GmbH, and as reservoir engineer in EOR research and project coordination before coming to NTNU. Her topics of research interest are reservoir engineering, experiments, SCAL analysis, enhanced oil recovery, surfactants/low salinity, wettability, streaming potential, convective flow CO_2 storage, and foams.

Ole Torsæter is Professor Emeritus in reservoir engineering at the Norwegian University of Science and Technology (NTNU). He is also a research associate at PoreLab, a Norwegian Centre of Excellence. Before this, he was Adjunct Professor at the University of Oslo. He has been a researcher or visiting professor with SINTEF, Phillips Petroleum, ResLab, New Mexico Tech, Texas A&M, University of Bordeaux, and A*STAR, Singapore. Torsæter has supervised 220 master's- and 25 PhD-candidates. He has published 200 research papers and edited 8 books. Ole Torsæter has a doctorate degree from NTNU with a thesis on water imbibition in the chalk formations in the Ekofisk Field. Torsæter received the Darcy Technical Achievement Award from the Society of Core Analysts in 2014, the Distinguished Achievement Award in 2016, and the Management and Information Award in 2018 from the Society of Petroleum Engineers. Torsæter is a member of the Norwegian Academy of Technological Sciences.

Preface

Understanding of flow of fluids in porous media is important for the evaluation of many industrial processes. This book concentrates on flow in rocks and the examples and laboratory exercises are related to groundwater flow, flow of hydrocarbon fluids, and storage of carbon dioxide and hydrogen.

With this book we aim to create an easily accessible document that can be used for students – preferentially with a background in geosciences or fluid mechanics or related topics – or for inexperienced staff in the laboratory, who shall learn to execute experiments for the characterization of rock and fluids, with the focus on characterization of reservoir rock and multiphase flow in porous media. The book covers more material than a one-semester course usually includes, and certain chapters may be selected to meet course requirements. It is also possible that the text can be the basis for an introductory course and afterwards the detailed description of measurement of fluids, rocks, and rock/fluid interaction parameters may be used in a laboratory course later in the studies. The book is useful for everyone interested in practical measurements of parameters needed for analysis of fluid flow in porous media.

A collection of notes from decades of teaching of a course covering basic reservoir engineering and reservoir laboratory experiments at the Norwegian University of Science and Technology (NTNU) can be regarded as the start of this book. The class notes were earlier collected in a handout workbook written by Ole Torsæter and Manoochehr Abtahi and were inspired by the report from a research project carried out at SINTEF by Odd Hjelmeland and Ole Torsæter on core analysis for the oil industry. The idea of making a textbook from this material came from Hassan Karimaie and he wrote the first draft. Antje van der Net became involved when she started teaching experimental analysis of flow in porous media at NTNU.

Many colleagues and students have contributed to data, examples, and illustrations and have given advice. We want to record our thanks to Manoochehr Abtahi and Odd Hjelmeland for allowing us to use the old manuscripts as the fundament for our work. Thanks to Shirin Safarzadeh for assisting on the preparation of illustrations and examples during her MSc studies. We appreciate the help in providing photo illustrations: Tomislav Vukovic, Mai Britt E. Mørk, and Knut Reitan Backe of NTNU; Ole Støren, Bård Bjørkvik, and Ane E. Lothe of SINTEF Industry Reservoir Laboratory; Gunn Anita Løvrød, Ragnhild Vinsnesbakk, and Tor Inge Wold of STRATUM, Trondheim; Lars Erik Mollan Parnas of SINTEF, Industry, Biotechnology and Nanomedicine.

The book has developed from the core analysis environment found in Trondheim and we are especially thankful to Department of Geoscience and Petroleum and PoreLab, Norwegian Centre of Excellence, at NTNU, for their support of the project.

Knowledge of core analysis of reservoir rock systems has been built up in the traditional oil and gas industry and is now increasingly applied in new developing interests like storage of CO_2 and hydrogen. This work shall be a basis for traditional reservoir characterization and shall give a motivation for research topics that are briefly introduced, to be further developed in the future.

1 Introduction

1.1 POROUS MEDIA

Porous media are materials which consist of pores that can contain fluids. This can be a sponge filled with gas and water or soil filled with ground water. If the fluids are contained such that they cannot move away, it is a reservoir. Lakes can be reservoirs for drinking water. There are also subsurface reservoirs, rocks destined for containment of liquid/gas within. Oil or gas reservoirs are an example.

Two important characteristics of a subsurface reservoir are, on a small, pore-level scale:

- Porosity: Volume to contain hydrocarbons/store CO_2 or hydrogen.
- Permeability: Connectivity of the pores to allow flow to a well from the reservoir or the reverse.

On a large scale it is important to know the volume and shape, concretely of importance is its geometry, especially regarding the gross volume, compartmentalization by faults, and the internal structure, for example, layering, fractures, and net to gross ratio, the parameter describing how much of the total volume can be considered contributing to the reserves/storage potential on a pore level.

Second, a large-scale factor that will have an effect on the reservoir rock quality is e.g. the distribution of properties – heterogeneity and the grain surface properties, determining the fluid distribution of multiple phases.

So far mainly the reservoir rock was addressed, though for containment of liquid or gas different components are needed, together called a reservoir system.

1.2 RESERVOIR SYSTEM

A reservoir system consists of:

- Reservoir rock.
- Seal rock (impermeable top layer) e.g. salt/shale layer.
- Trap (confining structure).

Additional necessary component in oil/gas reservoir:

- Source rock; for maturation of organic material with pressure/temperature.
- Migration path.
- Geological time-tectonics.

A schematic cross section of the underground with reservoir is shown in Figure 1.1. As seen in this figure, the hydrocarbons separate in the reservoir due to density

DOI: 10.1201/9781003382584-1

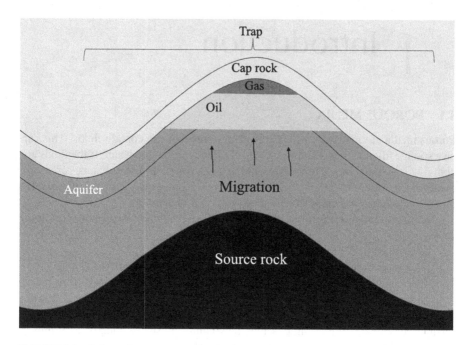

FIGURE 1.1   Schematic cross section of reservoir system containing oil and gas.

differences. Gas accumulates on top, then oil, and at the bottom water. As shown in Figure 1.1, the oil and gas system include a source rock, reservoir rock, trap, and seal or cap rock. In a commercial hydrocarbon project these elements must be studied in detail. In storage projects of, for instance, carbon dioxide or hydrogen, knowledge of the system of reservoir rock, trap, and seal are necessary. Reservoir rock systems consist of all the rocks in the reservoir systems, being the cap, reservoir, and source rock.

1.2.1   RESERVOIR ROCK

Reservoir rocks are usually sedimentary rocks deposited in basins, on the earth's surface, or at the sea bottom. These rocks consist of particles from eroded rocks, seashells, and rock salts, which after erosion were transported and deposited. Alternatively, by chemical solution and precipitation, sedimentary rocks were formed. In these deposits, today, oil and gas are looked for, or they are searched for suitable gas storage locations. The rock being built up from particles allows the rock to have pore space and makes it suitable as reservoir rock. Alternatively, rocks that obtained a dense fracture network can equally qualify as a reservoir rock when enough storage volume is present.

Sediments are deposited in subsided or still subsiding regions, basins. There are different kinds of depositional environments where sediments can be deposited, e.g. located a) on the continent; lacustrine (lakes) or fluvial (river) systems, b) coastal regions, with deltaic and beach deposits, and c) marine: tidal dominated, shallow, or deep marine environments. The depositional environment of the reservoir rock is important to understand as it affects the buildup of the internal structure of the reservoir, affecting its properties for storage and flow dynamics. The lithologies of main interest as reservoir rock are clastic in origin: sandstone, clay stone, and biochemical carbonate (organic shells). In Section 1.3 more details on these different reservoir rocks will be given.

After deposition, the sediments can undergo several processes that can further affect the properties of the reservoir rock as it is found presently. Under influence of increased pressure and temperature due to burial, the sediments will consolidate, a process called lithification. Lithification includes the compaction and cementation of the sediments and recrystallization of minerals and dolomitization on a pore level. On a larger scale also, structural deformation can affect the sediments and the reservoir geometry. Figure 1.2 shows two example of sandstone reservoir rock created in different depositional environments. Figure 1.2a shows a sandstone formation formed in a fluvial continental setting, forming a thick sand body, where Figure 1.2b shows tidal marine deposits with a clear layering structure, as consequence of differences in sedimentation. The two deposits both being sandstone will have different flow properties as a consequence of the deposition and lithification. The nearby field Wytch Farm produces from these formations.

FIGURE 1.2 Two sandstone reservoir rocks exposed at the coast of Southern England with different lithologies based on their depositional environment and diagenesis. a. Sherwood sandstone as outcrop located in Sidmouth. This is a fluvial deposition, with internal structures, as seen in b. c. shows Bridgeport sandstones shallow marine deposits, containing carbonate-cemented sands alternating with poorly cemented sands (Davies 1969; West 2020).

1.2.2 Cap Rock, Seal

As part of the containment, a seal rock is needed. This is a rock that prevents the gas or liquids, formed in the subsurface or injected, from flowing farther up as consequence of buoyancy forces. The caprock shall therefore, opposite to the reservoir rock, contain a limited number of pores or pores that are not or are minimally connected. This can be, for example, shales or salt layers.

For gas storage the caprock studies are more important and they focus on securing the seal for safe storage. This includes also understanding geomechanical properties, as reservoir pressures during injection will increase, which might lead to leakage of the seal or fracturing. During oil or gas production, pressures will rather decrease or remain constant.

1.2.3 Trap

Having a reservoir rock and seal is not enough to create a reservoir system. A confining structure is needed. This can be created structurally or stratigraphically. Examples of structural traps are anticline traps (as seen in Figure 1.3a), fault traps (see Figure 1.3b,) or salt dome traps. Stratigraphic traps are based on a change in lithology in the deposition, the lithofacies, e.g. a sand body formed by a river flowing

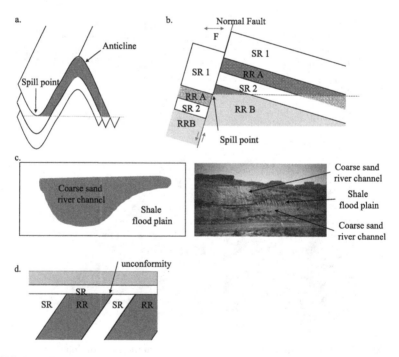

FIGURE 1.3 Examples of traps. a. Anticlinal structure and b. fault trap, two examples of structural traps and c. an example of a stratigraphic trap with a practical example from Huesca, Spain showing fluvial deposits where a river channel deposit as reservoir rock is contained in a sealing shale, the floodplain of the river, and d. a stratigraphic trap via an unconformity.

in a flood plain formed of shale, as illustrated in Figure 1.3c with an image from the Huesca region in Spain, where these structures can be observed at the surface. Finally, as drawn in Figure 1.3d, an unconformity acts as a trap.

1.2.4 Source Rock

Important for oil and gas reservoirs is its content. Oil and natural gas are the product of compression and heating of ancient organic materials over geologic time. Oil is formed from the preserved remains of zooplankton and algae that have been settled to the sea (or lake) bottom under anoxic conditions, see Figure 1.4. Terrestrial plants tend to form coal. Over geological time this organic matter is buried under heavy layers of sediments resulting in high levels of heat and pressure. The organic material in the so-called source rock will be chemically changed. First it will transform into kerogen as found in oil shales, and then with more heat into liquid and gaseous hydrocarbons. The temperature range in which oil forms is often called the "oil window". Below the minimum temperature of around 60°C (Press 1997) hydrocarbons remain trapped in the form of kerogen, and above this temperature the kerogen is converted to oil and natural gas through the process of thermal cracking. The "oil window" temperatures in 3–6 km subsurface are usually ~90–160°C. The origin of the organic material and maturation conditions determine the phase and type of hydrocarbon components created, affecting the pressure, volume, and temperature (PVT) behavior.

FIGURE 1.4 An example of a deep marine, type 2 kerogen source rock, a. Blue Lias shales and limestones located in southern England, Lyme Regis, showing b. bioturbation and c. a fossil ammonite.

The source rocks can themselves be an unconventional reservoir. The permeability is in general low and can be improved, for example by fracking, in order to produce from the rock.

1.2.5 Migration and Trapping

To create hydrocarbon-filled reservoirs, migration and trapping need to occur. Oil and gas have generally a lower density than water that originally fills the rock pores and may migrate long distances from the source rock due to buoyancy long distances from the source rock. When the hydrocarbon fluid reaches seal rock like a shale/trap, the fluid will be trapped and accumulate beneath the cap rock. It is important that in geological time the caprock and trap are formed before or during the migration phase, otherwise no trapping occurs. So, the sequence of deposition and geological timing is important to generate a hydrocarbon-filled trap. But even afterwards, due to tectonic events, traps can disappear. New leaking faults can appear within the reservoir. The fault trap presented in Figure 1.3b assumes a sealing fault, but if the fault is reactivated and/or compressed the trap might reduce in size and start leaking. An understanding of the events in the basin after deposition is important for exploration geologists to judge whether a structure can be a hydrocarbon trap in the present time.

The filling of the trap continues till migration into the reservoir. In geological times this could have happened e.g. due to structural changes such that the migration path changed or the maturation in source rock ceased, as the source rock tectonically moved out of the maturation window. Alternatively, the reservoir is full, and the so-called spill point is reached, the lowest point of the trap (see Figure 1.3 a and b for two examples).

1.2.6 Aquifer

In the subsurface reservoir other than the hydrocarbons, all porous layers, called aquifers, are filled with ground water and brine. There can be found trap structures, which are not filled with hydrocarbons but filled with ground water. Besides gas or oil reservoirs, these locations can be suitable for storage of hydrogen or sequestration of CO_2.

The sealing and trap properties of the cap rock then become more important to verify as there is no natural proof of trapping, as is in the case of hydrocarbon reservoirs.

For more information on geology of reservoirs, see the textbook by Press (Press 1997).

1.3 TYPICAL RESERVOIR ROCKS

Reservoir rocks containing oil and gas or rocks suitable for storage of CO_2 are mainly sedimentary rocks that are of three main types:

1. Clastic rocks formed by debris of eroded, transported, and deposited minerals of older rocks. The sediment accumulations are buried, compacted, and

finally consolidated in a process called diagenesis. The porosity and permeability of the clastic rocks are dependent on grain size, grain shape, sorting, degree of compaction, and the diagenetic process.

2. Rocks of organic origin are the result of deposition and diagenesis of remains of animals and plants, on the location where they originated. These underwent minimal erosion and transport compared to the clastic rocks. The distribution of animals and plants can vary considerably and therefore rocks of organic origin often are very heterogeneous.

3. Chemical rocks are formed by precipitation of salts from aqueous solutions, and some are interesting reservoir rocks even though they initially have no porosity. However, during folding and faulting these rocks may develop extensive fracture networks (on many scales) and thereby quite high permeability.

The main sedimentary rocks of interest in the context of fluid flow in porous media are clastic rock; sandstones, shales, limestones, and dolomites as carbonate reservoirs, formed from organic material.

SANDSTONES

Sandstones are defined based on the grain size, ranging from 63–2,000 μm (Press 1997). The grain size distribution, sorting, and shape depends on the nature of the eroded material, the transport media (ice, wind, water), distance of transportation, depositional environment, and many other factors and is a determining factor for porosity, permeability, and mechanical properties.

The most common grain type mineral in clastic sandstone rocks is quartz (SiO_2). The cementing material can be silica, calcium carbonate, or clay. Figure 1.5a. shows a thin section of a quartz sandstone, without much cement. Sandstones are often stratified in patterns or beds, and this is due to successive depositions by flowing water at the shoreline or in rivers.

SHALES

Clay minerals like kaolinite, illite, and montmorillonite are the main components of shale. The clay particles are very small (less than 1 μm in diameter), and they may be chemically active. Clay minerals have layered molecular structure of aluminum oxide and silica oxide, which might permit ion adsorption and water binding dependent on the salt concentration. The exchange of ions due to excess of cations on the surface of a clay leads to conductivity, measured as the cation exchange capacity (CEC). The ion sensitivity can cause volume changes, which can affect the permeability and porosity. When the clay is fissile and laminated the rock is called shale. Shales may occur as reservoir rocks, source rocks, and seal rocks. Production can come directly from shales where hydrocarbons maturated, oil, or gas shales. These reservoirs are accounted for as unconventional reservoirs, as permeabilities are low.

FIGURE 1.5   a. Thin section of sandstone and b. Scanning electron microscopy (SEM) picture of chalk (magnification: 5,000).

CARBONATE ROCK

Beside sandstone reservoirs, carbonate rock is a common reservoir rock. A carbonate mineral consists of an anionic complex CO_3^{2-} plus divalent metal ions e.g. Ca^{2+}, Mg^{2+}, Fe^{2+}. There are multiple carbonate minerals, with the commonest varieties being calcite ($CaCO_3$ crystalline system hexagonal), dolomite (Ca-, Mg-, CO_3 crystalline system hexagonal), and aragonite ($CaCO_3$; crystalline shape orthorhombic), which is not stable under normal reservoir conditions. Three kind of carbonate rocks can be distinguished:

> **Limestone** is the most important chemical rock with fluid storage capacity and permeability, which is mainly composed of the mineral calcite ($CaCO_3$). Limestones may also have organic origin like algae or reefs.
>
> **Chalk** is a carbonate rock containing > 80% calcite. Chalk is a bioclastic carbonate formed by small single-cell algae (coccoliths) with high porosity and low permeability; see Figure 1.5b showing a microscope picture of the coccoliths.
>
> **Dolomite** is a calcite rock with more than 50% by weight of the mineral dolomite (Ca Mg(CO_3)$_2$).

With some exceptions, there are distinct differences between siliciclastic, sandstone reservoirs, and carbonate reservoirs, (Moore 1989), which are important to understanding the fluid flow in the reservoir. Compared to siliciclastic sandstone, carbonates differ concerning:

- Fractures: Carbonate is more brittle and can fracture easily. The reservoir layer can be broken up in matrix or matrix blocks by fractures, creating a fracture system. This will affect the preferential fluid injection paths.
- Pore structure: Carbonates are generally more heterogeneous with the presence of secondary porosity like fractures due to tectonics or vugs due to dissolution. On the other hand, strong diagenesis in the form of cementation or

compaction can lead to closing of pores and low porosity and permeability (Choquehe and Pray 1970).

- Permeability: In a fractured system, there can be dual permeability: low matrix permeability in mD range and a fractured network with high permeability.
- Wettability: Carbonates are generally more oil wet (Roehl and Choquette 1985). This will make it hard to drain the matrix with brine, especially in the presence of fractures.

A detailed geological characterization of rocks is outside the scope of this book. In the recent years the focus of fluid flow description has changed to pore scale, where major advances in the field of pore type classification have been made brought about by the development of models for simulation of fluid flow on pore scale. Informative texts on pore type characterization are given in Anovitz and Cole (Anovitz 2015) and Lucia (Lucia 1999).

1.4   EXPLORATION AND FIELD DEVELOPMENT

Exploration geologists study the history of sedimentary basins, where accumulation of sediments occurred and the region where subsidence takes place, such that significant accumulation can occur. These can be interesting for the presence of natural reservoirs for oil and gas or storage. Seismic data allows for the creation of structure maps of the reservoir systems. Structure maps are a source of data at low resolution, but they cover a large scale (kilometers).

1.4.1   OIL AND GAS

Access to the reservoir will be obtained by well drilling, either injection or production wells. For oil and gas exploration, the first well that finds hydrocarbons is the exploration well. From the well's core, data can be obtained to perform detailed geological study but also to obtain petrophysical and dynamic flow properties performing experiments in the laboratory. Generally during the appraisal phase of the field, while drilling appraisal wells, core samples are taken that shall be sufficient to build up knowledge about the field performing a core analysis study (see Section 1.4) and respectively store them for studies in later field life. This requires that cores be stored appropriately in a core storage for decades; see Figure 1.6.

Well logging by petrophysicists can obtain further information of the reservoir rock and cap rock. For example facies, porosity (neutron density log), saturation (resistivity log), net/gross (spontaneous potential log), and well data can be correlated, enabling data interpolation between the wells.

After the appraisal phase, a field development plan will be made, determining a field production/injection strategy by understanding flow of fluids like gas, oil, and water through the pore space of rocks and techniques to recover/store them in an economically efficient manner and with a low carbon footprint. If agreed upon and if licenses are obtained, production and possible injection wells will be drilled. Injection concerns water or gas injection for pressure maintenance or

FIGURE 1.6 An example of core storage (courtesy of Stratum).

displacement using water or chemicals as displacement enhancement, for optimization of the field performance, called enhanced oil recovery methods (EOR) (Sheng 2011; Green and Willhite 1998). There can be different phases of drilling additional wells, allowing more samples to be obtained, depending on the economics. Finally, when the field does not provide economical production anymore, the field is abandoned and wells are closed. For more information on field life of oil and gas fields and reservoir engineering, see Ahmed 2010; Press 1997; Dake 1978; Zolotukhin and Ursin 2000.

1.4.2 CO$_2$ Sequestration and H$_2$ Storage

Oil and gas accumulations occur naturally, and with exploration they are to be found. CO$_2$ as a greenhouse gas and H$_2$ as an alternative fuel gas are considered for storage in subsurface reservoirs by gas capture and gas generation and injected into the reservoirs, thus reservoirs need to be found.

CO$_2$ is a greenhouse gas and storage of this gas is considered to limit the impact of CO$_2$ emissions. The storage of CO$_2$ is intended to be permanent, called CO$_2$ sequestration. Therefore, the natural reservoirs are attractive, enabling storage of large volumes, low risk of leakage, and a reduced effect on the surface. There are four main storage mechanisms considered (Members of the CO2 Capture Project 2009):

1. **Mechanical trapping**: Injection of CO$_2$ forming a gas cap under the cap rock, based on buoyancy forces, which requires good sealing properties of the caprock.
2. **Capillary trapping**: When CO$_2$ driven by buoyancy forces moves up though an aquifer, part of the CO$_2$ phase remains trapped in the brine/pore capillary structure based on the relative permeability (see, for the concept of residual saturation, Chapter 11).

3. **Dissolution trapping**: CO_2 is dissolvable in brine, which happens at the gas water contact. A CO_2 saturated brine denser than water which tends to move down, based on density difference, driving non-saturated brine upwards. This is the so-called density driven convective flow. Figure 12.5 in Chapter 12 shows an experiment visualizing the convective flow of CO_2 enriched water.

4. **Solidification trapping**: CO_2 can react with specific rock types, like basalt, forming carbonate salts, which precipitate in the rock formation, storing CO_2 as a solid phase.

The first three mechanisms are likely to occur in sedimentary reservoirs. The mechanical trapping happens, and capillary trapping starts during the injection phase (see Figure 1.7 for an example).

The other processes proceed rather in time ranges of decades. Different from hydrocarbon reservoirs, storing large quantities by a) safe quick injection, b) understanding the CO_2 behavior and interactions, and c) providing long-term safe storage are requirements for the field screening (Members of the CO2 Capture Project 2009). Safety includes no long-term leakage (via well or seal integrity) and no fault activation or mechanical fracturing. Mainly aquifers and potentially old oil fields are considered as storage locations. Different from oil and gas is that, after injection phase and the abandonment, long term monitoring is required to secure the safety.

Note that the engineering of the optimized gas storage is intensively studied, and it is referred to in literature for the latest development updates.

Similar reservoir screening criteria are to be considered for storage locations for hydrogen. Hydrogen is considered as an alternative fuel gas, storing renewable energy in a form that during combustion only forms water. As demand and supply might not always be predictable, temporary storage is needed. So, hydrogen is to be produced opposite to CO_2.

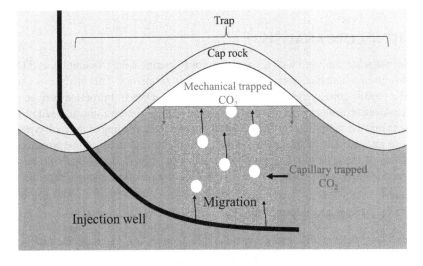

FIGURE 1.7 CO_2 storage, an example of the injection phase.

The difference between gas storage and hydrocarbon production is that storage costs money, so a limited budget is available for sampling, where hydrocarbon production will earn money, making it worth investing in sampling the reservoir. It is typical in an oil reservoir the aquifers are not sampled, balancing costs versus value. Decisions for gas storage therefore are based on limited data sets.

Hydrogen underground storage (HUS) is presently in the research stage and will only be mentioned briefly in this book. An interesting discussion of the feasibility of storing hydrogen in reservoir rocks is found in Ren et al. (2023). Core analysis aspects of CO_2 storage will be covered in several chapters in this book and a comprehensive presentation on how to store CO_2 underground is given in the book by Ringrose (Ringrose 2020).

1.5 CORE ANALYSIS

Knowledge of petrophysical and hydrodynamic properties of reservoir rocks are of fundamental importance in the evaluation of hydrocarbon reservoirs and CO_2 and H_2 storage reservoirs. Understanding the multiphase flow and storage of fluids in porous media is needed to optimize the activities. The data needed for this understanding of the reservoir flow properties is obtained from two major sources: core analysis and well logging. In this book details are presented about the fluid flow analysis of porous media like reservoir rocks; the main core analysis experiments are described and the quality of the experimental data is reviewed.

For the analysis of the reservoir rock, representative samples need to be available. These samples, being cores, can be obtained from the well during the drilling with a special diamond core drill bit, which is hollow to allow the rock consistency to remain intact. Chapter 3 will give more details on this coring process.

When the core arrives in the laboratory, core plugs are drilled usually 20–30 cm apart throughout the reservoir interval. On these core plugs a series of experiments is performed.

ROUTINE CORE ANALYSIS

All the plugs are analyzed with respect to porosity, permeability, saturation, and lithology, to obtain a statistical overview of the spread in properties. This analysis is usually called a routine core analysis laboratory (RCAL) test. Data from routine core analysis and from supplementary tests and the application of these data are summarized in Table 1.1. Note that the sampling can deviate depending on the spreading of the rock types and facies of interest. The results from routine core analysis are used in interpretation and evaluation of the reservoir. Examples are prediction of gas, oil and water production, definition of fluid contacts and volume in place, definition of completion intervals, etc.

SPECIAL CORE ANALYSIS (SCAL)

On reservoir rocks, additionally, a special core analysis (SCAL) is performed, which includes several measurements with the objective of obtaining detailed information about multiphase flow behavior. Table 1.2 provides a list of the various special core analysis tests. Special core analysis gives information about the distribution of

TABLE 1.1

Routine Core Analysis (RCAL) and Supplementary Measurements with a Reference to the Chapter of Relevance

Data	Application	Reference to
	Routine Core Analysis Laboratory Tests	
Saturations	Define the mobile hydrocarbons (productive zones and contacts), type of hydrocarbons	Chapter 3
Porosity	Storage capacity	Chapter 5
Permeability	Flow capacity; productivity/injectivity	Chapter 7
Lithology	Rock type and characteristics (fractures, layering, net to gross, etc.)	
	Supplementary Measurement	
Vertical permeability	Effect of coning, gravity drainage, etc.	Chapter 7
Matrix density	Calibrate the density log	Chapter 4/5
Oil and water analysis	Density, viscosity, interfacial tension, composition etc.	Chapter 4 Chapter 8
Core-gamma surface log	Identify lost core sections, correlate cores, and logs	

TABLE 1.2

Various Special Core Analysis Laboratory (SCAL) Tests

Tests/Studies	Data/Properties	Reference to
	Static Tests	
Compressibility studies	Permeability and porosity vs. pressure	Chapter 5
Petrographical studies	Mineral identification, diagenesis, clay identification, grain size distribution, pore geometry, etc.	
Wettability	Contact angle and wettability index	Chapter 9
Capillarity	Capillary pressure vs. saturation	Chapter 10
Electric tests	Input data for well logging interpretation	Chapter 6
Acoustic tests		
	Dynamic Tests	
Flow studies	Relative permeability and end-point saturations	Chapter 11
		Chapter 12
EOR-flow tests	Injectivity and residual saturation	

the fluid phases in the reservoir (capillary pressure data), residual saturation, and multiphase flow characteristics (relative permeability). Occasionally, measurements of electrical and acoustic properties are included in SCAL, and this information is mainly used in the interpretation of well logs.

The effect of pressure and temperature on rock and fluid properties is significant, and laboratory measurements should therefore be made at – or corrected to – reservoir conditions. Often included in SCAL is detailed petrographic analysis (grain size distribution, clay identification, diagenesis, etc.). Wettability analysis and special tests for enhanced oil recovery (EOR) are also often part of special core analysis.

The core analysis programs are set up such that multiple experiments can be done on a single core plug. This is done to save core material and some preparation steps can obtain required parameters. Single phase brine flooding, providing permeability (Chapter 7), is needed to prepare for the multiphase drainage experiment, measuring relative permeability (Chapter 11). For single phase brine permeability, the core needs to be saturated with liquid, which provides data to calculate the liquid porosity (Chapter 5). The program is planned in multidisciplinary teams, involving the subsurface experts and the geologists, geophysicists, production, and reservoir engineers to select the material and optimize the data obtained from the core material.

The traditional RCAL and SCAL analysis for oil and gas production needs suitable adaptions to the application of gas sequestration and storage, which is under development (Worden n.d.). In Table 1.3 a general overview is given on the main properties considered to be of importance for the characterization of reservoir and seal rock comparing the traditional target of oil and gas production versus CO_2 sequestration and H_2 storage in aquifers. The addressed topics will be discussed in detail in the following chapters.

1.6   NOTES ABOUT THE BOOK CONTENT

In this book basic theory and step-by-step procedures of the generalized experimental set-ups are presented, together with a detailed description of the data analysis. The scope is on the basic characterization of porous material for the purpose of multiphase flow, specifically consolidated natural materials. The presented methods are part of the special core analysis standardly executed within the oil and gas industry, with the extension to the upcoming topic of storage, complemented with occasional references to research-oriented equipment, to follow up for more information. Porous media are all around in daily life, and many of the principles discussed here are applicable on other porous media, e.g. polymer membranes for filtration, open polymer or metal foam, and organic materials like plant stems or sponges.

In this book the laboratory experiments supporting the description of flow in porous media will be presented, starting with paying attention to general routines to enable a safe working environment and precise working procedures, in Chapter 2. Sample creation including drilling, cleaning, and drying will be presented in Chapter 3. Fluid sampling and determination of density and viscosity as necessary fluid characterization for the flow description are presented in Chapter 4. As part of the routine core analysis, porosity is the volume of importance, containing the liquid and gas phases of interest, discussed in Chapter 5. A sidestep is made in Chapter 6 to a discussion on electrical properties to support the logging data, which provides porosity data as well. The volume in the pores shall be able to flow towards the well, the access to the reservoirs, where the basic resistance to flow is described by absolute permeability addressed respectively in Chapter 7. Moving to multiphase

TABLE 1.3

A General Overview on the Main Properties of importance for the Characterization of Reservoir and Seal Rock Comparing the Traditional Oil and Gas Production versus CO$_2$ Sequestration and H$_2$ Storage in an Aquifer

Property	Oil and Gas Production	CO$_2$ Sequestration	Hydrogen Storage
RCAL-SCAL	Oil -Gas Reservoirs	Aquifers	
Porosity	Storage capacity in place as found	Storage capacity created dynamically by aquifer displacement	
Initial water saturation	Volumetrics of original hydrocarbons in place, as found	Determined by microscopic displacement efficiency (at pore level).	
Permeability/relative permeability	Primary imbibition, secondary drainage, etc.	Primary drainage	Primary drainage, primary imbibition
	Production oil/gas; Injection water/gas for pressure support/ displacement	Injection	Injection; Back production
Capillary pressure	Reservoir rock: imbibition; Initial fluid distribution	Reservoir rock; drainage; fluid distribution affecting dissolution trapping; Cap rock: capillary entry pressure, sealing efficiency; drainage	
Wettability	Oil or water wet	Water wet (assuming no exposure to oil)	
	Displacement efficiency	Displacement efficiency drainage	Displacement efficiency drainage, imbibition
Fluid viscosity	Gas: Low; Oil: Depends on PVT	Low- detrimental for displacement efficiency	
Fluid density	Gas: Low; Oil: Depends on PVT	Depends on PVT; Gas: Low; supercritical comparable to brine	Low
Interfacial tension	Assumed immiscible (black oil model)	Solubility and miscibility phase behavior dependent	Solubility, phase behavior dependent
Geomechanical experiments	Maximum injection pressure (injectivity, fracturing risk)		
	Effect of reduction of reservoir pressure (subsidence risk)	Effect of increase of reservoir pressure	Effect of increase and reduction of reservoir pressure
	Sealing properties of the cap rock under mechanical stress and possibility of fault activation		

flow description, interfacial tension (IFT) is a necessary fluid characterization for the flow description addressed in Chapter 8. IFT, combined with pore fluid surface affinity, the so-called wettability, and capillary pressure determine the fluid distribution and flow behavior; the latter two respectively described in Chapters 9 and 10. Finally the extension of absolute single-phase permeability to the two-phase flow description in a porous medium using relative permeability is described in Chapter 11. Special attention is given in Chapter 12 to 2D and 3D flow visualization techniques, which can support the traditional core sample usage in core flooding. Out of scope are geomechanical experiments. The focus within the chapter lies on the description of the basic principles of the parameters and the common methods available to measure the respective parameter. The characterization focuses on the reservoir rock. Nevertheless, some experiments can be applied for the cap rock characterization. Detailed procedures are provided to enable performing the experiment in the laboratory if the equipment is available. The experiments and experimental exercises are considered at standard room conditions, rather than at reservoir conditions with possible high pressure and high temperature. A main HSE (health, safety, environment) guidance is given, but working in accordance with HSE regulations is one's own, personal responsibility and local guidelines should always be checked and followed.

The core analysis concerns experiments with samples at centimeter scale, representative for the matrix. Possible fracture system descriptions are not considered, and neither are effects on flow due to heterogeneities on a field scale. The rock and possible gas, oil, or brine will be considered immiscible and without occurrence of chemical reactions.

Practical exercises with solutions are included, for practicing theory and data analysis, mainly using standard international (SI) units.

REFERENCES

Ahmed, T. 2010. *Reservoir Engineering Handbook*. Elsevier.

Anovitz, L.M., Cole, D.R. 2015. "Characterization and analysis of porosity and pore structures." *Reviews in Mineralogy & Geochemistry* 80: 61–164.

Choquehe, P.W., Pray, L.C. 1970. "Geologic nomenclature and classification of porosity in sedimentary carbonates." *The American Association of Petroleum Geologists Bulletin* 54 (2): 207–250.

Dake, L.P. 1978. *Fundamentals of Reservoir Engineering*, 17th 1998. Elsevier Science, Shell Learning and Development. ISBN 0-444-41830-X.

Davies, D.K. 1969. "Shelf sedimentation: An example from the Jurassic of Britain." *Journal of Sedimentary Petrology* 39: 1344–1370.

Green, D.W., Willhite, G.P. 1998. *Enhanced Oil Recovery*. Henry L. Doherty Memorial Fund of AIME, Society of Petroleum Engineers.

Lucia, F.J. 1999. *Carbonate Reservoir Characterization*. Springer-Verlag.

Members of the CO2 Capture Project, edited by C. Cooper. 2009. *A Technical Basis for Carbon Dioxide Storage*. Berks: CPL Press. ISBN: 978-1-872691-48-0.

Moore, C.H. 1989. *Carbonate Diagenesis and Porosity*. Vol. Developments in Sedimentology 46. Elsevier Science Publications.

Press, F., Siever, R. 1997. *Understanding Earth*. W.H. Freeman.

Ren, B., Jensen, J., Duncan, I., Lake, L. 2023. "Buoyant flow of H2 vs CO2 in storage aqui-fers: Implications to geological screening." *SPE Reservoir Evaluation & Engineering*: 1048–1058.

Ringrose, P. 2020. *How to Store CO2 Underground: Insights from Early-Mover CCS Projects*. Springer Briefs in Earth Sciences.

Roehl, P.O., Choquette, P.W. 1985. *Carbonate Petroleum Reservoirs*. Springer Verlag.

Sheng, J.J. 2011. *Modern Chemical Enhanced Oil Recovery, Theory and Practice*. Elsevier Inc. https://doi.org/10.1016/C2009-0-20241-8.

West, I.M. 2020. *Geology of the Wessex Coast of Southern England: The World Heritage Jurassic Coast- and more by Ian West*. Vers. 22.07.2020. iSolutions. 22 07. Accessed 2023–2024. https://wessexcoastgeology.soton.ac.uk.

Worden, R.H. n.d. "Value of core for reservoir and top-seal analysis for carbon capture and storage projects." In *Core Values: The Role of Core in Twenty-first Century Reservoir Characterization*. https://doi.org/10.1144/SP527-2022-38.

Zolotukhin, A.B., Ursin, J.-R. 2000. *Introduction to Petroleum Reservoir Engineering*. Høyskoleforlaget. ISBN 9788276340655.

2 Working in the Laboratory

2.1 INTRODUCTION

The aim of this book is to understand the theory and procedures to perform the different laboratory methods for characterization of rock, fluid, and their interactions. For this purpose, discussion of some general approaches and laboratory skills are necessary. It is important that the experiments are performed safely, and data quality is sufficient. This needs a conscious approach of all three phases of the experimental work: the planning, execution, and data analysis.

In this chapter we will go into more details on what is to be considered when performing the experiments safely. We will address some general laboratory rules and regulations that can be found in laboratories. Everyone who shall work or follow practical hours in a new laboratory shall make themselves acquainted with the local laboratory rules and regulations. So, one should do this for all the specific activities in the laboratory. The information here does not replace local, lab-specific safety instruction or specific institute regulations that may be in place. It shall solely increase awareness of laboratory safety and organization in general.

Besides safety, there are also standard methodological skills as preparation that are useful for performing the experiments accurately. These include awareness of standards for measurement accuracy and general skills required for handling the laboratory equipment, like glass cleaning and sample labelling. For the execution of the specific experiments, we refer to the individual chapters. As part of the data analysis, estimating the use of units and uncertainties will be addressed.

2.2 LABORATORY HEALTH, SAFETY, AND ENVIRONMENT

During laboratory experiments there can be risks that can generally be grouped into the following categories: use of high/low pressure, high/low temperature, chemical danger, mechanical movement, electrical use, use of radiation, or biological-medical dangers. Laboratories are dedicated locations where these experiments can be performed and where general safety measures for experiments are already in place or can easily be implemented. In the laboratories, regulations are to be in place based on national legal regulations for safety of workplaces.

Safety in a laboratory is a responsibility for everyone. It is therefore important to familiarize yourselves with the regulations and safety standards of the laboratory. These regulations are the "health, safety and environment (HSE) regulations" (Arbeidstilsynet 2024), (IMO 2024). These are specific to each country and every laboratory is obliged to implement and follow them. In the laboratory this information is to be provided or requested and learned.

DOI: 10.1201/9781003382584-2

For each new experiment and set-up or adaptions to experiments, safety needs to be re-considered. Hereby the most important aspects are presented to consider while planning and performing experiments.

2.2.1   RISK ASSESSMENT AND PROCEDURE

In laboratory activity there is always a risk that harm can be done. Accidents that could result in harm are to be avoided. Harm can be present in different forms. First, harm to the persons in the laboratory leading to injuries shall be avoided. This is what "health" in HSE stands for. Besides personal harm, damage to the lab equipment is to be avoided as well as pollution to the environment (e.g., pollution of soil or groundwater or the sewage system). If a severe accident happens with injuries, fatalities, or significant environmental pollution, negative press can affect the reputation of the company, potentially resulting in fewer customers, reduced job applicants, or retraction of local support from politicians and government for activities and permits. So also, the protection of the reputation of the company or institute that the laboratory is part of is important. Besides damage to health, equipment, environment, and/or reputation, there is an economic impact as well, which ideally is all avoided, therefore prevention and mitigation is key. To minimize the risk of harming personnel, equipment, environment, or reputation, a risk assessment needs to be performed when an experiment is planned on existing experiment, adapted experimental set-ups, as well as newly designed experiments; see Figure 2.1 for an example. The objectives of the risk assessment are multiple:

- Identification of possible accidents that can result in harm to a person, an institute, equipment, or the environment.
- The risk probability and impact of the accident occurring.
- Action list to mitigate the identified risks or to minimize risks.
- Support in prioritization of the risks and the actions.
- In general, it shall increase awareness of the risks and the feeling of responsibility.

The weighting of risk of an event occurring depends on two factors. First, how likely it is for the event to occur, which is expressed in the probability of occurrence. Second, when an accident happens, its impact on the health, equipment, environment, and organization. The sum of the risk values is considered to prioritize the measures for risk reduction.

$$\text{Risk value} = \text{Probability of occurrence} \cdot \text{impact} \qquad (2.1)$$

For example, in daily life, the risk of the event where an airplane has motor problems in both motors is very small, but if it occurs, consequently, the impact of a crash to the passengers and airplane is enormous. On the other hand, taking a car might have a significantly higher risk of an accident occurring, but the impact, having a fatal accident or destroying the car, is much lower. The risk weighting, being the accident probability × accident impact, might be similar.

RISIKOANALYSE SOXHLET CLEANING								
Unit/Institute:								
Responsible line manager (name):								
Responsible for activities being risk assessed (name):								
Participants in the risk assement (names):								

Beskrivelse av den aktuelle aktiviteten, området mv.:/ Description of the activity, process, area, etc.:
Cleaning cores inside the fume hood with soxhlet, condenser, distillation flask with metanol or toluene heated to boiling point using a heat mantle. Boiling point for toluene is 110.6 deg C and for methanol 65 deg C. The whole apparatus is placed inside the fume hood.

Activity / process	Unwanted incident	Existing risk reducing measures	Probability (P) (1-5)	Consequence (C) (1-5)	Risk value (P x C)	Risk reducing measures - suggestions. Measures reducing the probability of the unwanted incident happening. Highest risk value should be prioritized.	Residual risk after measures being implemented (S x K)
General; use of glass	breaking glass	wearing safety glasses			18	Use necessary protective equipment, follow the user procedure and read information in the safety data sheet for methanol and toluene.	3 (S=1)
General; use of easy vaporizing liquids	vapor formation leading to inhalation of high quantities; eye, skin irritation, explosion danger	Activity in the fume hood, limited quantities used, no open flame in the fume hood			10		
Fill the distillation flask with methanol or toluene	spill of liquid	Perform in the fume hood, limited amount of 800ml. (80% of extraction flask volume) Wear PSE: glasses, safety shoes, lab coat			10		
Connect the soxhlet and the condenser	eye and skin damage due to fluid spill	Wear PSE: glasses, safety shoes, lab coat			9		
	overflooding of water in the sink	check that the tube goes to drain			4		
Start condenser fluid circulation	leaking of water from the tubes of the condenser down to the heat mantel	Slow start up of water flooding to reduce the quantities Procedure: stop the cleaning prosess and close the leak			8	check manual, whether heater can handle droplet spill volumes!	
Start/use of heater	skin damage due to heat (=boiling temperature of the solvent)	no dismantling with heater on. min 30 min waiting after heater stopped, before handling the glass and fluids. use of safety gloves for the heat			5		
leakage during experiment of solvents, condensing water		seals are to be checked at the start (procedure), experiment only in office hours. No use over night			4		
Dismantling	fluid spills	Procedure; only dismantle when extraction chamber is empty;			8		
	heat exposure	Procedure: wait 30 minutes minimum for the equipment to cool down.			10		
	vapour exposure	Procedure; only dismantle when extraction chamber is cooled down; use the condenser and running water til min 30 minutes after the heater was turned off			8		
Drying cores in heating cabinet	spill of liquid and the heat for 60 deg.C	glasses, safety shoes, lab coat and safety gloves for the heat			4		

FIGURE 2.1 An example of the risk assessment for Soxhlet cleaning (see Chapter 3 for the background of the method).

Similarly, in the laboratory, for example, the risk of the accident breaking of glassware might be high and it occurs frequently, but the impact is rather small. The likelihood of an artery being hit is small. The risk of a chemical explosion might be much less but much more detrimental in its impact.

Depending on the kind of experiments, different risks can occur. As mentioned previously, these can be categorized as followed: High/low pressure; high/low temperature, e.g. when considering performing experiments at reservoir conditions or a vacuum or freezing conditions; electrical hazards, e.g. using electrical devises in a wet environment; mechanical, e.g. in the form of rotating or moving motions; chemical hazards depending on the solids and fluids used; radiation when, for example, using laser or X-rays; and biological hazards e.g. in case of using bacteria. The last instance is rather seldom for routine core analysis laboratory (RCAL) and special core analysis laboratory (SCAL) experiments.

Besides the risk assessment, a procedure needs to be written if not available. The procedure should include a step-by-step description of how the experiment is to be executed, for clarity, repeatability, and safely.

Note that for some equipment the procedure is standardized, and industries have uniformed the procedures, such that all labs follow the same procedure and data can be compared between the labs. The procedure standardization is documented internationally in American Society for Testing and Materials (ASTM) standards or International Organization for Standardization (ISO) standards.

For every step in the procedure, risks and existing measures are to be evaluated for risk value (see the column on activity and process in Figure 2.2). Based on the level of risk, additional measures need to be implemented to reduce the risk to an acceptable level. Acceptability is based on legislation and internal policies and standards defined. The scaling of risk estimation is to be described to enable a more objective evaluation.

The description of the risks is accompanied with a scale of how to categorize the risk for each category (see Figure 2.1). The first two columns include the description of the planned activities and possible expected incidents. A listing of the existing measures shall be included. Based on this information, the risk value is to be evaluated on a judgement of probability of occurrence and consequence. Afterwards the risk values can be ranked, and risk-reducing measures, in case needed, shall be suggested and implemented. An estimation of the consequent risk reduction can be made accordingly.

The assessment of the risks should follow the experimental procedure and steps envisioned during the experiment. In general, a risk evaluation shall start when the first design of the experiment is made. It shall be updated with the development of

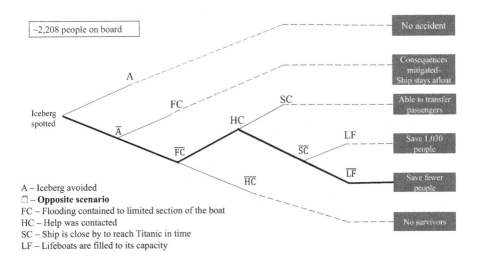

FIGURE 2.2   Event tree diagram of *Titanic* accident sequence. Modified from Haga (2013).

the experimental set-up and re-evaluated with every new experiment performed or new person working with the set-up.

2.2.2 Risk Mitigation

Considering a risk, mitigation is best, so eliminate the risk. It needs to be considered if this is a feasible option. e.g., no heat during the experiment, no pressure, or changing equipment. This can, however, have an effect on the experimental quality or change its objective. E.g., one of the risks of using a capillary viscometer is the breaking of the device made from glass. A mitigation would mean avoiding using glass. But if you manufacture the device from a polymer, the material might not be inert to the liquids to be used, or it might not remain transparent to trace the fluid interface. Then the decision needs to be made whether to accept the risk or minimize the risk instead. The latter, for example, can be done by using a stand to hold the capillary viscometer in place or having a procedure where the device is to be held over the table and walking through the lab with the device should be minimized.

During the risk assessment, as indicated, an evaluation is made to what extent risks can be avoided or minimized. Ideally the risk caused by personal mistakes is minimized, for example, using a choke of 10 bar on a gas bottle of 200 bar pressure, which has a maximum pressure that is below the pressure limits of the set-up 35 bar. The pressure regulated in person in this way can never be above the maximum pressure the choke can provide and will never approach the limiting system pressures. This is an example of a system risk mitigation measure, instead of relying on the user to not regulate pressure higher than allowed. Other examples are setting pressure limits on pumps or temperature or timer limits on ovens.

It might be that the systematic measures are not possible technically or they are not sufficient in risk reduction. The investment costs also play a role. Therefore, procedures shall be in place with details on use of the equipment, sequences to follow, and precautions to implement to avoid dangerous situations. The procedures shall be the guide to follow, describing all steps of the experiment from the set-up preparation to the execution to shut down or predictable experimental problems. This is all part of the experimental preparations.

2.2.3 Risk Reduction

2.2.3.1 Systematic Measures

Human mistakes are the main risk, which shall be reduced. From this point of view consideration of implementation of systematic measures is important, similar to risk mitigation. Also, one has to consider that avoiding situations is better than coping with the damage and consequences of activities that go wrong. E.g., the equipment in the oven can have limited temperature up to 80°C, where the oven itself can heat up to 400°C. A person shall set the temperature of the equipment. The scenario of the risk that someone will set the temperatures too high (> 80°C) can be dealt with several ways: You can install an alarm to warn for too-high temperatures, coping with the risk. But it might be avoided if oven temperature cannot be set higher than 80°C. The risk of a human mistake can then be avoided. For double security both measures can be implemented. One risk of temperature is in this way mitigated, but still a risk of 80°C temperature remains.

2.2.3.2 General Personal Safety Equipment

To reduce the risk to persons there is general personal safety equipment (PSE) to use. This can include safety goggles, a lab coat, and safety shoes. Depending on the set-up and chemicals to be used, this can be extended with gloves, masks, and radiation counters. In the risk assessment for the activity evaluated, the personal safety equipment to be used needs to be determined. If available, the handbook of the purchased equipment shall guide in deciding which safety equipment to wear for which activity, where for chemicals material safety data sheets (MSDS) are available to guide.

2.2.3.3 Impact Reduction

Large disasters can generally be explained by sequences of failures, like the sinking of the *Titanic* (Håvold 2013; Haga 2013) or the Deep Horizon oil spill (Makocha et al. 2019; BP 2010; Commission 2011). So, the different combinations of possible failures are important to consider as well. The different sequential events that led to the death of many people on the *Titanic* are shown in an event tree diagram (Figure 2.2), with the outcome at the right if the sequential events would have been resolved earlier or later. The bold black line is the actual sequence of events. Not hitting the iceberg would have mitigated all problems. But along the way there would have still been options to save all passengers. So, in the risk assessment, not only would the question of how to avoid collisions be asked but also what happens in case of a collision. Can we mitigate sinking? If not, how can we save the people?

Similarly, if you have a leakage; what can be the consequence? Can you avoid leakage? If not, how can you minimize its impact, e.g., by using the least amount of fluid needed, or confining leakage area? If that does not happen, what can happen? For example, slipping of people, suffocation by evaporation, creation of flammable vapors. To reduce these effects, measures can be to perform activities in ventilated areas, use good shoes or have other personal safe equipment available, or no open flames close to the set-up. If no safety gear is available or other methods of mitigation are not present, an evacuation procedure needs to be in place.

Be aware that the risk assessment shall contain existing and planned or future ideas for measures to reduced risks. Ideally they shall be implemented, improving the set-up directly e.g. including use of safety equipment like pressure release valves of the set-up, which releases the pressure or bursts in case the maximum pressure is reached or e.g. heat insulation. Allowance for use of the equipment can be given when all risks are minimized to an accepted level. With each new experiment performed, an update of the risk assessment needs to be made.

2.2.4 Manuals and Material Safety Data Sheets

Not does equipment purchased usually come with guidelines on personal safety measures, but a manual does also contain information on how to use the equipment safely. This includes descriptions of accepted risks and safety measures and procedures for safe handling. This shall be consulted and included when making a risk assessment for planned activities.

Similarly for chemicals, when a chemical is bought, the supplier needs to provide the product specific material safety data sheet (MSDS). This is a document describing the general product content, its physical properties, hazards, general measures for first aid, firefighting and accidental release (spills), a guideline on handling, storage, transport and disposal, and information on exposure control and consequent personal protection advised working with the chemical of interest. Related to this also information is provided on chemical stability, reactivity, and toxicological and ecological information for judgement on risk to human and environment. Material safety data sheets shall always be part of the risk assessment documentation.

2.2.5 APPARATUS CARD AND PROCEDURES

For unauthorized persons or personnel not trained, a minimum amount of information needs to be available to understand the main risks of a set-up in the laboratory and what to do in case of an emergency (for instance labels on a fume hood, Figure 2.3). Therefore, for each set-up an apparatus card is to be made (see Figure 2.4). Here the main information is given on e.g. the person responsible for the set-up, the main risks, and what to do in case of emergency. The latter can be, for example, the advice to make use of the emergency button or turn off a device by a switch-off button or unplugging the device.

In general, to inform on whether an experiment is ongoing, an additional indication is advised, as seen in Figure 2.4b.

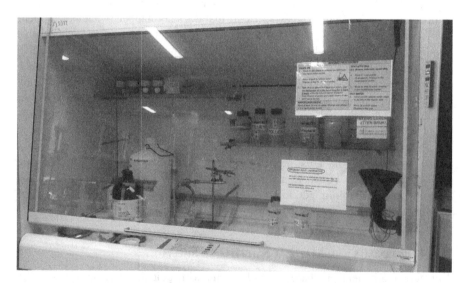

FIGURE 2.3 A fume hood, an enclosed location with an increased ventilation such that vapors and gasses are not released into the laboratory environment.

UNIT CARD

UNIT	
Helium Porosimeter_30	
UNIT RESPONSIBLE	
▮▮▮▮▮	
SAFETY HAZARDS	
Helium bottle – max choke pressure: 10 bar	
CHEMICALS	**QUANTITY**
Core plugs	
SAFETY RULES	
Lab coat , Safety glasses *Safety shoes and gloves*	
RESTRICTION	
Limitations in working hour: 08:00-15:45	
EMERGENCY SHUTDOWN	
Close the gas bottle valve.	
OTHER INFORMATION	
BELONGING	
IGP Reservoirlab	
SIGNATURE	**DATE**
▮▮▮▮	▮▮▮▮

FIGURE 2.4   a) An example of an apparatus card for the Helium porosimeter (Chapter 5). b) An experiment-in-progress sign.

2.2.6   FIRST AID AS IMPACT REDUCTION

For cases of an accident happening, there is general safety equipment present in the laboratories to reduce the impact of the accident. For each new lab you are going to work in, this shall be shown. Generally, for a chemical lab or workshop this includes, for example, fire extinguishers and blankets, emergency doors and routes, fire detectors, and alarm systems (Figure 2.5). For spills, the adsorption materials and use of fume hood shall be possible.

For personal injuries or exposure to chemicals this can include eye showers, full body showers, first aid kits, and first aid courses provided to the personnel. Depending on the event, a judgement needs to be made on what is the best approach, considering whether it is possible to resolve the consequences on your own or if professional help is needed and whether there is urgency or not.

2.2.7   INCIDENT REPORTING

With the measures in place, prevention and risk reduction are targeted. Where activities are performed there always remains a risk that something can go wrong. When unwanted incidents occur, it is to be reported and analyzed for the purpose of learning from mistakes. This is of great importance for severe accidents but more importantly with small-scale incidents that occur. It is not paramount to find the person

FIGURE 2.5 Emergency equipment available top from left to right: a. Fire extinguisher, b. fire alarm, c. emergency shower, d./e. first aid kits, f. an eye shower, and g. a defibrillator.

to blame but, as mentioned earlier, in general to find more factors and conditions that led to the accidents taking place that need to be analyzed. Learning must occur to improve such that more incidents can be prevented in the future. In aviation and car safety a saying is used: "Safety regulations are written in blood". In general, the preventing of small incidents in larger frequencies helps to minimize large impact events, as schematized in Figure 2.6.

There is a general correlation between the number of "almost accidents" happening with low impact but generally high frequencies and how many of these small incidents develop into catastrophic consequences. Reporting and making incidents discussable is therefore of great importance.

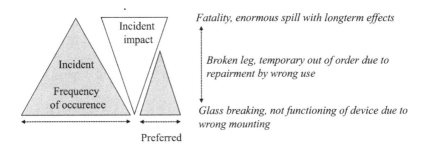

FIGURE 2.6   The higher impact events occur less frequently.

EXERCISE 2.1

1. What does the abbreviation HSE stand for?
2. Give four examples of possible personal safety gear that can be used to protect yourself from being harmed.
3. Which two factors determine the level of risk for an activity?
4. What risks can you identify for walking up the stairs with books and a telephone in your hands?
5. Describe one way to mitigate one of the risks.
6. Describe one way to reduce one of the risks.

SOLUTION

1. Health, Safety and Environment: Health focusses on the physical health of the person. Safety; hazards by use of tools, surroundings, location. Environment; impact on surrounding, pollution of air, water, waste.
2. Coat, shoes, glasses, gloves, ear plugs, helmet, breathing mask.
3. Probability of occurrence and consequence/impact of the occurrence.
4. Falling from the stairs, dropping equipment/books, tripping over dropped components.
5. Find alternative to stairs. For example, go by elevator or other solutions as long as the goal is still achieved.
6. Use of stairs in an improved way: For instance, free hands/put telephone and books in backpack etc., use of handrail, concentrate.

EXERCISE 2.2

1. In which document can you find how to handle, store, and dispose of the chemicals you want to use?
2. Which two documents do you need to prepare before working with (new) equipment?
3. What is the use of a fume hood?

SOLUTION

1. MSDS: material safety data sheet.
2. Risk assessment, experimental procedure (apparatus card).
3. Controlled ventilation with the objective of removal of vapors in the workspace.

EXERCISE 2.3

1. In which four categories can damage during activities in the lab be grouped? Which are part of the risk assessment evaluation?
2. What is the purpose of a risk assessment, and what can be two different outcomes/results from a risk assessment?
3. Which two factors are to consider when evaluating a risk to determine the risk value of an event occurring?

SOLUTION

1. Health of people, equipment damage, environment of the lab, environment of the water/air outside, name of the university.
2. Access the safety of the equipment. Improvements to the equipment to make it safer or if all is okay, acceptable continuation towards generation of procedure/apparatus card to perform the experiment.
3. Probability of occurrence, consequence/impact.

2.3   LABORATORY ORGANIZATION

Depending on the focus and function of the laboratory, the working environment can be organized differently. Note that the organization is partially based on functionality but also enshrined in legislation.

In a laboratory for rock characterization and flow in natural porous media, chemicals can be used, like oils, solvents, and salt solutions. In this section the chemical labelling and glassware cleaning will be briefly addressed.

2.3.1   Chemical Storage; Labelling and Safety Symbols

As part of the safe use and handling of chemicals, labelling of the containers is very important. The label shall give information on the content and its main hazards. Pictograms are used to indicate the hazards of the chemicals. There is a globally harmonized system of classification and labelling of chemicals (GHS) (UN 2021) that enables a worldwide applicable safety hazards system for classification and labelling of chemicals (Table 2.1). This shall be part of the information present on the label.

When chemicals arrive in the lab from the supplier, they shall already be labelled according to the legal standards. This shall be used as basis in case of re-bottling the chemical or use of the chemical. Some chemicals have an expiration date, which one needs to be aware of. Be aware that batch chemicals can vary in composition such that as part of the experimental documentation the product number or container notification needs to be noted, which can be found on the original product packaging.

TABLE 2.1

The Nine Hazard Classes and Hazard Pictograms and a Summarizing Description (UN 2021)

GHS Hazard Classes

Explosive
Explosive, self-reactive,
  organic peroxides

Flammable
Gas, aerosols, liquid, solid,
  self-heating, self-reactive,
  (pyrophoric)

Oxidizing
Liquids or solids

Compressed gas
Gases under pressure

Corrosive
Corrosive to human skin, eyes,
  metals

Toxic
Acute toxicity
(Cat. 1-3)

Health hazard
Respiratory or skin sensitization
Toxicity by single/repeated
  exposure, aspiration, may
  cause cancer or genetic effects

Environmental hazard
Hazard to aquatic life (short
  and long term)

Harmful
Hazard to public health
  (in contact, swallowed, inhaled,
  or in contact with skin) to the
  environment (e.g. ozone layer)

Before labelling, attention shall be given to the following points:

- Choose an appropriate container:
  - Consider material compatibility, sunlight exposure, size, closing mechanism.
- Avoid contamination!
  - Clean the container properly.
  - Clean the tools you use to transfer the chemical into the new container.
  - Have appropriate adsorption material closely available to reduce possible spills.
- Perform the rebottling at the proper location (fume hood, sink).

An overview of possible labels:

A. Chemical reception

Label design for when chemical is received (reception) and stored in its original package in the lab; see Figure 2.7. Its date of opening can help to keep trace of first oxygen or humidity exposure. There might be an official expiration date given by the manufacturers that can be stated on the label, and if the product is meant for someone specific this can be added additionally.

B. Label long-term storage of created samples; new, mixtures, dilutions, to be stored > 1 day

Figure 2.8 shows an example of the label for long term storage of a made sample, where long-term storage indicates storage longer than 1 day. An appropriate safety hazard class sign should be added to the label, based on the material safety data sheet and sample preparation. Here an example of the danger label for acute toxicity is added.

The local chemical storage organization, depending on the storage advice in the MSDS, needs to be consulted to appropriately store the chemical.

Summary on making your own samples:

- Dependent on the size of the glassware/container, use the appropriate label size.
- Based on the MSDS add the safety hazard stickers on the glassware.

```
Reception_____

Opened_____

Expiration_____

Designated to_____


```

FIGURE 2.7   An example of the label for chemical reception.

Name_____(State the sample creators name)_____   **

Content_____ (Describe the content) _____

Date___ (Use the date of making the sample or filling the container) ____

Supervisor_____ (in case the sample creator works no longer in the lab) _

Sample #_____ (Sample identification label for own purposes) _____

FIGURE 2.8 An example of the label for long-term (> 1 day) storage of a prepared sample. ** Shall be adapted according to the content.

- Identify where the sample needs to be stored.
- Know your sample expiration date!
- Clean out when finished with your project or hand it over to another student/ supervisor.

2.3.2 GLASSWARE CLEANING

Using clean glassware for your experiments is important. Depending on the kind of chemicals used, different cleaning procedures are to be applied and routines in the lab need to be consulted. The basic principle of cleaning is the removal of the components adhering to the glass using the right solvent. The solvent is suitable when these components will preferably dissolve in solvent compared to the dirt. Polarity of the solvent is leading there. The commonly used cleaning solvents are given in Table 2.2.

Be aware of the safety hazards of the solvents via the MSDS to determine the personal safety gear to use and handling and disposal procedures. The cleaning procedure with these volatile solvents listed, except for pure water, needs to be performed in the fume hood. A rinse with acetone and use of an air pistol to evaporate the acetone can be used to dry the glassware quickly. An alternative is to dry it in a dedicated oven. Paper use shall be avoided as this leaves microparticles on the glass.

Procedure for manual glass cleaning to remove crude oil:

1. Rinse 2× with toluene. Dispose of everything in the liquid waste bucket.
2. Rinse 2× with methanol. Dispose of everything in the liquid waste bucket.
3. Rinse at least 3× with DI water. Dispose in the liquid waste bucket.
4. Take off your gloves and have your gloves and the equipment sit in the fume hood for at least 2 hours while the toluene fumes dissipate. Then dispose of the gloves and paper towels in the solid waste bucket or dedicated waste bucket as organized in the laboratory.

Speaking from experience, the quality of the de-ionized cleaning water is critical when using the glassware for precise analytical measurements like the pendant drop

TABLE 2.2

Common Glass Cleaning Solvents Listed Based on Polarity

Polar solvents	Deionized water	-
	Methanol	Miscible in water and toluene
	Alcohol [ethanol]	-
	Isopropanol	
	Acetone	For quick evaporation as final solvent
Non-polar solvents	Toluene	-

method for interfacial tension measurements. Freshly prepared de-ionized water is to be used. Water purification methods of reverse osmosis, distillation, or filtering and oxidation can be used. The containers for transport and tools to prepare samples then need to be cleaned properly. Use gloves to prevent natural hand grease from polluting the glass when handling the cleaned glassware. When glassware is very dirty, you can soak it overnight in detergent specific for laboratory glassware or rinse it in the detergent water. Complementing the manual cleaning procedures is the use of a laboratory glassware dishwasher.

2.4   EQUIPMENT DESIGN

In the laboratories, set-ups might be available to perform standard experiments. When performing an experiment, the equipment has a measurement range that needs to match the experiments planned.

In case of design of a new experimental set-up, the needed measurements and measurement ranges determine which set-up components are needed and which settings can be used, e.g. for a flooding experiment the pump design, choice of pressure transducer, or flow rate controller depend on the system size and flow rates intended. Besides measurement range or sensitivity (lowest quantity that can be measured) also preferred accuracy, material compatibility, and temperature resistance are important to consider for the experimental design. Exercise 2.4 shows how the range of pressure transducer influences the range of permeabilities that can be measured.

EXERCISE 2.4

In a core flooding setup, three different pressure transducers can be used, with ranges of 0–200 psi, 0–50 psi, and 0–5 psi, with an error accuracy of 1% of the range. Flow rates used in the set-up shall represent field rates, (Darcy velocity) from 0.5 to 1.5 ft/day. The cores used are 4.57 cm in length and 3.82 cm in diameter. The viscosity range of the fluids used is between 1.0335 and 1.16 cp.

What is the permeability range of the rock this set-up can cover for a single-phase experiment?

SOLUTION

See Chapter 7 for the application of the Darcy equation for single-phase flooding:

$$Q = \frac{kA\Delta P}{\mu L} \qquad (2.2)$$

$$k = \frac{Q\mu L}{A\Delta P} \qquad (2.3)$$

$$v = \frac{Q}{A} \qquad (2.4)$$

with k = permeability, A = cross-sectional flooding area, ΔP = pressure difference, L = core length, μ = viscosity, and v = Darcy velocity

Minimum permeability: Combine lowest flow rate, lowest μ with highest pressure drop.

Maximum permeability: Combine highest flow rate, highest μ with lowest pressure drop, the 1% error in lowest dP meter.

$$1 \text{ feet} = 0.3 \text{ m}, 1 \text{ bar} = 14.5 \text{ psi} = 1 \cdot 10^5 \text{ Pa}, 1\text{D} = 9.869 \cdot 10^{-13} \text{ m}^2$$
$$\Delta P_{highest} = 200 \text{ psi}, \Delta P_{lowest} = 1\% \text{ of } 5 \text{ psi} = 0.05 \text{ psi}$$

$$k_{lowest} = \frac{0.5 \cdot \dfrac{0.30}{86400} \cdot 1 \cdot \dfrac{0335}{1000} \cdot 0.0457}{\dfrac{200}{14.5} \cdot 10^5} = 6.0 \cdot 10^{-13} \, m^2 \approx 0.060 \, mD$$

$$k_{highest} = \frac{1.5 \cdot 0. \dfrac{30}{86400} \cdot \dfrac{1.16}{1000} \cdot 0.0457}{\dfrac{0.05}{14.5} \cdot 10^5} = 0.8 \cdot 10^{-12} \, m^2 = 0.8 \mu m^2 \approx 0.8 \, mD$$

The variables of the experiment need to be clearly defined, as well as what the parameter you measure is dependent on. These parameters need to be set, known, or measured during the experiment. For instance, in a flooding experiment without temperature control, the laboratory temperature is to be determined such that all fluid properties can be measured or found for the room temperature during the flooding experiment.

Preparations before the experiment:

1. Design experimental set-up.
   a. Cover objective of experiment, including all parameters to be measured.
   b. MSDS collection.
   c. Equipment manual collection.

2. Perform a risk analysis.
3. Write a procedure.
4. Make a unit/apparatus card.

2.5   EXPERIMENT ORGANIZATION

Besides risk assessment and procedure, it is important to plan the experiment precisely, step by step. Table 2.3 is a typical approach for the preparation, execution, and closing down of the laboratory experiment.

Before any experiment it needs to be clear what is to be done and what is needed for that, not only in materials but also in data collection. Ordering or collection of rock material and chemicals needs to be done. It needs to be listed what to be measured, such that all relevant characterization measurements before and during the experiment are measured. For example, during viscosity or density experiments the temperature needs to be measured as the experiments are temperature dependent. A clear experimental plan/procedure shall be made. Have a hypothesis of what data you expect from the experiments. If you expect e.g. the water weight to be higher

---

TABLE 2.3

Laboratory Preparation and Execution Approach

	✓
Before experiment	
Read through the procedure.	☐
Make a plan of approach considering:	☐
What is the objective of the measurement?	☐
Which parameters do you know before the measurement?	☐
Which parameters are needed for the data analysis?	☐
What materials/fluids are needed to perform the experiment? If not available, how does one prepare them?	☐
What are the assumptions of the theory? Can they be met/how does one fulfill them?	☐
What settings will you choose?	☐
How many data points are needed to get a reliable mean?	☐
What is your hypothesis on the results/measurement values to expect?	☐
During experiment	
Follow your plan.	☐
Use a log book; have a laboratory notebook and make notes/drawings/pictures/etc.	☐
Document unexpected events with notification of time, what happened, and anything else of potential importance for further data analysis.	☐
Consider uncertainties/errors as assumptions.	☐
After experiment	
Clean up, store the materials properly.	☐
Report:	☐
Data analysis, including uncertainty and error.	☐
Conclusion and lessons/recommendation.	☐

than for oil in a similar volume and the results are different, you might want to redo the experiment one more time to confirm the results.

During the experiment, the plan shall be followed to make sure you perform all steps and gather all the data you need. Notes and pictures shall be made in a laboratory logbook with proper dating of the information. This concerns the planned events, but just as important are the unexpected events and/or observations that shall be recorded as they can help you to understand the recorded data afterwards.

The experiment needs to be properly finished by cleaning up and leaving the equipment in a state that it either can be used again or stored properly before being used again. The data from all sources needs to be collected to perform a data analysis and to draw conclusions from the results. And possibly additional experiments after the experiment can be done, to confirm data.

For example, in a displacement experiment, where water displaced a volume of oil from a core, recordings of oil and water production have been made. A material balance calculation can be performed to derive how much oil was displaced. An additional resistivity measurement or weighing the core can help to confirm that the results from the mass balance calculations are correct. This is possible if information on saturation and resistivity are known or densities of rock, oil, and water are known.

2.6 DATA ANALYSIS

2.6.1 GENERAL UNITS

Measurement values or quantities always come with a unit and these shall always be stated. The use of units used here in this book will be both SI units and non-SI units.

The SI units are standard units of measurements defined by the international system of units, which are universally recognized. These are stated in the Table 2.4. These can be used to derive all other units, for example, speed is the amount of distance moved in a time period, which will have as an SI unit [meters/seconds]. So-called prefixes (BIPM 2022) can be used to get appropriate orders of magnitudes to display based on a metric system. Some of the most common ones used are deci-10^{-1}, centi-10^{-2}, milli-10^{-3}, micron-10^{-6}, kilo- 10^{+3}, and mega-$10^{+6.}$

Non-SI or field units are not based on a metric system but are still frequently used in the society, like hour or minute. Also, in the discipline of characterization of rocks and fluids they are still used, for example length in feet or inches, volumetric quantification as barrels, centipoise for dynamic viscosity. To familiarize the reader with the order of magnitude of parameters in various unit systems and the need for unit conversions, both SI and non-SI units are used.

A third unit system is the Gaussian or centimeter–gram–second system of units (CGS units). This system is based on the three base units of centimeter, gram, and second.

2.6.1.1 Dimensional Analysis and Conversion

Units can be used in equations to derive whether the equation is correct or unit conversion is performed properly.

TABLE 2.4

The Seven Base SI Units (BIPM 2022)

Measure of	Unit	Symbol	Non-SI Unit
Temperature	Kelvin	K	
Length	meter	m	feet
			inches
Time	second	s	minute
			hour
			day
Current	ampere	A	
Weight	kilogram	kg	pounds
			ounces
Amount	mole	mol	
Luminous intensity	candela	cd	

$$\text{shear stress } \tau = \text{viscosity } \mu \cdot \text{shear rate } \frac{dv}{dz} \qquad (2.5)$$

The unit for shear stress is force (Newton, N) acting on a surface (meter²) expressed as Newton per meter² (Pascal, Pa). Force = mass • acceleration, such that one Newton is $N = kg \dfrac{m}{s^2}$. Substituting this in Equation 2.5 gives a unit for viscosity:

$$Pa = \left[\frac{N}{m^2}\right] = \left[\frac{kg\,m}{m^2 s^2}\right] = [\text{viscosity } \mu] \bullet \frac{\left[\dfrac{m}{s}\right]}{[m]}$$

$$Pa = \frac{kg}{ms^2} = [\text{viscosity } \mu] \bullet \frac{1}{s}$$

$$\text{viscosity } \mu \text{ unit} = \frac{kg}{m\,s} \text{ or } Pa\,s$$

EXAMPLE

Velocity; conversion ft/day to m/s

$$v = 1\left[\frac{ft}{day}\right] = \left|\frac{0.33\dfrac{m}{ft}\,ft}{24\dfrac{hr}{day} \cdot 60\dfrac{min}{hr} \cdot 60\dfrac{sec}{min}\,day}\right| = \left|\frac{0.33\,m}{864000\dfrac{sec}{day}\,day}\right| = 3.82 \cdot 10^{-6}\frac{m}{s}$$

2.6.2 Error and Uncertainty

When performing the experiments, the data acquired shall be as accurate as needed. Higher accuracy equipment costs more, so a balance needs to be found between necessity and economics. This can be influenced by the experimental design. Ultimately in the data analysis, the experimental uncertainty or error shall be determined to quantify the relevance for comparison and applicability.

In data acquisition, accuracy and precision are two different parameters to define the deviation or error from the actual true value, called uncertainty. Accuracy defines how far the measured value is off compared to the true value. The precision defines the scattering of the data points around an average. In real life we do not know the true value. Repeatability with the same set-up and comparison with other, independent measurement methods can be used to make a judgement on the uncertainty.

There are two kinds of errors: Systematic errors, which depend on the set-up and are repeatable, and random errors, which are unpredictable and not repeatable.

The systematic errors are, for example, wrong taring of a balance or wrong calibration of a scale. They are hard to determine and are avoided by e.g., calibration or measurement in another device or experimental approach. The latter is for example comparing porosity measurements done in the helium porosimeter compared to CT scanning or liquid porosity measurements. Random errors are fluctuations in the measurement due to random events e.g. due to varying draft in the lab, changing atmospheric pressure, and temperature fluctuations. Human judgements reading from an analogue scale are part of random error.

As measurements are never perfect, it is important to make an estimation of uncertainty (or other terms used; the error range or measurement range) that the true value must lie in. For example, the length of a table is 102 cm ± 0.1 cm. So, the form is $X\Delta X$, with X as the measured value or average (102 cm) and ΔX the estimated error (±0.1 cm).

The number of digits or the finesse of the scale defines the uncertainty of each individual measurement.

The uncertainty of a digital scale is ±50% of the last digit, unless the reported accuracy is given differently (Check the equipment manual for this information). So, if the balance measures and displays three digits behind the decimal (0.000 gram), the uncertainty is ±0.0005 gram.

For example, a measured value of 0.567 gram can actually be between 0.56650 gram rounded up or 0.56749 gram rounded down. The value reported is 0.5670 ± 0.0005 gram.

For an analog scale the rule of thumb for uncertainty estimation is 10% of the smallest scale. This is more accurate than digital scale as it is expected that the human eye can judge better than 50% between the scale marks. So, for example, on a glass tube the finest marking is marks every 0.1cm³. The estimated error is then 0.01cm³, and the reading can be also in that accuracy. So, the liquid interface is between the markings 2.1 and 2.2 cm³; then the person can make their own estimation between the 0.1 and 0.2 e.g., 0.16. The uncertainty is then 2.16 ± 0.01 cm³. See another example in Figure 2.9.

To define the random error, data statistics is performed. The approach of defining the error varies depending on the parameter measured and methodology used, e.g., the possibility to perform multiple experiments to get statistical data.

FIGURE 2.9   What is the oil volume here? Top interface-bottom interface = 2.45 ± 0.01 − 1.48 ± 0.01 = 0.97cm³ ± 0.02 (image courtesy of Stratum).

When determining the diameter of a core with a caliper it is advised to repeat the measurements at different locations and average the value. The spread in the obtained data can be used to define the uncertainty using the standard deviation.

$$\sigma = \sqrt{\frac{\sum_{i=1}^{n}\left[\overline{x} - x_i\right]^2}{n}} \tag{2.6}$$

with σ being the standard deviation, n the total number of measurements, and $\overline{x}$ the average of the data set. The data spread in this case is in general larger than the uncertainty in the individual measurement.

Note that errors can also be reported as percentage of the reported value, e.g. 43 cm ± 12%.

Not all errors can be captured based on scale or statistics. Judgement on whether steady state has been reached to make a reading can be an example or where to draw the contact lines to measure contact angle. In general, it is important to have a hypothesis of what data to expect and whether the data obtained is to be trusted and can be realistic.

Data measured can be input for models to compute a specific parameter using mathematical equations. Multiple variables with uncertainties can be needed, with all their individual uncertainty. There are rules on how to estimate the uncertainty of a variable using parameters with uncertainty. This shows the basics necessary mathematical rules to work with; more can be found in the reference (Taylor 2022).

- Addition, subtraction; errors are added.

$$Z = X + Y;$$

$$error\ \Delta Z = \Delta X + \Delta Y \tag{2.7}$$

- Multiplication, division; the error fraction of the calculated variable is the sum of the error fractions of each parameter.

$$Z = X * Y;$$

$$\text{error fraction } \frac{\Delta Z}{Z} = \frac{\Delta X}{X} + \frac{\Delta Y}{Y} \tag{2.8}$$

- When using equations, there are different rules, depending on whether the input variables are dependent or independent relative to the source of error. For example, when all parameters are measured with the same device, they are dependent. The easiest, however, is to find the largest and smallest calculated values using the highest/lowest values possible of the input data.
- Data plotting and manually obtaining gradients requires that the error bars are plotted and can be used to find the extreme gradients to define the uncertainty of the gradient obtained using the average value without uncertainty.
- Note: Round your data appropriately. As the digits represent accuracy, the number of digits shall not be increased as a consequence of the calculations. ASTM standards D6026-21 can be used as a guide (ASTM n.d.).
- Note the averaging is dependent on whether parameter is a scalar or vector/tensor. This is beyond the scope of this book to further discuss.

2.6.3 DATA QUALITY CONTROL

Data quality control is a crucial aspect of data interpretation. Raw data usually needs to be reconstructed to represent the "physics and theory" behind the experiment. Filtering and removing noises are essential to reconciling measured data. Raw data should be corrected for: a) general trend, b) monotonicity, c) smoothing, d) noise filtering.

As mentioned previously, laboratory-measured data often contain errors results from various sources, which needs to be taken into account. Despite the errors, trends need to be checked, as they are expected to happen based on the theory. For example, according to Darcy's equation the flow rate is linearly corrected to the pressure differences of the core. When there is no flow pressure, data shall be zero. With data interpretation a linear trend might be fitting that does not go through zero. This needs to be considered. Or when production data is collected the mass balance needs to be correct. The flow into the core, e.g. based on the pump settings, needs to correspond to the mass rate produced at the outlet. If this is not the case, either the production data contains errors or the pump settings do not match the actual rates produced.

It is not always too easy to find out whether data is of good quality. Obtaining similar data from other sources to compare with is a possibility. Sources can be literature data, if that is appropriate. For rock properties order of magnitudes can be used, but as each rock is unique, values need to be obtained using a different source, e.g. comparing liquid permeability with air permeability or liquid porosity with helium porosity. Comparing data from different sources should all give similar results, within the error margins of the methods. Liquid data as viscosity or density is easier to compare with literature data. Here it is important to compare the data at equal measurement conditions such as composition, pressure, and temperature, so knowledge of the dependencies of the parameters is necessary.

Repeatability is another option to obtain confidence in the data measured. This shall be part of the experimental plan, to obtain some repeated data, within one experiment or performing more than one experiment, if possible. Availability of material and time/availability and expenses of performing the experiments can limit that option.

2.7 A TYPICAL LABORATORY PLANNING SCHEME

The following is a typical laboratory planning scheme.

TABLE 2.5
Typical Laboratory Planning Scheme

LAB_____   Date _____

Objective: _____

Approach (Same as Table 2.3)

Before experiment   ✓

Read through the procedure.

Make a plan of approach considering:

  What is the objective of the measurement?

  Which parameters do you know before the measurement?

  Which parameters are needed for the data analysis?

What materials/fluids are needed to perform the experiment? If not
  available, how does one prepare them?

  What are the assumptions of the theory? Can they be met/how does
    one fulfill them?

  What settings will you choose?

  How many data points are needed to get a reliable mean?

What is your hypothesis on the results/measurement values to expect?

During experiment

Follow your plan.

Use a log book; have a laboratory notebook and make notes/drawings/
  pictures/etc.

Document unexpected events with notification of time, what happened,
  and anything else of potential importance for further data analysis.

Consider whether uncertainties/errors/assumptions are met.

After experiment

Clean up, store the materials properly.

Report:

  Data analysis, including uncertainty and error.

  Conclusion and lessons/recommendation.

Equations applicable:

_____

Assumptions:

_____

Parameters to measure/to know beforehand:

_____   _____

_____   _____

Materials needed:

- _____

- _____

- _____

REFERENCES

Arbeidstilsynet. 2024. *Laboratoriearbeid.* www.arbeidstilsynet.no/tema/kjemikalier/laboratoriearbeid/. Accessed 01 07, 2024.

BIPM. 2022. *The International System of Units (SI).* Bureau International des Poids et Mesures. www.bipm.org/en/publications/si-brochure.

BP. 2010. *Deep Water the Gulf Oil Disaster and the Future of Offshore Drilling Report to the President.* BP, 1–192. www.bp.com/content/dam/bp/business-sites/en/global/corporate/pdfs/sustainability/issue-briefings/deepwater-horizon-accident-investigation-report.pdf. Accessed 01 08, 2024.

Commission, National. 2011. *Deep Water. The Gulf Oil Disaster and the Future of Offshore Drilling. Report to the President.* Government.

Haga, R.A. 2013. "Reexamining the Titanic with Current Accident Analysis Tool. Multidisciplinary Eduation and System Safety Primer for Engineering Students." *IEEE Global Engineering Education Conference (EDUCON)*, 1032–1041. IEEE.

Håvold, J.I. 2013. "Conference: Safety, Reliability and Risk Analysis: Beyond the Horizon: Proceedings of the European Safety and ReliabAt: Amsterdam." *European Safety and Reliability.* Amsterdam.

IMO, International Maritime Organisation. 2024. *IMO.* wwwcdn.imo.org/localresources/en/OurWork/Safety/Documents/TITANIC.pdf. Accessed 01 24, 2024.

Makocha, I.R., Ete, T., Saini, G. 2019. "Deepwater Horizon Oil Spill: A Review." *International Journal of Technical Innovation in Modern Engineering& Science (IJTIMES)*, vol. 5.

Taylor, J.R. 2022. *An Introduction to Error Analysis. The Study of Uncertainties in Physical Mesurements*, 3rd ed. Sausalito: University Science Books.

UN. 2021. *Globally Hormonized System of Classification and Labelling of Chemicals (GHS)*. United Nations. https://unece.org/transportdangerous-goods/ghs-pictograms. Accessed 01 08, 2024.

3 Core Preparation and Fluid Saturation

3.1 INTRODUCTION

To describe flow in a reservoir, the core samples used for the experiments come from the reservoir. Specifically routine core analysis (RCAL) and special core analysis (SCAL) shall be performed on reservoir rock. Reservoir cores are actual "point data" from the field, but they are expensive to drill. As an alternative, if no core material is available or there is too little, model or analog cores can be used to study how the fluids interact within a porous medium. These model cores come from outcrops, where formation is dry. The model cores are cheap to obtain and can be used for screening studies like initial assessment of enhanced oil recovery or effect of CO_2 on carbonate rock dissolution. The most commonly used sandstone model cores are Berea, from Ohio in the USA and Fontainebleau, France; Bentheimer or Gildehaus from Germany; and Clashach from Scotland. Model carbonate rocks are Austin chalk, USA and Stevns Klint chalk, Denmark and limestones from Ainsa, Spain and Angola. Caprock shales can come from Kimmeridge in the south of England or Mont Terri rock, Switzerland. The disadvantage of the model cores is that the rock is consolidated, such that the minerology is fixed. In case mineralogy is important to control, packing can be used, like sand or glass bead packs, which are most common, but any unconsolidated minerals can be mixed in the grain size and grain size distribution preferred. The mineralogy can be chosen but permeability and porosities are generally high and not representative.

In this chapter we will briefly introduce the process to obtain the original reservoir core samples. Care should be taken during this process to preserve the original properties of the reservoir rock as much as possible. This chapter will address how the core material is obtained by drilling, including which tools are used in the well to recover the core material and thereafter the preparation and handling of the drilled core at the well site. Once transported to the laboratory, the core material is processed. Core preparation procedures will be dependent on rock types, considering sandstones or carbonates and types and volume of clays present in the rock. The discussion includes a detailed description of the cleaning process of the cores, removing remains of the drilling fluids and possible precipitants due to the changed conditions of temperature and pressure. Recovery of the reservoir fluids during the cleaning process can give an indication of the phase saturations from the original core. This is one of the main parameters in fluid flow in porous media and is important in the laboratory studies of reservoir rocks.

3.2 CORING

Rock samples are recovered from the reservoirs or horizons of interest during a coring operation, brought to the surface and transported to a core laboratory.

DOI: 10.1201/9781003382584-3

The coring processes are classified into four main categories according to the principle applied:

1. Conventional coring.
2. Wireline coring.
3. Continuous coring.
4. Sidewall coring.

The first three methods are based on obtaining the core while drilling the well, whereas four is performed in an open but already existing hole. In the following, only a brief overview of the coring methods is given. More details are given in the textbooks on drilling (Ashena and Thonhauser 2018; Bourgoyne et al. 1991).

3.2.1 CONVENTIONAL CORING

Core drilling is a special operation during the drilling of a well. The well is drilled using a drill string with a drill bit at the end. During conventional drilling, the drill bit crushes the rock into pieces called cuttings, which are removed from the drill hole by circulation of a drilling fluid. The drilling fluid, also called drilling mud, has additional function of cooling the drill bit, keeping a hydrostatic head in the well to avoid fluid from the formations flowing into the wellbore and to form a mud cake at the wall of the drill hole to minimize drilling fluid flowing into the formations. The drilling fluid consists of oil or brine as a base fluid, clay to increase its density, and chemicals e.g. corrosion inhibitors or friction reducers. For drilling a core a special drilling bit is needed; see Figure 3.1 for two examples compared to a conventional drill bit.

FIGURE 3.1 Examples of drill bits. a. and b. show coring drill bits, compared to c. the conventional drill bit. d. Shows the core catcher, holding the core in the drill string while it's being pulled up.

The construction to collect the core is called a core barrel, which is a part of the complete drill string. When drilling the core from the formation, the drilled core shall not rotate with the drill string. There are, depending on the core quality, different types of core barrels used in conventional coring:

1. Single-tube core barrels.
2. Double-tube rigid core barrels.
3. Double-tube swivel core barrels.
4. Rubber sleeve core barrels.
5. Pressure core barrels.

Today, almost all petroleum related coring is done using diamond core bits and double tube swivel core barrels, (see Figure 3.2). The barrel system includes an inner and outer barrel. The barrels are separated by ball bearings and thereby it is possible to keep the inner barrel with the core stationary while the outer barrel is rotating. The inner barrel has a liner of fiberglass, plastic, or aluminum of 9.1 m length. The liner's

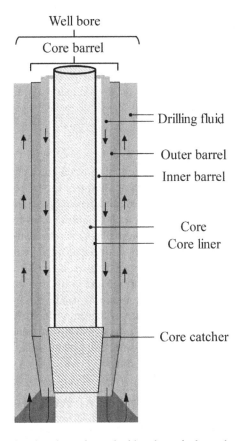

FIGURE 3.2   Conventional coring using a double tube swivel core barrel. The shaded components do not rotate.

purpose is to support the core during drilling, and it is also the package of the core when transported to storage or laboratory.

To retrieve the core, the complete drill string needs to be pulled up the drill hole. To ensure the core does not drop out, a core catcher is installed to hold the core in the inner barrel; see Figure 3.1d for an example.

3.2.2   WIRELINE CORING

It is a disadvantage in conventional coring that the core needs to be retrieved by pulling up the complete drill string. This is time consuming and costly. Wireline coring systems provide added utility to conventional coring in situations where continuous coring over long intervals is desired. The inner barrel is detachably mounted in the outer barrel and can be retrieved to the surface through the inside of the drill pipe on a wireline. The main advantage is that the drill string does not have to be removed from the hole to recover the core. A disadvantage is the small core diameter (2¾"), which is limited by the inner diameter of the drill string. Wireline cores generally provide only marginally adequate samples for proper formation evaluation and will not be discussed any further here.

3.2.3   CONTINUOUS CORING

Coring while drilling is the idea behind various concepts of reversed circulation drilling methods. These methods have never obtained commercial application and will not be further discussed here.

3.2.4   SIDEWALL CORING

Sidewall samples are taken by a wireline tool lowered down an existing open hole to a prescribed depth. In principle the two possibilities for the recovery of cores from the hole wall exist: Percussion procedures and drilling procedures.

Percussion sidewall coring is performed by shooting a series of hollow bullets that are loaded with explosive charges. The core plugs are retrieved from the formation to the tool by a chain, (comparable to a cork wine bottle opener) and the tool is returned to surface (Figure 3.3 The plugs are small (approx. 2.5 cm in diameter and length) and they are used for lithological description and grain size analysis.

Rotary side wall tools are presently the most used. This is also a wireline tool with a series of rotary coring bits which is driven by the drilling mud or an electric motor. The samples obtained from rotary drilling are better than samples from percussion. The plug size is similar to those from percussion, but more trustworthy porosity and permeability are often obtained.

3.3   PREPARATION AND HANDLING AT WELL SITE

The coring process/handling and analysis of the core sections should be done in close cooperation between the engineers and geologists. Before the core is brought to the surface, it is important to know exactly what procedures will be followed during removal of the core from the core barrel.

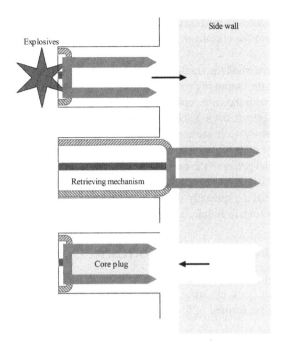

FIGURE 3.3 Sidewall coring by percussion, sequence from top to bottom.

The following procedure is advised once the core is retrieved from the drill hole:

1. When removing the core from the core barrel all parts shall be kept in their original orientation. All core materials that come out of the core barrel should be retained. Loose pieces should be collected in depth-labelled plastic bags. Such core material may be fitted together later in the laboratory and provide useful information concerning lithology and vertical sequence.
2. The core should be laid out and fit together at the well site. This will make reassembling of the core in the laboratory much easier.
3. The core depths can be measured, from the top of the core down. A downhole orientation can be made while the core is laid out by marking parallel lines down the entire length of the core. (See the marking on the cores in Figure 3.5.)
4. An initial general lithologic description of the core sequence shall be made, including a recording of the general condition of the core. Further geological description will be done at the core laboratory. There might be core material losses because of a) shaley material being dropped out of the core barrel, b) grinding of the core at "spin-off" points, or c) loss of fragments during extraction from the core barrel; this can result in the core parts not fitting together. Collecting this data can provide an estimate of percent recovery from the cored interval and will help in later correlation of the core with wire-line logs and other subsurface data.

5. The core shall be packed carefully for transport to the core analysis laboratory to prevent the core from breaking during transport.

When the core has reached the rig floor it is important to minimize further alteration of the rock. Presently, the common procedure is to leave the core in the liner. The standard 9-meter-long liner with the core could be sent to the laboratory, but usual practice is to cut the liner in 1-meter sections. Each section is capped or sealed to avoid drying out on its trip to the laboratory. Core sections meant for special core analysis can be preserved more carefully than others. However, the best practice today is to keep the core in the liner and remove and preserve the core in the laboratory. The special precautions taken at the rig could be to dip the whole liner section in wax, but sometimes the core inside the liner must be stabilized, especially if it is weak and unconsolidated. Such stabilization can be done by freezing, resin injection, gypsum injection, or foam injection.

3.4 PROCESSING IN THE LABORATORY

The processing in the lab consists of scanning the core, drilling core plugs for routine and special core analysis, cleaning, and preservation, for short term purpose till the experiment can be performed or for the long term (years).

3.4.1 SCANNING

In the lab the core is usually gamma ray scanned while it is in the liner. Figure 3.4 shows an example. These scans are used for assessment of core recovery, depth correlations from various sources (gamma ray logs, driller), and decisions on selecting good locations for plugs. Dense material absorbs X-rays and appears white, while

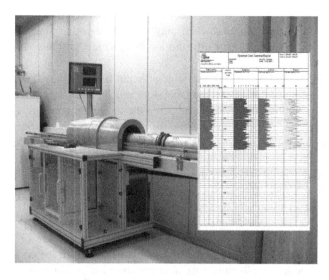

FIGURE 3.4 The core in the liner being gamma ray scanned (courtesy of Stratum).

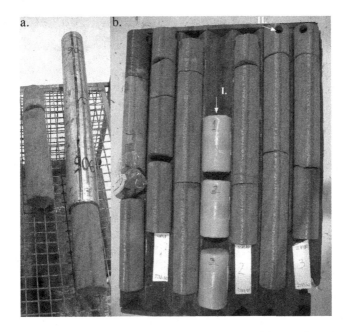

FIGURE 3.5   a. The core taken out of the liner; b. the core with the orientation markings (two lines) where I. shows the in wax preserved core pieces and II. top core plugging (courtesy of Stratum).

less-dense material appears darker in the scans. Afterwards the core can be removed from the liner and laid out on a viewing table, marked, and then wrapped with plastic film. At this stage, a core analysis program is determined, and the preservation is done accordingly for parts of the core. Figure 3.5a shows the core taken out of the liner and preservations (I) and initial top coring (II).

3.4.2   CORE SLABBING AND PLUG DRILLING

Routine core analysis (RCA) and special core analysis (SCA) tests are usually performed on plugs taken from the whole cores. Depending on the test, the orientation shall be chosen (perpendicular or along the axis or along the layer orientation.). The standard plug diameters are either 1 in (2.54 cm) or 1.5 in (3.81 cm). The fluid used for plug drilling is generally tap water, assuming limited interaction with the plugs. The locations of typical core plugs taken from the whole cores are shown in Figure 3.6.

Geologists will use part of the core to perform a thorough lithological study. Therefore, the core is slabbed along the vertical axis of the core. The core is always slabbed with maximum exposed bed dips to enable studying the layering and to ensure further core plugs obtained are parallel to the bed strike and not perpendicular to the bed strike. Figure 3.7 shows an example of slabbing for an exploration well. One-third (A, B) are for geological studies, where B is for the epoxy resin sampling, and C and D can be used for coring if that was not performed before or for permeability studies and/or for sharing with partners.

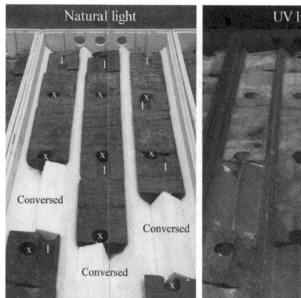

FIGURE 3.6 A core in natural light (left) and UV light (right; Part A of Figure 3.7). The locations of the core plugs are marked; x shows the horizontally drilled plug locations and the I shows plugs drilled along the axis of the core (courtesy of Stratum).

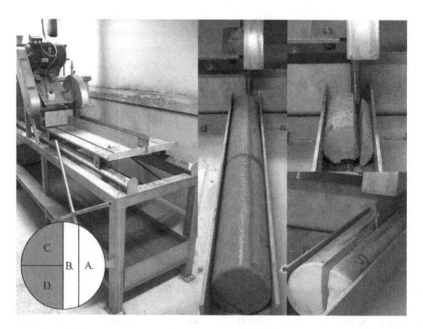

FIGURE 3.7 Core slabbing. The inlay shows an example of slabbing for a sample from an exploration well (courtesy of Stratum).

Coring operation, core handling, and preparation in a proper way require a lot of detailed practical procedures not included here. Details can be found in Wisenbaker (1952), McPhee et al. (2015), and American Petroleum Institute RP 40 (API 1998).

3.4.3 PRESERVATION OF CORE AND CORE PLUGS

Both whole core and plugs can be preserved, and the two main methods are "wet" or "dry" preservation. The objective is to avoid exposure to oxygen, avoid evaporation, and avoid fluid ingress/egress.

Several methods of preservation are available:

3.4.3.1 Wet

In this method the core is submerged in simulated, deoxygenated formation brine for water zone cores or kerosene for oil zone cores. Samples could additionally be preserved under nitrogen blanket at pressure (~30bar).

3.4.3.2 Dry

The core material is wrapped in vapor and watertight material. This can be done as wax coated or commonly called seal-peal, for all SCAL purposes and especially wettability and residual oil saturations. Here the core first wrapped in plastic and aluminum, is dipped in hot wax or plastic, see Figure 3.5 I. Other barrier foil laminates can be used, where many layers of plastic, nylon and aluminum are used and finally the package is air evacuated and heat sealed. Alternatively, the core plugs can be wrapped in cling film and frozen in solid CO_2 for fluid saturation measurements. This can be used for unconsolidated core specifically.

The decision of what type of preservation technique to use is based on several factors. The rock nature, the nature of transport from field to lab, length of time between storage and analysis, and objectives of the coring program all must be considered. Hard and well consolidated cores should not require special care. However, soft and consolidated core should be provided with adequate support and protection to prevent damage or alteration. The use of glass jars, easily deformable plastics, paper cartons, and other nonrigid containers should be avoided, if the sample are to be shipped in such a way as to encounter rough handling.

There are a number of acceptable practices for preserving cores during transportation and storage. These methods are clearly described in the American Petroleum Institute Publication API RP40 (API 1998). See for some examples Table 3.1.

3.4.4 CORE CLEANING

Core cleaning is needed prior to petrophysical measurement of the core such as porosity and permeability. The objective of the cleaning is to remove pollutants and all original components in the pore space and from the grain surfaces, so the removal of oil, water and salts. The pollutants are introduced into the core while drilling using drilling fluid. Precipitates from the oil or brine need to be removed. Traditionally the absorbed oil components are also removed, relevant for reservoirs

TABLE 3.1

Several Types and Variations of Preservation Methods

Type of preservation	Variations
Containment	Sealing in air-tight cans
	Sealing in Air-tight Steel, Aluminum or Plastic Tubes
	Sealing in Plastic Bags
Fixating	Freezing in Dry Ice
Coating/wrapping	Wrapping in Metal Foil and Plastic Tape
	Plastic Coatings
	"Seal-peel" cores

that have seen hydrocarbons. The cleaning is based on dissolution of the pollutants/pore fluids in solvents using different techniques of exposure, or evaporation of the components using temperature.

The American Petroleum Institute (API) presents in their recommended practice for core analysis (RP 40) (API 1998) detailed cleaning procedures, and the following is based on this recommendation. The textbook by Amyx, Bass and Whiting is also presenting core cleaning methods and apparatus in a concise manner (Amyx et al. 1960).

3.4.5 SOLVENTS

For the case of using solvents as cleaning phase, the number of cycles or kind and amount of solvents which must be used depends on the nature of the hydrocarbons being removed and the solvent used. Often, more than one solvent must be used to clean a sample. The solvents selected must not react with the minerals in the core. The commonly used solvents are given in Table 3.2.

As nonpolar solvents, toluene and benzene are most frequently used to remove oil, and methanol being polar is used to remove water.

The cleaning procedures used are specifically important in special core analysis tests, as the cleaning itself may change wettability (Amott 1959; Anderson 1986). See Chapter 9 for a discussion on how hard and soft cleaning affects wettability, respectively removing or maintaining the absorbed oil components on the grains.

3.4.6 SOXHLET EXTRACTION

A Soxhlet extraction apparatus is the most common method for cleaning samples and is routinely used by most laboratories. Toluene and methanol are usually used for removing oil and water from the core sample, respectively. As shown in Figure 3.8, toluene is brought to a slow boil in a Pyrex flask; its vapors move upwards, and the core becomes engulfed in the toluene vapors (at approximately 110°C). The

TABLE 3.2
Solvents for Core Cleaning and Their Use (API 1998)

Solvent	Boiling Point, °C	Solubility
Acetone	56.5	oil, water, salt
Chloroform/methanol azeotrope (65/35)	53.5	oil, water, salt
Cyclohexane	81.4	oil
Ethylene Chloride	83.5	oil, limited water
Hexane	49.7–68.7	oil
Methanol	64.7	water, salt
Methylene Chloride	40.1	oil, limited water
Naphtha	160	oil
Tetrachloroethylene	121	oil
Tetrahydrofuran	65	oil, water, salt
Toluene	110.6	oil
Trichloroethylene	87	oil, limited water
Xylene	138–144.4	oil

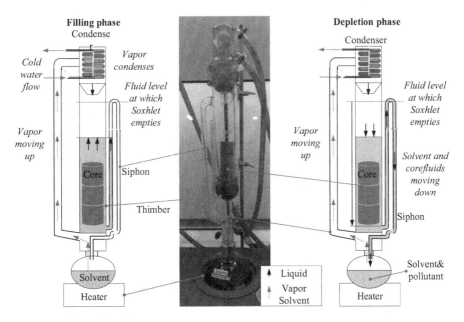

FIGURE 3.8 Schematic diagram of Soxhlet, left showing the filling phase of the thimber, right the depletion phase (picture courtesy of Stratum).

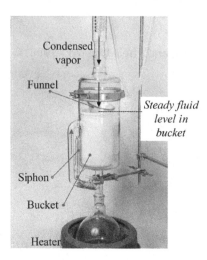

FIGURE 3.9 Total immersion Soxhlet cleaning set-up, where the bucket avoids the liquid level to drop, and the funnel leads the condensing solvent to the bottom of the bucket (courtesy of Stratum).

toluene vapor enters the inner chamber of the condenser and the cold water circulating outside the inner chamber condenses the toluene. Recondensed toluene falls from the base of the condenser onto the core sample. The toluene soaks the core sample and dissolves any oil with which it comes into contact. When the liquid level within the Soxhlet tube reaches the top of the siphon tube arrangement, the liquids within the Soxhlet tube are automatically emptied by a siphon effect and flow into the boiling flask (Figure 3.8, left). The toluene is then ready to start another cycle. Note that the toluene needs a lower boiling point than the pollutants, such that it evaporates as pure vapor once more. As toluene has a higher boiling point than water, water can evaporate from the core, condense and it collects at the bottom of the thimble. It however evaporates with the toluene once it reaches the boiling flask, and the core is not cleaned from water. A complete extraction may take several days to several weeks in the case of low high-density crude oil or the presence of heavy residual hydrocarbon deposits within the core. Low permeability rock may also require a long extraction time. The core sample achieves complete cleaning when no change in the color of the toluene in the Pyrex flask is observed. Toluene will then be replaced by methanol to clean the water content from the core sample.

When the thimble is emptied, the gas liquid interface moves through the core. This can damage fragile minerals/clays. The alternative is total immersion Soxhlet cleaning, where the fluid level in the thimble always stays above the cores. An overflowing container is used, where the condensate is led to the bottom of the thimble. This setup can also be used for unconsolidated rocks. Chloroform/methanol azeotrope could be used such that the temperature can be reduced.

Procedure for Soxhlet cleaning:

General: Personal safety gear; wear (heat) gloves, glasses, safety shoes and lab coat. Check the material safety sheet data for the solvent whether gas masks are needed additionally for handling the soaked cores. Check the boiling point for the maximum temperature of the system.

Start cleaning procedure

1. Put the cores careful in the extraction chamber, maximum height shall lie below the top of the siphon.
2. Use a funnel and fill up the 1000 ml boiling flask with 800ml with the first solvent (A).
3. Place the boiling flask in the heating mantel and connect the extraction chamber to the boiling flask.
4. Place the condenser on the top of the extraction chamber, fasten it carefully and check whether the water tube ends in the drain.
5. Turn on the water slowly on till maximum flow rate, check for leakage while increasing the flow rate.
6. Set the heating of the heating mantel, such that the solvent will boil gently and adjust the rate of boiling so that the reflux from the condenser is a few drops of solvent per second.
7. Close the glass-door of the fume hood and put the experiment in progress notification with name and date on the fume hood door.

Pause cleaning procedure

a. **In case of continuation with the same solvent after the pause**
   • Turn off the heating mantle when the fluid level in the extraction chamber stands _above_ the sample surfaces.
   • Stop the water flow after waiting minimal 30 min.
   • Start up the cleaning by turning on the water slowly on till maximum flow rate and turning on the heating mantle again (step 5 onwards).

b. **In case of continuation with a different solvent after the pause**
   • Turn off the heating mantle when the fluid level in the extraction chamber stands _below_ the sample surfaces.
   • Stop the water flow for after waiting minimally 30 min.
   • Continue the cleaning procedure at step 2, starting with another boiling flask filled with the new solvent (B), connecting the Soxhlet with the cores exposed to the initial solvent (A).

Closing down the cleaning procedure

1. Turn the heating mantel to zero directly after the extraction chamber emptied.
2. Disconnect the electricity from the heating mantel.
3. Let the glassware cool down 30 minutes before turning off the water.

4. Disconnect the condenser from the Soxhlet and the extraction chamber from the glass-container.
5. If there is some methanol or toluene in the extraction chamber and connecting glass tubes, use a funnel and put it back into the boiling flask.
6. Clean the boiling flask.
7. Dry the cores in the oven.

3.4.7   DIRECT INJECTION OF SOLVENT

The method is based on core flooding and injection of a proper solvent into the core sample while being placed within a fume hood or having a closed system. Figure 3.10a shows an example of a flooding rig assembly, where the color of the fluid production is monitored (Figure 3.10b) according to a color scheme 'cleaning card'.

The following steps can be used in this method:

- Place the rock sample in a rubber sleeve inside a core and apply confining pressure of at least 7 bar more than the inlet pressure to avoid leakage between sleeve and core (see Chapter 7, Figure 7.7 for an example of a core holder).
- Start with toluene injection to remove any hydrocarbon inside. The injection rate should not be higher than 0.5 cm^3/minute to prevent any fines migration inside the core sample.
- Inject three pore volumes and more if it is needed. In fact, the process should last until there is no color change in the fluid on the production side of the core.
- Replace toluene with methanol and repeat step two and three to clean the sample from water content.

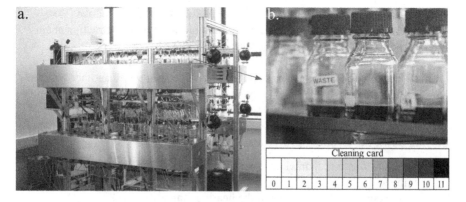

FIGURE 3.10   Direct solvent injection set-up a., the produced fluids b. and monitoring of its color according to the cleaning card (courtesy of Stratum).

3.4.8 CENTRIFUGE FLUSHING

Detailed procedure for running centrifuge experiment is given in Chapter 10. Samples are placed in either drainage (to remove water content) or imbibition buckets (to remove hydrocarbon oil) and surrounded with proper solvents. The centrifuge for the purpose of cleaning is equipped with a special fluid spray device that sprays warm solvent onto the sample. The centrifugal force then moves the solvent through the sample. The solvent used can be collected and recycled. The speed of rotation may vary from a few hundred to several thousand rpm, depending on the permeability of the samples.

3.4.9 GAS-DRIVEN SOLVENT EXTRACTION

The sample is placed in a pressurized atmosphere of solvent containing dissolved gas. The solvent fills the pores of the sample. When the pressure is decreased, the gas comes out of solution, expands, and drives fluids out of the rock pore space. This process can be repeated as many times as necessary.

3.4.10 CLEANING CONSIDERATIONS

To determine the approach, method, solvent used, temperature consider:

1. The presence of fragile clay and minerals.
2. Wettability.
3. Erosion/fines migration.
4. Temperature effects.
5. Efficiency/duration of cleaning process.

The direct-injection method is effective, but slow. The method of flushing by using centrifuge is limited to plug-sized samples. The samples also must have sufficient mechanical strength to withstand the stress imposed by centrifuging. However, the procedure is fast. The gas driven-extraction method is slow. The disadvantage here is that it is not suitable for poorly consolidated samples or chalky limestones. The distillation in a Soxhlet apparatus is slow but gentle on the samples. The procedure is simple and multiple cores can be cleaned at once.

3.4.11 CORE DRYING

The core sample is dried for the purpose of removing connate water from the pores, or to remove solvents used in cleaning the cores. When hydratable minerals are present, the drying procedure is critical since interstitial water must be removed without mineral alteration. Drying is commonly performed in a regular oven or a vacuum oven at temperatures between $50°C$ to $105°C$. A vacuum can be added to increase the speed of drying (vacuum oven drying). If problems with clay are expected, drying the samples at $60 °C$ and 40% relative humidity will not damage the samples, called humidity oven drying. Prior to porosity and permeability measurement, the core sample should be completely dried.

Alternatively for very delicate structures present at the rock surface or in the pores, a much more sophisticated technique like critical point drying can be used, where the supercritical CO_2 is used to transition from a liquid phase to a gas phase. The solvent in the sample shall be soluble in liquid CO_2, which will displace the solvent first. Here after the CO_2 will be brought into its critical state (T > 31°C, P > 73,8 bar), before the pressure will be reduced, so that the CO_2 turns from the supercritical state into a gas phase, not creating any interfaces such that interfacial tension cannot damage the fragile structures. This is also applied to fabricate aerogels.

In Table 3.3 some core drying methods are listed for different rock types:

TABLE 3.3
Core Sample Drying Methods (API 1998)

Rock Type	Method	Temperature, °C
Sandstone (Low clay content)	Conventional oven	116
	Vacuum oven	90
Sandstone (High clay content)	Humidity oven (40% relative humidity)	63
Rock with fragile clays	Critical point drying	>71
Carbonate	Conventional oven	116
	Vacuum oven	90
Gypsum bearing rock	Humidity oven (40% relative humidity)	60
Shale or other high clay rock	Humidity oven (40% relative humidity), Conventional vacuum	60
Unconsolidated	Flow-through drying	

Note that drying times may vary from sample to sample. The check point for completely dried samples is that each sample should be dried until the weight becomes constant.

3.5   FLUID SATURATION MEASUREMENTS

In hydrocarbon reservoirs, the saturation shows how much of the reservoir's pore space is occupied by hydrocarbons. In petroleum reservoirs, the present fluids are hydrocarbon oil and gas plus aqueous brine. In connection with CO_2 storage, only water and CO_2 are present in the system.

In core plugs it is possible to determine fluid saturation directly. Before measuring porosity and permeability, the core samples are cleaned of residual fluids and thoroughly dried. In this cleaning process we can directly measure the fluid volumes in the pore space and thereby determine the fluid saturation. Also, after experiments like wettability, capillary pressure or relative permeability tests, saturations can be determined during the core cleaning to obtain an independent measurement of saturations at the end of the experiment.

Definition: Before introducing the core cleaning procedures and saturation determination methods it is necessary to define the concept of fluid saturation. Fluid saturation is defined as the ratio of the volume of fluid in a core sample to the pore volume of the sample as:

$$S_o = \frac{Volume\ of\ oil}{Pore\ volume} = \frac{V_o}{V_p} \qquad (3.1)$$

$$S_g = \frac{Volume\ of\ gas}{Pore\ volume} = \frac{V_g}{V_p} \qquad (3.2)$$

$$S_w = \frac{Volume\ of\ water}{Pore\ volume} = \frac{V_w}{V_p} \qquad (3.3)$$

The saturation of each individual phase ranges between zero and 100%. Therefore, by definition, the sum of saturations is 100%, and:

$$S_w + S_o + S_g = 1 \qquad (3.4)$$

Where V_w, V_o, V_g, and V_p are water, oil, and gas pore volumes respectively and S_w, S_o, and S_g are water, oil, and gas saturations. Note that fluid saturation may be reported either as a fraction of total porosity or as a fraction of effective porosity. Since fluid in pore spaces that are not interconnected cannot be produced from a well, the saturations are more meaningful if expressed relative to the effective porosity. The mass of water collected from the sample is calculated from the volume of water by the following relationship:

$$m_w = \rho_w V_w \qquad (3.5)$$

Where ρ_w is water density in g/cm³. The mass of oil removed from the core may be computed as the weight of liquid minus weight of water:

$$m_o = m_L - m_w \qquad (3.6)$$

where m_L is the mass of liquids removed from the core sample in gram. Oil volume may then be calculated as m_o / ρ_o. Pore volume V_p is determined by a porosity measurement, and oil, gas, and water saturations are calculated by Eqs. 3.1, 3.2, and 3.3.

While cleaning, there is a possibility to quantify fluids extracted from the core, which come from the reservoir. Depending on the wettability of the rock and drilling fluid used (oil- or brine-based), drilling fluid could have imbibed the core, changing the saturations that are measured. This occurs when *water-based mud* is used for a *water wet core* at initial water saturation, where oil-based mud would not imbibe. Oil-based mud could imbibe in an oil wet core, but as it can be presumed the reservoir is found at initial water saturation, all mobile water is displaced already. Water wet mud would not want to imbibe in an oil wet core. So, in the case of water-based mud used for a water wet core, the data on saturations might not represent the reservoir saturations. For the background we refer to Chapter 9, Wettability and Chapter 10, Capillary Pressure.

DEAN-STARK DISTILLATION

The Dean-Stark distillation provides a direct determination of water content. The schematic of the apparatus is given in Figure 3.11. In this method, the water and solvent are vaporized, recondensed in a cooled tube in the top of the apparatus, and the water is collected in a graduated tube, initially filled with solvent. The lighter solvent remains on top, overflows, and drips back over the samples. The oil removed from the samples remains in solution in the solvent.

Oil content is calculated by the difference between the weight of water recovered and the total weight loss after extraction and drying.

PROCEDURE

1. Weigh a clean, dry thimble. Use tongs to handle the thimble.
2. Place the cylindrical core plug inside the thimble, then weigh the thimble and sample.
3. Fill the extraction flask two-thirds full of toluene. Place the thimble with sample into the long-neck flask.

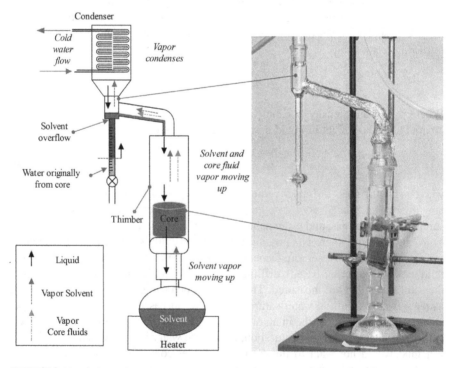

FIGURE 3.11   Schematic diagram of Dean-Stark apparatus (picture courtesy of Stratum).

4. Tighten the ground joint fittings, but do not apply any lubricant for creating tighter joints. Start circulating chilly water in the condenser.
5. Turn on the heating jacket or plate and adjust the rate of boiling so that the reflux from the condenser is a few drops of solvent per second. The water circulation rate should be adjusted so that excessive cooling does not prevent the condenser solvent from reaching the core sample.
6. Continue the extraction until the solvent is clear. Change solvent if necessary.
7. Read the volume of collected water in the graduated tube. Turn off the heater and cooling water and place the sample into the oven (from 105°C to 120°C) until the sample weight does not change. The dried sample should be stored in a desiccator.
8. Obtain the weight of the thimble and the dry core.
9. Calculate the loss in weight, W_L, of the core sample due to the removal of oil and water.
10. Measure the density of a separate sample of the oil.
11. Calculate the oil, water, and gas saturations after the pore volume, V_p, of the sample is determined.

The following is an outline of the detailed calculation procedure for obtaining saturations:

1. The volume of extracted water accumulated in graduated tube is a direct measure of the sample's water content.
2. A simple material balance in the core sample is given here:

$$m_{sat} - m_{dry} = M_o + M_g + M_w \qquad (3.7)$$

Where m_{sat}, m_{dry}, M_o, M_g, and M_w are saturated core mass, dry core mass, and mass of oil, gas, and water, respectively. The mass of the phases can be expressed in terms of volume and density, $V_o\rho_o$, $V_g\rho_g$, and $V_w\rho_w$.

3. Pore volume of the sample can also be written as:

$$PV = V_o + V_g + V_w \qquad (3.8)$$

Where V_o, V_g and V_w are volume of oil, gas, and water in the core sample.

4. Equations (3.7) and (3.8) can be rearranged as:

$$V_g\rho_g + V_o\rho_o = m_{sat} - m_{dry} - V_w\rho_w \qquad (3.9)$$

$$V_g + V_o = PV - V_W \qquad (3.10)$$

5. By knowing $PV, V_W, m_{sat}, m_{dry}, \rho_w, \rho_o, \rho_g$ the previous equations can be solved to obtain V_o.

6.
$$V_o = \frac{(m_{sat} - m_{dry} - V_w \rho_w) - (PV - V_w)\rho_g}{(\rho_g + \rho_o)}$$
(3.11)

With ρ_g very small, Equation (3.11) can be simplified to $V_o = \dfrac{m_{sat} - m_{dry} - V_w \rho_w}{\rho_o}$

7. The fluid saturations for each phase are calculated by:

$$S_w = \frac{V_w}{PV}$$
(3.12)

$$S_o = \frac{V_o}{PV}$$
(3.13)

$$S_g = \frac{V_g}{PV} = \frac{PV - V_o - V_w}{PV}$$
(3.14)

Where S_o, S_g, and S_w are oil saturation, gas saturation, and water saturation.

Alternatives to the Dean Stark method are methods like vacuum distillation and retort distillation.

Vacuum distillation is often used for full diameter cores because the process is rapid. Vacuum distillation is also frequently used for poorly consolidated cores since the process does not damage the sample. The sample is placed in a vacuum while being heated to a maximum temperature of 230°C. Liquids in the sample are vaporized and passed through a condensing column cooled by liquid nitrogen (Koederitz et al. 1989) where their quantity can be measured.

Retort distillation works in a similar way, except that no vacuum is used, and temperatures are usually 700–1,000°C. Due to the high temperatures, some oil will become solid and will deposit, resulting in loss of oil. Collected water will be a function of heating time. First the free water will be collected; thereafter it will take some time (20–30 minutes) before the clay lattice water is produced.

EXERCISE 3.1

During a Dean-Stark extraction 1.4 ml water is recorded. Determine the saturation of oil, water, and gas given the following data:

Porosity = 0.2
Bulk volume = 25.0 ml
Density of the oil = 0.88 g/ml
Density of water = 1.0 g/ml
Mass of saturated core = 57.0 g
Mass of dried core = 53.0 g

SOLUTION

$$Pore\,volume = porosity \cdot bulk\,volume = 0.2 \cdot 25 = 5.0\,ml$$

$$Water\,saturation = S_w = \frac{Water\,volume}{Pore\,volume} = \frac{1.4}{5} = 0.28 = 28\%$$

$$Oil\,saturation = S_o = \frac{mass\,of\,saturated\,core - mass\,of\,dry\,core - mass\,of\,water}{oil\,density\cdot pore\,volume}$$

$$S_o = \frac{57 - 53 - 1.4 \cdot 1}{0.88 \cdot 5} = 0.59 = 59\%$$

$$S_g = 1 - S_o - S_w = 1 - 0.28 - 0.59 = 0.13 = 13\%$$

3.6　SOME NOTES ON THE CORE SAMPLE USE

In a coring operation, the core undergoes changes when taken from in-situ, down-hole conditions to the well-site, such as alteration of rock wettability by contact with mud filtrate; release of overburden pressure, which may alter porosity and permeability; the flushing of the sample by mud filtrate, which can strip reservoir fluids; and the expulsion of natural fluids in the core by expansion of trapped gases as the core is brought to the surface. In addition, other changes may occur because of the coring and subsequent handling; deposition of heavy fractions of the oil, such as asphaltenes, decrease in temperature and oxidate upon contact with air. All these changes should be kept in mind in the subsequent handling and treatment of the cores in the laboratory but also afterwards when performing the experiments and interpreting the data.

- Note that cores are reservoir samples, but they are not in the reservoir any-more. This affects the results and adds uncertainty to the data obtained from all core experiments.
- Before coring/processing and cleaning:
  - A plan shall be made to obtain as representative core material as possible.
  - RCAL and SCAL program is to be worked out representative for the data needed.
- After working with core data:
  - Information shall be requested on the history of the cores and experi-mental methods to judge the value of the data.
  - Consider how representative the cores are.

This core material will be the basis for the core experiments described in the chapters to follow. The samples used will be further addressed as *cores*, instead of core plugs as they were named in this chapter, to make the distinction from the core drilled from the well.

3.7　EXPERIMENT 3.1: SATURATION DETERMINATION, DEAN-STARK DISTILLATION METHOD

Saturation determination, Dean-Stark distillation method: Experiment 3.1

　Objective: The experiment's objective is to determine the oil, water, and gas saturation of a core sample from a Dean-Stark cleaning process.

Data and calculations (see Table 3.4):

TABLE 3.4

Saturation Determination Calculation Table for a Sample with Porosity of 17.41%

Measured								Calculated			
m_{org} [g]	m_{dry} [g]	ρ_w [g/cm³]	ρ_o [g/cm³]	V_w [cm³]	V_p [cm³]	m_o [g]	V_o [cm³]	S_o	S_w	S_g
156	147	1.05	0.842	2.1	10.717	6.789	8.062	0.752	0.195	0.051

Where:

m_{org}: Mass of original saturated sample

m_{dry}: Mass of desaturated, dry sample

EXERCISE 3.2

1. Determine the saturation of oil, water, and gas given the data presented in Table 3.4 under "measured".
2. The core is water wet and was drilled with water-based mud. The reservoir pressure was above bubble point. Discuss the results.

SOLUTION

1. $\quad Water\,saturation = S_w = \dfrac{Water\,volume}{Pore\,volume} = \dfrac{2.1}{10.717} = 0.195 = 19.5\%$

$Oil\,saturation = S_o = \dfrac{mass\,of\,saturated\,core - mass\,of\,dry\,core - mass\,of\,water}{oil\,density\cdot pore\,volume}$

$$S_o = \frac{156 - 147 - 2.1\cdot1.05}{0.842\cdot10.717} = 0.752 = 75.2\%$$

$$S_g = 1 - S_o - S_w = 1 - 0.195 - 0.752 = 0.053 = 5.3\%$$

2. During the drilling, water-based mud could have entered the water wet core at initial water saturation, displacing part of the oil/gas. With the fluid pressure dropping below bubble point during the drilling and core retrieving, the gas coming out of the solution might partly have displayed the fluids in the core, either mobile oil and/or mobile drilling fluid. The results of this method therefore need to be considered with caution, as they likely underestimate the oil saturation and overestimate gas and water saturation.

REFERENCES

Amott, E. "Observations relating to the wettability of porous rock." *Transactions of the AIME 216*, 1959: 156–162.

Amyx, J.W., Bass Jr., D.M., Whiting, R.L. *Petroleum reservoir engineering.* McGraw-Hill, 1960.

Anderson, W.G. " Wettability literature survey—part 1: Rock/oil/brine interactions and the effects on core handling on wettability." *Journal of Petroleum Technology 38*, 1986: 1125–1144.

API, American Petroleum Institute. *Recommended practices for core analysis, RP40*. API Publishing Services, 1998.

Ashena, R., Thonhauser, G. *Coring methods and systems*. Springer, 2018.

Bourgoyne, A.T., Millheim, K.K., Chenevert, M.E., Young, F.S. *Applied drilling engineering*. SPE Textbook Series, Vol. 2, Richardson, Texas, USA, 1991.

Koederitz, L.F., Harvey, A.H., Honarpour, M. *Introduction to petroleum reservoir analysis; laboratory workbook*. Gulf Publ. Co., 1989.

McPhee, C., Reed, J., Zubizaretta, I. *Core analysis. A best practice guide in "Developments in petroleum science, Volume 64"*. Elsevier, 2015.

Wisenbaker, J.D. *Process for treating core samples*. United States Patent 2,617,296, Nov. 1952.

4 Fluid Properties

4.1 INTRODUCTION

In natural reservoirs the pore space can be filled with brine, gas, or oil. For storage or production purposes, the description of flow in the porous media of these phases is of interest. Forces acting on the fluids while flowing in the reservoirs are viscous forces, gravitational forces, and capillary forces. The evaluation of fluid flow in porous media will therefore require fluid properties like viscosity, density, and interfacial tension. The first two parameters will be discussed in this chapter.

Density differences between fluids can cause separation, like oil floating on water. A helium balloon moves up because helium is lighter than air or heated air is lighter than air at room temperature of 20°C, such that a hot air balloon can move up while heating the air inside the balloon. In a reservoir, gas being lighter than oil or brine and will form a gas cap, where denser CO_2-saturated brine will move down in time, creating convective flow. The density can additionally be used to convert mass in quantities of volume and vice versa. Depending on what is easiest to be measured, the other can be calculated.

Viscosity describes the "thickness" or resistance of the fluid flowing as describing the effect of viscous forces. Honey or syrup does not flow as easily as water, having a higher viscosity. Viscosity changes or differences can have a large effect on flow processes; this can be in industry processes, as in flow in pipes but also in nature, like landslides or quick clay and flow in natural reservoirs. Viscosity and viscosity ratio are important for multiphase flow. Viscosity has an effect on flow rates for a given pressure drop in the field, between the well bore pressure and the reservoir pressure. If viscosity differences are high, the easier flowing phase will bypass the higher viscous phase, called viscous fingering.

In this chapter we will discuss apparatus and procedures for measuring these parameters, mainly for liquids. During flow in the reservoir, the pressure and temperature may change, affecting the phase density and viscosity, which will be addressed. The measurement methods presented here will be mainly for liquids in ambient conditions. Gas properties will be discussed briefly at the end of this chapter. Gas viscosity and density measurements are more challenging, and common calculation methods for density and viscosity are discussed. The gas theory will be briefly presented and specifically the properties of hydrocarbon gas methane and the storage gasses like CO_2 and H_2 will be presented. However, complete coverage of pressure-volume–temperature relationships for gas/oil mixtures is outside the scope of this book.

4.2 FLUIDS IN NATURAL RESERVOIRS

4.2.1 Oil

Natural oils that can be found in the reservoirs are mixture of different hydrocarbons in the liquid phase with small amounts of non-organic compounds like sulfur,

 DOI: 10.1201/9781003382584-4

oxygen, nitrogen, CO_2, and nonorganic compounds like salts and metals. Density and viscosity differences are caused by:

- Physical parameters based on phase behavior: Temperature (T), pressure (P), and volume (V).
- Composition-molecular interaction depending on:

  - Different kinds and quantifications of oil components.
  - Gas/oil ratio.

Depending on the composition, pressure, and temperature, the fluids (oil/brine) can contain dissolved gas, which results in density and viscosity changes depending on the phase the mixture is in. This is described in phase behavior of the mixture. In Figure 4.1 the phase behavior for the single component CO_2 and a general sketch of a phase diagram for multi components are pictured showing the phase of the system dependent on pressure and temperature. The shape of the transition lines can vary and is dependent on the composition of the mixture.

Due to this phase behavior, it is important to decide on what kind of samples the density and viscosity measurements will be based on, as it depends on the composition and therefore much on the sampling conditions of pressure and temperature. Oil can be sampled downhole, but pressure there might be below bubble point where in the field it might be above. Also upstream in the separator samples can be taken, where gas has been separated completely. There are four kinds of samples that can be used for measurements of oil characteristics:

- Dead oil: no dissolved gas.
- Live oil: oil with dissolved gas as obtained from the field.
- Composite oil: dead oil enriched with gas from the separator.
- Model oil: 1–2 component systems chosen generally to match the viscosity of the live oil.

Depending on the purpose and sample material available, a choice needs to be made. In this book, for illustration purposes the commonly used model oil Exxsol D60 will be used.

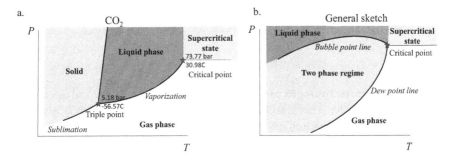

FIGURE 4.1 Phase behavior P-T correlation a) for a single component phase CO_2 and b) a general sketch for a multicomponent system.

4.2.2 BRINE

The reservoir contains brine if no hydrocarbons were trapped. Brine is water with dissolved gas and salts. The brine composition can be described by the salinity content of both the total dissolved solids (TDS) and different salt components and to a lesser extent the gas/oil ratio.

Brine can be sampled as a reservoir sample as produced. Here there is a risk of salt loss due to deposition with changing temperature and pollution due to drilling fluid invasion. Unnatural tracers might help to identify that fraction. Alternatively, once the composition is known, the brine can be composed, adding ions to pure water in the right ratio.

In case no reservoir-specific brine is to be used, model brine can be used. To avoid clay swelling, no pure water shall be used. Generally, a sodium chloride solution is used then, e.g. 3wt% NaCl solution, which will be used in the examples here as well.

4.2.3 GASSES

4.2.3.1 Natural Gas

Like oil, natural gases are mixtures of short hydrocarbon molecules, which are in the reservoir and surface conditions in the gas phase. Mainly they contain methane, but butane fractions can be present, with additional components of CO_2, N_2, H_2S, H_2, or noble gases. Heavier components can be available, which at low temperature and pressure can drop out as condensates.

4.2.3.2 Storage Gasses H_2, CO_2

Both hydrogen (H_2) and carbon dioxide (CO_2) are currently gases considered to be stored in natural subsurface reservoirs. In view of reducing presence and the emission of greenhouse gases in the atmosphere, capturing and storage of the main contributing gas CO_2 as combustion gas is under development. The storage of CO_2 is intended to be permanent, called CO_2 sequestration. To store CO_2, it needs to be captured first. This can be done when it is formed during combustion, using exhaust filtering facilities. This might be applicable for industrial processes but becomes more challenging for example during transport when using gasoline or diesel oil. Therefore, ideally an alternative energy source can be used, which does not form CO_2 when combusted or does not need combustion at all. For the latter the use of rechargeable batteries based on solar energy is an example, for the former, hydrogen gas is considered. When combusting hydrogen, only water is created. Therefore, hydrogen is considered as alternative fuel gas:

$$2H_{2(g)} + O_{2(g)}- > 2H_2O_{(l)}$$

The process to create hydrogen can be very clean using solar energy and electrolysis to form H_2 from water. This is still under development and alternatively it is formed involving a reaction of natural gas with water, which forms hydrogen and CO_2:

$$CH_{4(g)} + 2H_2O_{(l)} \rightarrow 4H_{2(g)} + CO_{2(g)}$$

In this process CO_2 can be captured and stored. Independent of how it is created, hydrogen is considered attractive as alternative fuel. As demand and supply might not always be predictable, temporary storage is needed. So, after injection, hydrogen shall ultimately be back produced for consumption, comparable to seasonal natural gas storage.

Therefore, for the natural reservoirs the main focus lies on hydrocarbon gas, CO_2, and H_2. CO_2 and H_2 can dissolve in brine, which needs to be considered for the experimental procedure. HSE measures for H_2 as flammable gas are important to consider as well as the corrosive effect of the CO_2, which reduces the pH when dissolved in the brine. $CO_{2(g)} + 2\,H_2O_{(l)} \leftrightarrow HCO_3^-{}_{(aq)} + 3\,H^+{}_{(aq)}$ or $CO_{2(g)} + 2\,H_2O_{(l)} \leftrightarrow CO_3^{-2}{}_{(aq)}, + 4\,H^+{}_{(aq)}$.

4.3 LIQUID DENSITY

4.3.1 DEFINITIONS

Density (ρ) is defined as the mass of the fluid, m, per unit volume V.

$$\rho = \frac{m}{V} \tag{4.1}$$

Officially the density is measured in a vacuum, but often the weighting in air is used, where buoyancy of air reduces the measured weight. Then the apparent density is measured but this is often taken as the density. Here the densities considered are based on weight measurements in air.

In general, density varies with pressure and temperature and fluid composition. The dimension of density is kg/m^3 in SI or lb/ft^3 in the English system.

Specific gravity (γ) is defined as the ratio of the weight of a volume of liquid to the weight of an equal volume of water at the same temperature.

$$\gamma_T = \frac{\rho_f}{\rho_w} \tag{4.2}$$

The specific gravity of liquid in the oil industry is often measured by a hydrometer that has its special scale. The American Petroleum Institute (API 1998) has adopted a hydrometer for oil lighter than water for which the scale, referred to as the API scale or API gravity, is:

$$^\circ API = \frac{141.5}{\gamma} - 131.5 \tag{4.3}$$

Note: When reporting the density, the units of mass and volume used at the measured temperature must be explicitly stated, e.g., grams per cubic centimeter (g/cm^3) at T($^\circ$C). The standard reference temperature for international trade in petroleum and its products is 15°C (60°F), but other reference temperatures may be used for other special purposes. Figure 4.2 shows the density and API gravity for different oil compositions.

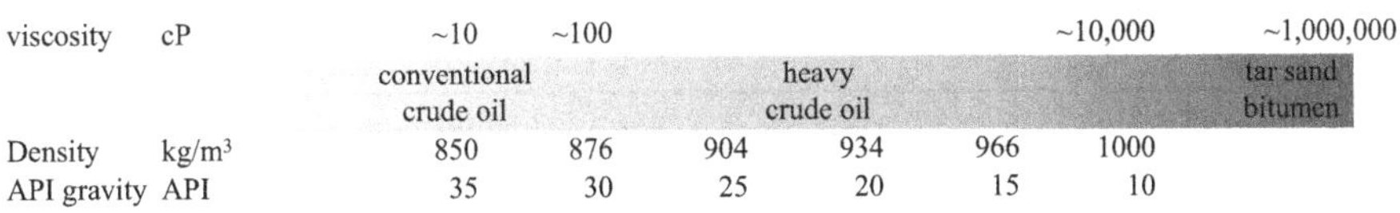

viscosity	cP	~10	~100			~10,000		~1,000,000
		conventional crude oil			heavy crude oil			tar sand bitumen
Density	kg/m³	850	876	904	934	966	1000	
API gravity	API	35	30	25	20	15	10	

FIGURE 4.2 Viscosity–density relationship. Modified from Speight (2001).

For an emulsion or solid mixture, the density can be determined using:

$$\text{Average density} \left(\rho_{avg} \right) = \frac{\sum_{i=1}^{n} V_i \rho_i}{\sum_{i=1}^{n} V_i}$$

with n as the number of components. It is required that the volumes of the different components are known, V_i, as well as their densities.

4.3.2 OIL DENSITY

Figure 4.3 shows how the density of phases in a hydrocarbon reservoir can change depending on pressure. Note that here the composition changes with pressure. Below the bubble point (P_b) the oil phase would contain lighter gas components with an increasing pressure, generally reducing its density, till P_b and maximum dissolution are reached. With pressure increase above the bubble point the fluid with constant composition will compress and increase density slightly.

The liquid compressibility describes this. Temperature dependencies generally show a decrease in density with an increase in temperature. This dependency is described in PVT studies and is further out of the scope of this book.

4.3.3 BRINE DENSITY

Brine is water with dissolved gas and salts. Density differences depend on physical parameters based on phase behavior like for oil, temperature, and pressure, where density decreases with increasing temperature and reducing pressure. Beside the

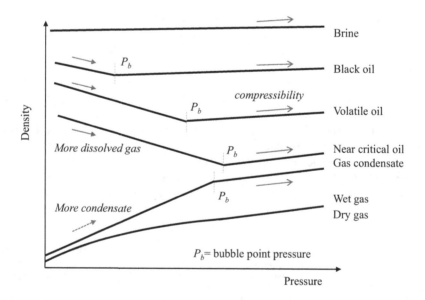

FIGURE 4.3 Density trends of different oil kinds and gas types as a function of pressure.

conditions, the composition-molecular interaction heavily affects the density of the brine, based on the salinity of both the total dissolved solids (TDS) and different salt components and to a lesser extent the gas/oil ratio.

4.4 MEASUREMENT OF LIQUID DENSITY

The most-used methods for determining density or specific gravity of a liquid are the Westphal balance and API hydrometer, based on the Archimedes principle and a pycnometer or bicapillary pycnometer, based on measuring the weight of a known volume. In-situ density measurements can be performed using an oscillating U-tube or rod.

4.4.1 Archimedes Principle

The Westphal balance and specific gravity balance are based on the principle of Archimedes: a body immersed in a liquid is experiencing a buoyancy force up, equal to the weight of the liquid it displaces. A weight with a known volume V is weighed in air and within the liquid for which the density shall be determined for.

$$\text{Weight in air: } F_1 = m \cdot g$$
$$\text{Weight in the fluid: } F_2 = m \cdot g - m_{\text{displaced liquid}} \cdot g$$

The difference in weight relates to buoyancy on a known volume V of the displaced fluid:

$$\Delta F = m_{\text{displaced liquid}} \cdot g = V \cdot \rho_{\text{fluid}} \cdot g \qquad \rightarrow \qquad \rho_{\text{fluid}} = \frac{Dm}{V}$$

with g as the gravitational acceleration. Note that here the buoyancy force in air is assumed to be negligible.

The API hydrometer is usually used for determining oil gravity in the oil field. When a hydrometer is placed in oil, it will float with its axis vertical after it has displaced a mass of oil equal to the mass of hydrometer (Figure 4.4a). The higher the density, the more buoyancy force so the less deep the hydrometer will float in

---

TABLE 4.1
Common Methods for Density Measurements of Liquids and Gasses

Principle	Methods	Phase	References
Archimedes principle	Westphal balance, specific gravity balance API hydrometer	Liquid	
	Magnetic suspension balance	Gas	(Yang et al. 2023) (Kuramoto et al. 2004)
Volume and weight	Pycnometer Bicapillary pycnometer	Liquid	
Resonance frequency	Oscillating U-tube	Gas/Liquid	

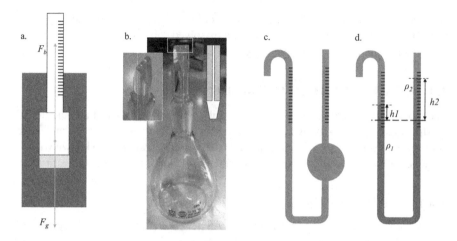

FIGURE 4.4 Schematic illustration of hydrometer (a), pycnometer (b), and bicapillary pyc-nometers (c, d).

the liquid. The scale allows us to read the density directly. Note that this scale is calibrated for a specific temperature and volume of the hydrometer. The hydrometer can be used at atmospheric pressure or at any other pressure in a pressure cylinder. ASTM standards are available for standardized procedures on the use of the hydrometer for different products (ASTM n.d.).

4.4.2 VOLUME AND WEIGHT

The pycnometer (Figure 4.4b) is a flask, which can accurately be filled with a known volume of liquid. The cap has a capillary that is used to reduce the filling error to a minimum. A variation to the pycnometer's geometry is the possibility to add a thermometer that can measure the fluid temperature in situ.

The liquid specific gravity is defined as the ratio of the weight of a volume of the liquid to the weight of an equal volume of water at the same temperature. Both weights should be corrected for buoyancy (due to air) if a high degree of accuracy is required. The ratio of the differences between the weights of the pycnometer filled with liquid and empty weight, to the weight of the flask filled with distilled water and empty weight, gives the specific gravity of the unknown fluid. The water and the liquid must both be at the same temperature.

Procedure for pycnometer:

1. Weigh the clean, dry pycnometer with the glass top.
2. Record the volume of the pycnometer.
3. Fill the pycnometer with liquid using a funnel, in the fume hood.
4. Gently put the glass top on the pycnometer. The liquid and the air will come out of the tube via the glass top.
5. Dry the exterior surface of the filled pycnometer.

6. Weigh the pycnometer filled with the fluid.
7. Clean the pycnometer with the appropriate solvents in the fume hood.
8. Leave the pycnometer to dry. The pycnometer must be completely dry before being used again.

Error sources can be that the pycnometer volume is not the same as the calibration volume any longer due to e.g. damage to the cap of the pycnometer. Air bubbles in the liquid shall be removed. The volume measured can be incorrect if excess liquid is present on the outside of the device. The balance should be calibrated and tared correctly.

The pycnometer can also be used to measure the density of a solid or the grain density. The sample needs to be crushed for this purpose. The weight of the pycnometer must be measured as dry empty (A) and filled with dry crushed solid (B). With the dry crushed solid in the pycnometer, the pycnometer shall be filled up till the calibrated volume with a fluid where the density is known and weighed (C). Air needs to be removed by vacuum or heat. See for further details the standard ASTM D854 (ASTM n.d.). The difference in mass between A and B will give the mass of the solid. The difference in mass between B and C is the mass of the liquid added, which can be converted to volume of liquid added knowing its density. By knowing the total volume of the pycnometer, the volume of the solid can then be calculated. Alternatively, when the samples are too precious to be crushed, a helium porosimeter can be used to determine the grain volume of a consolidated medium, which is discussed in Chapter 5, Section 5.2.2. By knowing the dry mass, the average grain density can be determined.

The bicapillary pycnometer (Figure 4.4c) is a variation on the normal pycnometer. The density of the liquid sample drawn into the pycnometer is determined from its volume and weight. The volume of fluid to use is flexible so can this be used to measure temperature dependent densities, where fluid shrinkage or expansion is measurable for a constant fluid mass.

Alternatively, filling the pycnometer with a fluid with a known density, followed by a lighter fluid in one of the capillaries, the height difference can be correlated to the difference in density pycnometer (Figure 4.4d) as $\rho_2 = \frac{\rho_1 \cdot h_1}{h_2}$, with h measured from the lower interface of liquid 2.

4.4.3 Forced Oscillation Method

The oscillating method, resonance, or U-tube density meter is based on the measurement of the resonance frequency of a U-shaped tube, which is dependent on the content of the tube (Enoksson et al. 1995; Kratky et al. 1969; Langdon 1985).

The U-tube is excited at different frequencies, and the corresponding resonance frequency and amplitude of oscillation of the U-tube is used to determine the density of the U-tube. The higher the density, the more inert the U-tube becomes and the lower the resonance frequency. The correlation between density ρ and resonance frequency f is:

$$\rho = A \frac{1}{f^2} + B \tag{4.4}$$

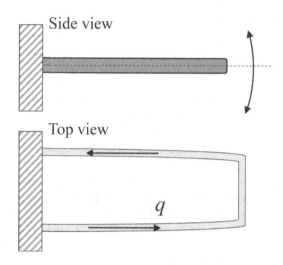

FIGURE 4.5   Resonant U tube.

with A and B constants to be found to calibrate the system using fluids with a known density (Stemme et al. 1983; Paar 2024). More information can be found in the ISO standard (ISO-a 15212-1:1998).

This method can be applied in-line to log the density of an effluent over time. Varieties of this principle have been developed e.g. where an object like a cylinder or rod is exited to its own natural frequency in a flow line or chamber filled with the sample of interest. The resonance frequency will be dependent on the density of the surrounding gas (Suzuki 2000).

4.4.4   EXPERIMENT 4.1: FLUID DENSITY

Fluid density using the pycnometer method:

> Determine the density of distilled water, 3%wt NaCl solution, Exxsol D60, and Zalo (a commercial detergent) solution with the pycnometer and compare the results.

The personal protection equipment needed for this experiment are lab coat, gloves, and safety glasses. Check the MSDS for the liquids to determine whether additional measures are needed when handling the fluids used.

Materials needed:

> Distilled water, 3%wt NaCl solution, Exxsol D60, Zalo solution.

Equipment needed:

> Balance, funnel and pycnometer, paper to remove spills.

Calculation and report:

1. Calculate the liquid density (ρ) and the average density based on your data by dividing mass by volume.
2. Calculate the absolute error for each measurement.
3. Calculate the specific gravity.
4. Perform an error source analysis of the pycnometer method.

$$m_L = m_{full\ pyc} - m_{dry\ pyc}$$
$$\rho_L = m_L / V_{pyc}$$

where m_L = weight of liquid, $m_{full\ pyc}$ = weight of full pycnometer, $m_{dry\ pyc}$ = weight of dry pycnometer, ρ_L = density of liquid, V_{pyc} = volume of pycnometer

See Table 4.2 and Table 4.3 for data and example calculation.

TABLE 4.2

Density of Water at Different Temperatures at Atmospheric Pressure

Temperature [°C]	Density [kg/m³]	Temperature [°C]	Density [kg/m³]	Temperature [°C]	Density [kg/m³]
18.0	998.5934	18.5	998.4995	19.0	998.4030
19.5	998.3070	20.0	998.2019	20.5	998.0973
21.0	997.9902	21.5	997.8805	22.0	997.7683
22.5	997.6536	23.0	997.5363	24.0	997.2944

TABLE 4.3

Fluid Density Using the Pycnometer Method Calculation Table

Fluid	Pycnometer mass [g]	Pycnometer + liquid [g]	Pycnometer volume [cm³]	Density, ρ [g/cm³]	Specific gravity, γ
Brine at 20°C	31.375	82.810	50.528	1.0180	1.0200
	31.321	82.802		1.0189	1.0209
	31.367	82.790		1.0177	1.0198
Average*				1.018	1.020
Absolute error. E_a [g/cm³] **				0.00046	
Std [g/cm³]				0.0009	
ρ_{avr} = 1.0182 + −0.0005 [g/cm³] γ_{avr} = 1.0120 + −0.0005 [g/cm³]					

*Average Density $\left(\rho_{avg}\right) = \dfrac{1}{n}\sum_{i=1}^{n}\rho_i$, with n as the number of experiments performed.

** Absolute error is E_a = | (Average Density) − (Measured Density) |

4.5 LIQUID VISCOSITY

Viscosity is defined as the internal resistance of fluid to flow. The basic equation of deformation is given by:

$$\tau = \mu\dot{\gamma} \tag{4.5}$$

where τ is shear stress, $\dot{\gamma}$ is the shear rate defined as $\partial v_x/\partial_y$, and μ is the viscosity. The term τ can be defined as F/A where F is force required to keep the upper plate moving at constant velocity v in the x-direction and A is area of the plate in contact with the fluid. By fluid viscosity, the force is transmitted through the fluid to the lower plate in such a way that the x-component of the fluid velocity linearly depends on the distance from the lower plate (Figure 4.6).

It is assumed that the fluid does not slip at the plate surface. Newtonian fluids, such as water and gases, have shear-independent viscosity and the shear stress is proportional to the shear rate.

Many systems like mixtures containing unsymmetrical particles will not have linear correlation between the shear stress τ shear rate $\partial v_x/\partial_y$. In these cases, the fluids are denoted as "non-Newtonian", and we may have many types as seen in Figure 4.7.

Bingham fluids have a yield stress that needs to be overcome to start flowing, like toothpaste. For shear thinning fluid or pseudoplastics like ketchup or shear thickening (or dilatant) fluids like corn starch solution (Oobleck), the viscosity decreases and respectively increases with increasing shear. The internal alignment of the molecules changes, which leads to the changing viscosity with increasing shear stress e.g. with alignment of polymer chains in a fluid, viscosity can decrease whereas when particles in a fluid jam, viscosity can increase. Combinations of these behaviors are also possible. A separate class of materials

Consider 2 plates, one stationary, one moving

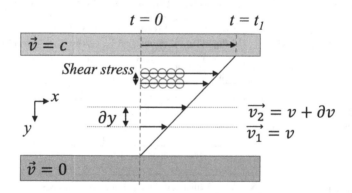

FIGURE 4.6 Steady-state velocity profile of a fluid entrained between two flat surfaces.

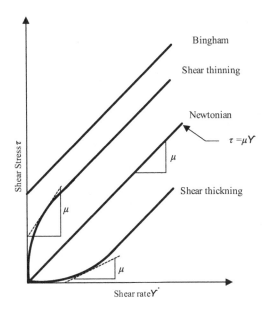

FIGURE 4.7   Shear stress vs. shear rate for a Newtonian and non-Newtonian fluid.

has a time-dependent viscosity, or viscoelastic materials show an elastic behavior, after shearing stops. For fluid design, e.g. paints or optimizing of industrial processes involving liquids, this is a large topic of research called rheology. Drilling fluids or EOR chemicals can show non-Newtonian behavior, but the fluid found in the fields shows Newtonian behavior. Here we will focus on the measurement and understanding of Newtonian fluids only.

Dynamic viscosity (μ) is the ratio between the applied shear stress and the rate of shear and is called coefficient of dynamic viscosity μ. This coefficient is thus a measure of the resistance to flow of the liquid; it is commonly called the viscosity of the liquid. Kinematic viscosity (υ) is the ratio μ/ρ where ρ is fluid density. Table 4.4 gives units and dimensions for kinematic and dynamic viscosity.

TABLE 4.4

Units and Dimensions for Kinematic and Dynamic Viscosity

	Symbol	cgs Unit	SI Unit	Dimension
Kinematic viscosity, υ	1 mm^2/s = 1 cSt	cm^2/s	m^2/s	L^2/T
	1 m^2/s = 10^6 cSt			
Dynamic viscosity, μ	1 Dyne·s/cm^2 = 100 cP	Dyne·s/cm^2	Newton·s/m^2	(M/L)T
	1 Newton·s/m^2 = 10^3 cP		(= Pa.s)	(FT/L^2)

where cSt = centistokes, cp = centipoise
$1cP = 10^{-3}\ Pa.s$, $1cSt = 10^{-6}\ [m^2/s]$

4.5.1 EFFECT OF PRESSURE AND TEMPERATURE ON VISCOSITY

Viscosity of fluids varies with pressure and temperature. For most liquids the viscosity is rather sensitive to changes in temperature but relatively insensitive to pressure until rather high pressures have been attained. For oils with a changing composition with pressure, the viscosity changes based on the pressure accordingly; see Figure 4.8. Before bubble point (P_b) oil becomes increasingly saturated with lighter gas components, reducing the viscosity till maximum saturation at P_b, while for gas condensate saturation can increase, increasing the viscosity till P_b. After P_b, viscosity increases slightly due to compressibility of the saturated phase.

The viscosity of pure liquids usually rises with pressure at constant temperature. Water is an exception to this rule; its viscosity decreases with increasing pressure at constant temperature. For most cases of practical interest, however, the effect of pressure on the viscosity of pure liquids can be ignored.

Temperature has different effects on viscosity of liquids and gases. An increase in temperature causes the viscosity of a liquid to reduce, increasing the distance between molecules, reducing density, whereas for gases it will increase due to an increase in Brownian motion and therefore molecular collisions. The effect of molecular weight on the viscosity of liquids shows generally that the liquid viscosity increases with increasing molecular weight.

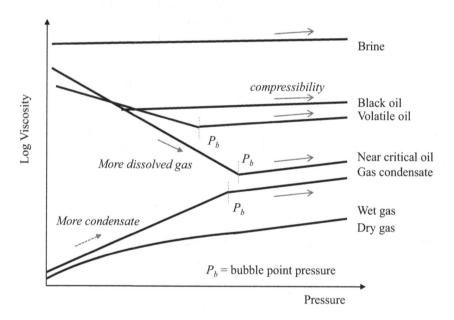

FIGURE 4.8   The change of viscosity of brine and different oils, gas types with increasing pressure.

4.5.2 Viscosity Modification

Injection of fluids is used to improve oil production, displacing the fluid in place with the injection fluid. Similar for storage of CO_2 or H_2, a good displacement of the liquid in place creates more storage volume. In both cases two or more phases are flowing. For optimization of two-phase flow, viscosity of the injection fluid is preferably similar to or slightly larger than the fluids in place. This is expressed in the viscosity ratio or mobility ratio (see Chapter 11, 11.2.3, Eq. 11.35). If the viscosity ratio $\dfrac{\mu_{inj}}{\mu_{in\,place}}$ is smaller than 1, viscous instabilities will arise, where the injection fluid has the tendency to find the quickest path though the in-place fluid, bypassing it and having a bad displacement efficiency. Gas injection in a liquid-filled reservoir will always have a viscosity ratio below one. To improve these, different solutions are possible. Either the viscosity of the injection phase is increased e.g. by polymer or foam, or the viscosity of the fluid in place is reduced, using the dependency of viscosity on temperature or composition. Table 4.5 shows the main methods and the underlying principle.

4.6 MEASUREMENTS OF VISCOSITY

There are several methods available to measure viscosity of fluids. They are based on creating a shear in the fluid, creating a velocity gradient moving fluid or moving boundaries. In Figure 4.9 the three common fundamental types of viscosity measurements are shown.

The first type (a) is based on describing fluid flow through a tube applying Hagen-Poiseuille's law, where the flow speed and respective pressure difference in relation to the tube geometry can be used to derive the viscosity. Instead of letting the fluid flow, the motion of a known object through the fluid in a fixed tube can be measured to derive fluid viscosity (b). Third, shear between two plates can be simulated by an inner

TABLE 4.5
Methods to Reduce the Viscosity of Injection Fluid or Fluids in Place

Parameter Modified	Method	Principle
Brine viscosity (injection)	Polymer	Increase internal friction by the presence of polymer chains (Sheng 2010)
Gas viscosity (injection)	Foam	Gas phase is made discontinuous by foam films resisting gas flow (Sheng 2010)
Oil viscosity (in situ)	Steam injection	In-situ temperature increase (Lake et al. 2014)
	Hot water injection	In-situ temperature increase
	CO_2 and hydrocarbon flooding	Gas dissolution in the oil reducing the oil viscosity (Jarrell et al. 2002)

a. Capillary type viscometers
Flow through pipe

b. Falling ball viscometers
Flow around object

c. Rotational viscometers
Rotational flow

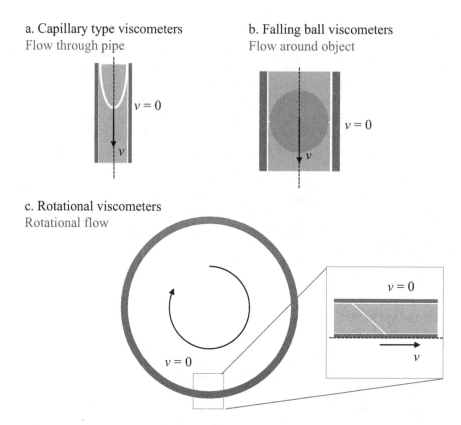

FIGURE 4.9   Common type of viscometers for bulk liquid viscosity measurements.

and outer cylinder, moving relative to each other, at a fixed distance; the rotational viscometers (c). Inline vibrational rod type viscometers are used (Dias et al. 2014).

4.6.1   CAPILLARY TYPE VISCOMETER

Viscosity of liquids is determined by instruments called viscosimeters or viscometers. One type of viscometer for liquids is the Ostwald viscometer or viscosimeter (Figure 4.10).

In this method the Poiseuille's law for a capillary tube with a laminar flow regime is used:

$$Q = \frac{V}{t} = \frac{\Delta P \pi r^4}{8 \mu l} \qquad (4.6)$$

where t is time required for a given volume of liquid V (k1 and k2 in Figure 4.10) with density of ρ and viscosity of μ to flow through the capillary tube of length l and radius r by means of pressure gradient ΔP. See Chapter 7 for the derivation. Applying this law, it is assumed the flow is laminar, steady state with no acceleration, the fluid

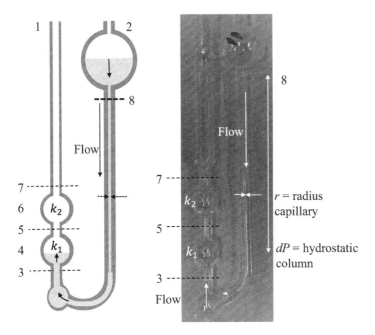

FIGURE 4.10   One type of Ostwald viscometer, measuring downward flow through a small capillary.

is Newtonian, and temperature (so density) is constant. The driving force ΔP at this instrument is $\rho\, g\, l$. Then:

$$\mu = \frac{\pi r^4 \rho g t}{8V} = K \cdot \rho \cdot t, \text{ with } K = \frac{\pi r^4 g}{8V} \qquad (4.7)$$

The capillary constant K is determined from a liquid with known viscosity and is provided for each device. As $\nu = \dfrac{\mu}{\rho}$, the kinematic viscosity can be written as the product of the measured flow time and the calibration constant of the viscometer.

$$\nu = K \cdot t \qquad (4.8)$$

Equation 4.8 needs a correction for extra pressure loss, the so-called Hagenbach correction. Several physical effects are included in this correction as wall friction, parabolic velocity profile and unsteady-state flow (fluid acceleration) (Wilke et al. 2000), which are not taken in account in the Hagen-Poiseuille law. Equation 4.8 becomes then:

$$\nu = K\,(t - \vartheta) \qquad (4.9)$$
$$\mu = \rho_{avg}\,\nu \qquad (4.10)$$

where
  ν = kinematic viscosity [cSt (centi Stoke)]
  K = calibration constant [cSt/s]
  t = flow time [s]
  ϑ = Hagenbach correction factor (see Table 4.6).
  μ = dynamic viscosity [cp (centipoise)]
  ρ_{avg} = average density [g/cm^3]

The manufacturer provides equipment specific time corrections, depending on the capillary tube and time; see Table 4.6 for Type Cannon-Fenske (ASTM n.d.), (ISO-b 3104:2023). The time correction reduces with longer measurement times, and with larger capillaries. The larger capillaries are used for higher viscosities (ISO-b 3104:2023).

TABLE 4.6
Hagenbach Correction Factor for Capillary Viscometer (Type Cannon-Fenske); the Grey Marked Time Corrections, where $t > 1.8s$, Indicates the Measurement Is Inaccurate and Another Capillary Shall Be Used

Capillary Number	25	50	75	100	150
Inner Diameter of Capillary [mm]	0.31	0.42	0.54	0.64	0.78
Flow time to fill given volume [s]	ϑ[s]				
50	4.61	2.11	0.66	0.24	0.06
60	3.20	1.46	0.46	0.16	0.04
70	2.35	1.07	0.34	0.12	0.03
80	1.80	0.82	0.26	0.10	0.02
90	1.42	0.65	0.20	0.07	-
100	1.15	0.53	0.16	0.06	
110	0.95	0.43	0.14	0.05	
120	0.08	0.37	0.11	0.04	
130	0.68	0.31	0.10	0.03	
140	0.59	0.27	0.08	0.03	
150	0.51	0.23	0.07	0.02	
160	0.45	0.21	0.06	-	
170	0.40	0.18	0.06		
180	0.36	0.16	0.05		
190	0.32	0.15	0.04		
200	0.29	0.13	0.04		
220	0.24	0.11	-		
240	0.20	0.09			
260	0.17	0.08			
280	0.15	0.07			
300	0.13	0.06			
350	0.09	0.04			
400	0.07	0.03			
>400	-	-			

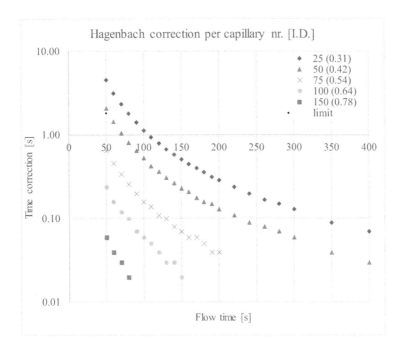

FIGURE 4.11   The Hagenbach correction plotted for time, for different capillaries. The data from Table 4.6 plotted.

PROCEDURE

The procedure described here is based on ASTM D445-23, Standard Test Method for Kinematic Viscosity of Transparent and Opaque Liquids (and Calculation of Dynamic Viscosity) (ASTM n.d.).

The numbers correspond to labels given in Figure 4.10.

1. Select a clean, dry calibrated viscometer having a range covering the estimated viscosity (i.e., a wide capillary for a very viscous liquid and a narrower capillary for a less viscous liquid). The flow time should not be less than 200 seconds.
2. Perform the experiment in the fume hood in case of use of harmful chemicals; check the MSDS.
3. Charge the viscometer: to fill, turn the viscometer upside down. Dip the tube, number (2) into the liquid to be measured while applying suction to tube (1) using a Peleus ball until liquid reaches mark (8). After inverting to the normal measuring position, close tube (1) before liquid reaches mark (3). Place the viscometer perfectly vertical in a lab stative.
4. Temperature equilibration: Allow the charged viscometer to remain long enough to reach the room temperature. Read the calibration constants directly from the individual viscometers' information.
5. Measuring operation: Open tube (1) and measure the time it takes the liquid to rise from mark (3) to mark (5). Measuring the time for rising from

mark (5) to mark (7) allows viscosity measurement to be repeated to check the first measurement. If the two measurements of viscosity agree within required error (generally 0.2–0.35%), use the average for calculating the reported kinematic viscosity.
6. After use: clean the viscometer in the fume hood with acetone and distilled water. Dry it in the oven at 60°C. Only when clean and dry can the next measurement be performed.

4.6.2 FALLING (ROLLING) BALL VISCOMETER

Another instrument commonly used for determining viscosity of a fluid is the falling (or rolling) ball viscometer (Figure 4.12), which is based on Stoke's law for a sphere falling or rolling in a fluid under effect of gravity (Yang and Lin 2008). The instrument is most suitable for experimental viscosity measurement of single-phase reservoir fluids and has the advantage of the capability of repeating the experiment very easily. The viscometer consists of a measuring barrel inside a heavy wall, stainless steel pressure housing in which a polished ball is placed. A polished steel ball is dropped into a glass tube of a slightly larger diameter containing the liquid, and the time required for the ball to fall at constant velocity through a specified distance between reference marks is recorded by an electronic timer. In fact, the ball inside the barrel starts its travel by breaking off the electrical contact and the contact made when the ball reaches the end of its travel stops the timer. The equipment has the advantage that human error to record the timing is eliminated. The equipment is able to measure viscosity of liquids in high-pressure, high-temperature conditions similar to petroleum reservoirs.

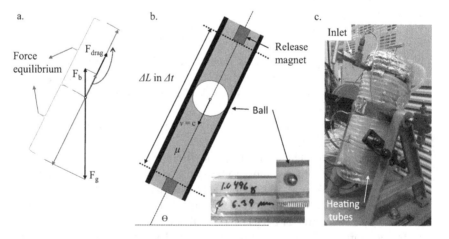

FIGURE 4.12 Schematic diagram of the falling ball viscometer with a ball of 6.39 mm (courtesy of SINTEF Reservoir Laboratory).

In the rolling-ball viscometer, viscosity calculation is based on the time it takes a metal ball to roll from one end to the other and mathematically can be expressed as:

$$\mu = \frac{2\left(\rho_B - \rho_F\right)\sin\theta\ g\ R^2}{9\ v_\infty} \tag{4.11}$$

where, μ = dynamic viscosity, ρ_B = density of the ball, ρ_F = density of the fluid, g = acceleration due to gravity, R = diameter of the ball, θ = the tilt angle of the measuring barrel, and v_∞ = velocity normal to the earth in the downward direction, defined as distance over time (d/t)

Considering that g, R, and θ remain constant, the previous equation can be expressed as follows

$$\mu = K\,t\,(\rho_B - \rho_F) \tag{4.12}$$

where K is a constant unique to the set of tests and density of the ball can be easily calculated from mass and volume. The equipment must be calibrated using a fluid with known viscosities and densities, e.g., hexane to measure ball constant K. The ball constant K is not dimensionless but involves the mechanical equivalent of heat. From the previous equation for a constant ball and fluid density, the viscosity is directly proportional to the ball roll time (t).

The rolling ball viscometer will give good results if the fluid flow in the tube remains in the laminar range, fluid is Newtonian, and density is constant, according to the assumptions of the Hagen-Poiseuille law. In some instruments of this type, both pressure and temperature may be controlled. The set-up can be used under pressure and temperature, as well for gasses, where the set-up is barely tilted to reduce the gravitational acceleration.

Force balance rolling ball

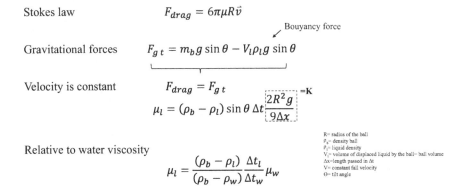

Stokes law $F_{drag} = 6\pi\mu R\vec{v}$

                                                          Bouyancy force

Gravitational forces $F_{g\,t} = m_b g \sin\theta - V_l \rho_l g \sin\theta$

Velocity is constant $F_{drag} = F_{g\,t}$

                          $\mu_l = (\rho_b - \rho_l)\sin\theta\,\Delta t\dfrac{2R^2 g}{9\Delta x}$ =K

R= radius of the ball
ρ_b= density ball
ρ_l= liquid density
V_l= volume of displaced liquid by the ball= ball volume
Δx=length passed in Δt
V= constant fall velocity
Θ= tilt angle

Relative to water viscosity

                          $\mu_l = \dfrac{(\rho_b - \rho_l)}{(\rho_b - \rho_w)}\dfrac{\Delta t_l}{\Delta t_w}\mu_w$

FIGURE 4.13   The force balance of a rolling ball experiment.

Procedure rolling ball experiment:

1. Clean the barrel and ball with toluene, methanol, and hexane prior to each experiment.
2. Evacuate the equipment by opening the inlet valve and connect it to the vacuum pump.
3. After reaching a complete vacuum, the vacuum valve should be closed.
4. Charge the sample to the viscometer and rock the tube to obtain a homogeneous sample.
5. Set the clock to zero.
6. Release the ball by toggling the drop button. The timer will start and record travel time for the ball to reach the end of the barrel.
7. Clean it as described in point 1.

4.6.3 Viscosity Measurement by Capillary Tube (Coil)

The principle of viscosity measurement by coil or capillary tube is also based on the Hagen-Poiseuille equation applied on a long narrow capillary tube. In this method, viscosity is measured by determining the pressure drop across the capillary tube while a fluid is flowing with a certain rate in the form of incompressible, laminar, steady-state flow. Figure 4.14 shows the schematics of an experimental set-up and the coil to the left used as long narrow tube. This can be applied to liquid as well as gasses under temperature and pressure. In a capillary tube with a length of L and internal diameter of D, pressure drop of ΔP and flow rate are

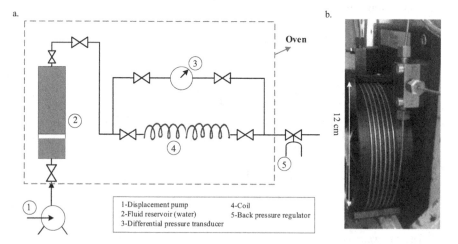

a.

b.

Oven

12 cm

1-Displacement pump   4-Coil
2-Fluid reservoir (water)   5-Back pressure regulator
3-Differential pressure transducer

FIGURE 4.14 Schematic of viscometer by capillary tube in a flow rig set up (courtesy of SINTEF Reservoir Laboratory).

related through the Hagen-Poiseuille equation as shown in Equation 4.6. Viscosity can then be expressed as

$$\mu = \frac{1}{K}\frac{\Delta p}{Q} \tag{4.13}$$

where constant K (capillary tube constant) can be defined as:

$$K = \frac{128L}{\pi D^4} \tag{4.14}$$

However, flow in the coil normally deviates from the ideal flow described by the Hagen-Poiseuille equation. One aspect is that, if the tube is not straight, then the flow path is not equal for each point parallel to the flow direction.

Therefore, it is recommended to obtain a capillary tube constant through a series of tests by measuring ΔP versus different flow rate using a fluid with a known viscosity (as a reference fluid).

Equation 4.14 clearly shows that capillary tube constant K can be determined from measurement of ΔP for several flow rates for a reference fluid with a known viscosity; plot it, obtain the slope, and divide by viscosity of the reference fluid.

For non-Newtonian fluid flowing through porous media, changing flow rates in the pores can lead to different in-situ viscosities than measured in the bulk. Apparent viscosities can be derived from core measurements. Polymer viscosity measurements in cores will be discussed in Chapter 7, Section 7.6.

EXERCISE 4.1

Experimental data for calibration of a capillary tube is given in Table 4.7: The experiment carried out using hexane as a fluid with known viscosity at the condition of 100 bar and 22°C. on a capillary tube with internal diameter of 0.25 mm, length 2,150 mm, and coil diameter of 100 mm. Calculate capillary tube constant. ($\mu_{Hexane} = 0.397\,mPa\,s$).

TABLE   4.7

Pressure Drop Versus Rate for Viscosity
Measurement by Hexane at 100 Bar and 22°C

Flow Rate Q [ml/min]	Pressure Drop ΔP [mbar]
0.098	61.24
0.204	125.60
0.311	192.23
0.417	256.83
0.524	324.49

SOLUTION

A graph of Q versus ΔP for the experiment is shown in Figure 4.15 in which the slope of the curve or $\dfrac{\Delta P}{q}$ is equal to 617.93 (mbar)/(ml/min) = 617. 93 · 60s/min·10^{11} = 3707. 58 · 10^{12} mPa s/m³

Then capillary-tube constant (K) can be calculated from Equation 4.10 as follows:

$$K = \frac{1}{\mu}\frac{\Delta p}{Q} \rightarrow K = (1/0.351) \cdot 3707.58 \cdot 10^{12} = 10562.90 \cdot 10^{12}\ m^{-3}$$

After determining capillary tube constant (K) with a reference fluid, one can determine viscosity of another fluid from similar measurement of ΔP for several flow rates and use Equation 4.13 to calculate viscosity of that fluid.

PROCEDURE

Schematic of the apparatus for viscosity measurement using capillary tube is shown in Figure 4.14 which consists of:

- Capillary tube with a known length, internal diameter, and coil diameter.
- A pump for injection.
- One-piston cylinder for delivery of the measurement fluid.
- Digital differential pressure transmitter to measure pressure drop across the capillary tube.

Prior to each experiment, the fluid should be prepared inside a cylinder with a movable piston. This kind of piston-cylinder can increase or decrease the volume of the process side due to the movement of the piston inside the cylinder.

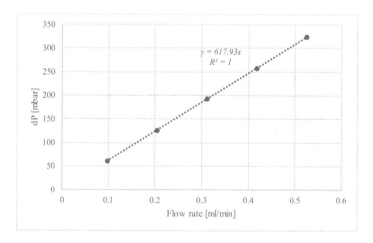

FIGURE 4.15 Capillary-tube constant determination using hexane for a capillary tube with internal diameter of 0.25 mm, length 2,150 mm, and coil diameter of 100 mm.

EXPERIMENTAL PROCEDURE

1. The piston-cylinder, capillary tube, and lines should be cleaned with proper solvents such as hexane, toluene, and methanol to remove any residual fluid from previous experiment.
2. The pressure differential meter (ΔP) should be filled with liquid that is compatible with the device. Typically, water is used.
3. The system should be vacuumed to remove any trapped air inside flowlines, fittings, and connections. Wait for 1 hour to reach the proper vacuum along the capillary tube and tubes.
4. Start injecting fluid through the capillary tube.
5. Measure ΔP for at least five different flow rates.
6. Viscosity can then be calculated by plotting ΔP versus flow rate using Equation 4.13.

EXERCISE 4.2

Experimental values for rate versus pressure drop by the same capillary tube in the previous example for liquid CO_2 at 180 bar and 20°C is given in Table 4.8. Calculate CO_2 viscosity (theoretical value = 0.097 cp). Use a capillary tube constant from the previous example.

TABLE 4.8

Pressure Drop versus Rate for Viscosity Measurement by CO_2 at 180 Bar and 22°C

Flow Rate Q [ml/min]	Pressure Drop ΔP [mbar]
0.25	42.30
0.40	67.61
0.52	87.86
0.75	126.67
1.02	172.23

SOLUTION

Plot of flow rate versus pressure drop is given in Figure 4.16. $\dfrac{\Delta P}{q}$ is equal to 167.86 (mbar)/(ml/min), which is equal to $K \cdot \mu$ according to Equation 4.13:

$$\mu_{CO2} = 1/(10562.90 \cdot 10^{12}) \cdot (167.86 \cdot 60s/min \cdot 10^{11}) = 0.095 \text{ mPas}$$

Figure 4.17 shows theoretical data of the viscosity for pure CO_2 (liquid state) at 0, 22, 50, and 100°C for different pressures. The figure clearly shows acceptable experimental value with deviation of less than 5% from theoretical values, being 0.099 mPas.

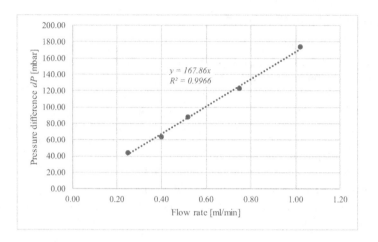

FIGURE 4.16 Plot of flow rate versus pressure drop.

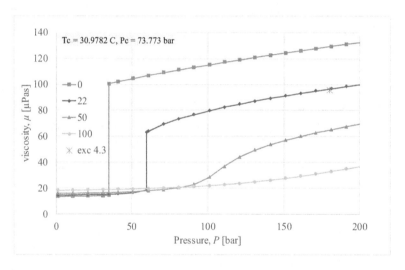

FIGURE 4.17 Experimental viscosity measurement (180 bar, 20°C) by capillary tube versus theoretical values for CO_2 (Lemmon et al. 2024).

4.6.4 ROTATIONAL VISCOMETER

Other frequently used viscometers especially for non-Newtonian fluids are the rotational viscometer or called rheometer, consisting of two concentric cylinders, with the annulus containing the liquid whose viscosity is to be measured (Figure 4.18). Geometries other than a cylinder or bob spindle can be used, like cones, plates, or a spindle. Here the bob is used as an example. Either the outer cylinder or the inner one (bob) is rotated at a constant speed, and the rotational deflection of the cylinder becomes a measure of the liquid's viscosity.

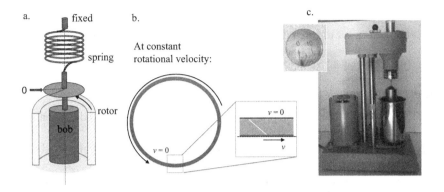

FIGURE 4.18   a. Schematic drawing of the rotational viscometer using a bob where the rotor moves c. Shows a lab example where the bob rotates and the torque is measured.

When the distance between the cylinders d, is small, we can define the viscosity gradient for laminar flow regime as:

$$\frac{dv}{dr} = \frac{\omega R}{d} \tag{4.15}$$

where R is radius of the inner cylinder (bob) and ω is angular velocity of the outer cylinder (rotor) defined by $\omega = 2\pi n$. When the rotor is rotating at a constant angular velocity ω and the bob is held motionless, the torque from the torsion spring on the bob must be equal but opposite in direction to the torque on the rotor from the motor. The effective area of the applied torque is $2\pi \cdot R \cdot h$ where h is length of the cylinder. The viscous drag on the bob is $k \cdot \theta \cdot R$, where k is the torsion constant of the spring and θ is angular displacement of the instrument in degrees. Then:

$$\frac{F}{A} = \frac{k\theta R}{2\pi Rh} = \mu\frac{dv}{dr} = \mu\frac{\omega R}{d} \tag{4.16}$$

which gives:

$$\mu = \frac{k\theta d}{2\pi h\omega R} = \frac{K\theta}{\omega h} \tag{4.17}$$

where K is the instrument's constant, which is determined by calibration.

PROCEDURE

The procedure will be equipment dependent. Read the local manual for specified operational and safety details. Here a general procedure for a rotational viscometer is given using a bob, where the torque on the bob is measured setting the rpm (shear rate). A heating mantle can be used to control the temperature.

   PPE: Lab coat, loves, safety glasses, safety shoes

Note:

- Use spill trays to catch possible spills from the measuring cup.
- Consider the necessity of doing the experiment in the fume hood based on the MSDS of the liquid.

Procedure, at room temperature

1. Switch on the device. Before attachment of the bob a zero adjustment is initiated automatically, to determine the torque without bob.
2. Mount the bob when the zero adjustment is finished by holding the top fixed and rotating the bob.
3. Fill liquid in the measuring cup in the fume hood. The filling volume depends on the bob used: Bob 00: = 20 ml, bob 18 = 10 ml, and for bob 31 and 34, 20 ml is used. All these spindles use the same measuring cup, which is placed in the center of the heat jacket. Here one can also use temperature during the measurements if desired.
4. Lower the bob into the measuring cup by placing the steering wheel in the viscometer stand. Be careful that the spindle is placed in the middle of the measuring cup.
5. Register in the device the kind of bob used, needed for the calculation of the viscosity.
6. Select the speed for the bob and start such that the bob will start to rotate in the cup.
7. In the display the following parameters are given, viscosity, RPM, torque, and temperature.

A torque between 10 and 30% often gives good viscosity measurements.

After use:

1. Turn off the viscosimeter switch and unplug the power cord.
2. Remove cup and bob. There are special green pliers for picking up the measuring cup that is mounted in the heat jacket.
3. When using different types of oils, the used container and measuring cup are to be cleaned in the fume hood using acetone/methanol and then rinsed with distilled water.

4.6.5 VIBRATIONAL VISCOMETER

A rod, tuning fork, or sphere is oscillating in the liquid or gas of interest, which is dampened by the inertia forces of the liquid, depending on the viscosity of the liquid. In diverse ways the dampening effect can be measured. The energy it costs to keep the vibration constant is a measure; the dampening or frequency phase change between resonance frequency and excitation frequency can be monitored (Viswanath et al. 2007).

4.7 GAS PROPERTIES

Like for liquids, gas densities and viscosities are also important for the description of flow of gases in porous media. Phase behavior of fluids dependent on pressure and temperature determines whether a fluid is in the gas or liquid phase. And when in the gas phase, pressure has a more significant effect as gasses are more compressible. Densities and viscosities are much smaller for gases than liquids and are therefore more challenging to measure. Alternatively correlations can be used to calculate parameters. In this section we will present real gas law and discuss the properties of hydrocarbon gases, CO_2, and H_2 and their most important properties: Gas density, gas critical pressure and temperature, gas viscosity, compressibility factor, and gas formation volume factor.

Changes in pressure, volume, and temperature conditions affect the physical properties of gas. Natural gas in the reservoir is therefore very different from the produced gas on the surface. To calculate the pressure gradient of a gas well or the amount of gas in a reservoir, the relationship among pressure, volume, and temperature (PVT) and properties such as density, compressibility factor, and viscosity must be known.

4.7.1 GAS DENSITY – THE REAL GAS LAW

An ideal gas is defined by the fact that there is no interaction between the molecules in the gas. The relationship between pressure, volume, and temperature of ideal gases is given by Boyle-Gay Lussac's law, which is often called the ideal gas law:

$$pV = n\,RT \tag{4.18}$$

where p = pressure, V = volume, R = universal gas constant, n = number of moles ($n = m/M$), T = absolute temperature, m = gas "total mass", and M = molecular weight of the gas. The units for the variables are given in Table 4.9. With $n = m/M$ and $\rho = m/V$, Equation 4.15 can be rewritten as

$$p = \frac{\rho RT}{M} \tag{4.19}$$

such that density can be expressed as:

$$\rho = \frac{p}{MRT} \tag{4.20}$$

TABLE 4.9
Gas Law Parameters in Different Unit Systems

Units	p	V	T	n	R	Temperature Units
Darcy	atm	cm³	K	gmole	2.057	K = °C + 273
Oil Field	psi	ft³	°R	lbmole	10.720	°R = °F + 460, °C = (°F − 32)$\frac{5}{9}$
SI	Pa	m³	K	gmole	8.314	K = °C + 273

There is no gas that follows the ideal gas law at all pressures and temperatures, and at reservoir conditions, natural gas shows large deviations from this law. One way to correct this discrepancy is to modify the ideal gas law by introducing a correction factor, z. The *real* gas law then becomes:

$$pV = z \cdot nRT \qquad (4.21)$$

The parameter z is also called compressibility factor or deviation factor, which depends on the composition, pressure, and temperature of the gas and is determined in the laboratory or from empirical diagrams. Figure 4.19 shows for different gasses the change of z with pressure, respective to temperature.

Generally, at low pressures, z is close to 1 and the behavior of the gas is almost ideal (Figure 4.19, left). The interactions between molecules are fewer at low pressures than at high pressures because each molecule has a greater available volume. At moderate pressures, the deviation from ideal states is less when the temperature is high (Figure 4.19, right). This is because the kinetic energy of the molecules, which is proportional to temperature, is "counteracted" by the attracting forces between the molecules. A given mole quantity of a particular gas will therefore have a larger volume at a high temperature than the same gas at a lower temperature. In general, deviations from ideal gas behavior increase with increasing molecular weight. At given pressure and temperature, a gas consisting of heavy molecules will occupy a smaller volume than a gas with light molecules, due to greater forces of attraction between the molecules. Z-diagrams are available for very few components, and to determine z for gases where such curves do not exist or for gas mixtures, the law of corresponding states must be applied. CO_2 is special as it changes phases. Awareness of the phase diagram and critical temperature and pressure (Figure 4.1) is therefore important. For further details PVT literature shall be consulted.

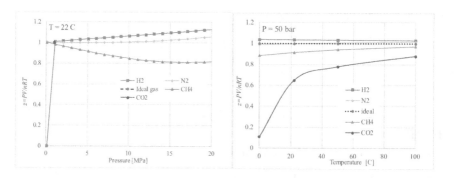

FIGURE 4.19   The development of z for hydrogen, nitrogen, methane, and CO_2 dependent on pressure for $T = 22°C$ and the temperature dependence at 50 bar (Lemmon et al. 2024).

4.7.2 Gas Viscosity

Viscosity is a measure of the internal resistance of fluid to flow. The viscosity of a gas is dependent on pressure and temperature and affects flow rates and potential pressure decline in the well. The viscosity of gases can be described from Newton's law of viscosity and the presented methods of capillary tube and vibrational viscometer can be suitable for gas phases. But due to viscosities ~3 orders of magnitude smaller than for liquids, it is difficult to measure the viscosity of a gas, especially at high temperatures. It is therefore common to make use of correlations between viscosity and easily measured parameters such as pressure (p), temperature, (T) and molecular mass (M).

In kinetic theory (molecular theory), temperature T expresses kinetic energy, pressure p expresses the distance between the molecules, and the molecular mass M is a measure of forces of attraction between the molecules. Figure 4.20 illustrates the qualitative variation of viscosity with pressure and temperature. In Figure 4.20a, the molar weight is constant. It can be seen that the viscosity increases with increasing temperature since increased kinetic energy results in more collisions between molecules. Note that this is the opposite from liquids, where viscosity reduces with increasing temperatures. In Figure 4.20b, the temperature is constant, while pressure and molecular mass vary. The sketch shows that higher pressure results in an increased viscosity, which can be explained by the fact that the distance between the molecules decreases and more collisions occur.

As previously mentioned, it is customary to use correlations to determine the viscosity of a gas, because measurements are complicated. Most used are the correlations of Carr et al. (1954).

4.7.2.1 Hydrogen

The viscosity of hydrogen is, as shown in Table 4.10 and Figure 4.21, very low compared to nitrogen and has a minor variation with pressure (ranging from 9 to 12 μPas

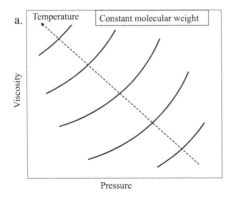
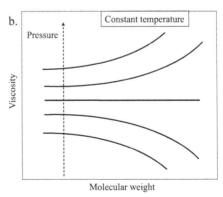

FIGURE 4.20 Qualitative variation of gas viscosity with a. pressure and temperature and b. with pressure and molecular weight.

TABLE 4.10

Viscosity of Hydrogen versus Temperature and Pressure (NIST 2024; Muzny et al. 2013)

Viscosity [μ Pas]	Pressure (MPa)					
Temperature [°C]	0.1	1	5	10	30	50
0	8.3969	8.4205	8.5245	8.6581	9.3275	10.077
25	8.9153	8.9369	9.0321	9.1533	9.7491	10.430
50	9.4193	9.4393	9.5270	9.6380	10.173	10.797
75	9.9102	9.9287	10.010	10.113	10.597	11.172
100	10.389	10.407	10.483	10.578	11.020	11.553
125	10.858	10.874	10.945	11.034	11.440	11.935

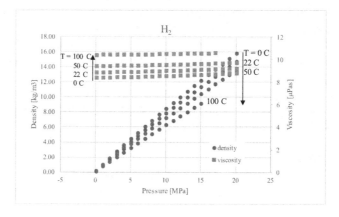

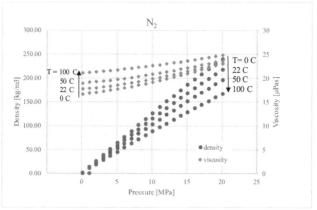

FIGURE 4.21 The density and viscosity of hydrogen dependent on temperature and pressure compared to nitrogen (Lemmon et al. 2024).

at pressure 0.1 to 50MPa and temperature from 25°C to 100°C). This low hydrogen viscosity should indicate high hydrogen mobility, negatively affecting the areal sweep. The viscosity increases with temperature, as gas molecules collide more, increasing more internal friction.

The low viscosity of hydrogen is helpful to suppress the water coning. However, the low viscous hydrogen will not displace water uniformly. Viscous fingering will most likely reduce the degree of recovery of hydrogen. The small variation of hydrogen viscosity indicates that it is unnecessary to use complicated equation of state calculations in multiphase flow calculations with H_2 and formation brine.

4.7.2.2 Carbon Dioxide

Different from hydrogen and methane or nitrogen, the critical temperature and pressure of CO_2 lie between surface conditions and reservoir conditions. The critical point is at 30.9782°C (Tc) and 73.773 bar (Pc). At reservoir conditions, it is likely that the CO_2 is in a liquid or supercritical state. This complicates the density and viscosity description of CO_2.

The sweep efficiency of CO_2 flooding is associated with the viscosity contrast between CO_2 and the fluids in the reservoir. A large database on experimentally measured viscosity of CO_2 is available in the literature. A publication by Laesecke and Muzny (2017) summarized all these experimental data and investigated available computed data for the viscosity of carbon dioxide. They developed a new reference correlation based on equation of state calculations and the result is given in Table 4.11, which covers data for viscosity of CO_2 for a wide range of temperatures and pressures. Figure 4.21 shows the development of the CO_2 density and viscosity. The abrupt jumps in viscosity and density at 0 and 22°C are due to a phase change from gas to liquid. Beyond the critical point the gas is supercritical. For CO_2 in the liquid phase, supercritical state densities are relatively high compared to methane (Figure 4.23) and in the order of magnitude of water. CO_2 viscosities compared to methane are higher but remain in the order of magnitude of 100 μPas, so lower than water or oil, more compatible to gas viscosities.

TABLE 4.11
Viscosity of CO_2 in mPas (Laesecke and Muzny 2017)

Pressure [MPa]	Temperature [K]			
	300	400	500	600
0.1	0.015 (g)	0.019 (g)	0.024 (g)	0.028 (g)
20	0.094 (l)	0.031	0.029	0.031
40	0.122 (l)	0.057	0.04	0.038
60	0.144 (l)	0.075	0.053	0.046
80	0.164 (l)	0.089	0.065	0.055
100	0.184 (l)	0.103	0.075	0.063

Bold numbers show values where CO_2 is in the supercritical state; (g) stands for the gas phase, (l) for the liquid phase

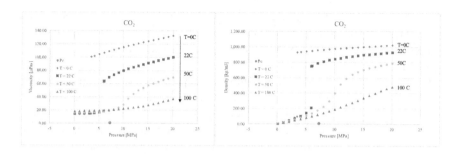

FIGURE 4.22 The viscosity and density of CO_2 dependent on pressure and temperature (Lemmon et al. 2024). Note that the sharp jumps correspond the phase changes, dependent on the vaporization line and critical point T_c and P_c.

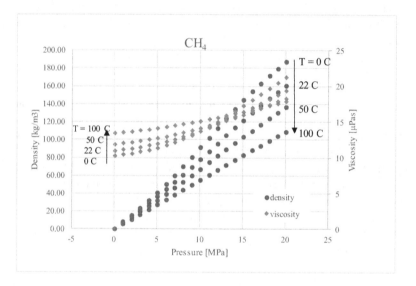

FIGURE 4.23 The density and viscosity of methane, as main component of natural gas, dependent on temperature and pressure (Lemmon et al. 2024).

4.8   EXPERIMENT 4.2: LIQUID VISCOSITY

Liquid viscosity measurement using capillary-type viscometer.

OBJECTIVE

Determine the dynamic viscosity of distilled water, 3%wt NaCl solution, Exxsol D60, and Zalo solution with the viscometer (Figure 4.10).

Special Notes

- The personal protection equipment needed for this experiment are lab coat, gloves, safety glasses, and safety shoes. Check the MSDS for the fluids to decide whether the experiment shall be performed in the fume hood.

- Use capillary no. 50 for both saltwater and Exxsol D60. Remember that each viscometer has its own correction factors (K values), so check that the viscometer matches the box it was packed in.
- If $t < 400$ seconds, the Hagenbach correction factor (ϑ) should be included according to Table 4.6. If $t > 400$ seconds, $\vartheta = 0$.

EQUIPMENT NEEDED

Viscometer, Peleus ball, glass work, and timer.
   A detailed procedure is given previously.

Calculation and report:

1. Calculate the kinematic viscosity υ from the measured flow time t and the instrument constant by means of the Equation 4.8. Calculate the viscosity μ from the calculated kinematic viscosity υ and the density ρ.
2. Report test results for both the kinematic and dynamic viscosity. Calculate the average dynamic viscosity.

EXERCISE 4.3

1. What is viscosity?

   With the capillary viscosimeter as shown in Figure 4.10, the viscosities of two fluids were measured.
   The following data was generated:
   Data for fluid one: $\rho = 0.867$ g/cm^3. Capillary 75 was used, with $K = 0.008$ mm^2/s^2:

   $$t_1 = 7 \text{ minute and } 35 \text{ s}, \ t_2 = 8 \text{ minute and } 2 \text{ s.}$$

   Data for fluid two: $\rho = 1.26$ g/cm^3. Capillary 100 was used, with $K = 0.015$ mm^2/s^2:

   $$t_1 = 150 \text{ sec}; \ t_2 = 130 \text{ s.}$$

2. What force drives the fluid? Based on which law is this device functioning, and what assumptions are used for the interpretation of the results?
3. Which fluid has the highest dynamic viscosity μ?
4. Name two potential sources of error for this measurement.
5. Derive which paramaters the constant K in Equation 4.9 is dependent on. How is the K value determined for a device?

TABLE 4.12
Kinematic Viscosity Calculation Table

Sample	Constant C, (cSt/s)	Time (s)	Hagenbach Factor, ϑ	Kinematic Viscosity, υ (cSt)	Density, ρ_{avr} (g/cm^3)	Dynamic Viscosity, μ (cP)
Brine 22°C	0.003667	261.81	0.08	0.96	1.01	0.97

$\mu_{avg} = 0.97$ cP

SOLUTION

1. Viscosity describes the internal resistance to flow; the gradient of shear stress t versus shear rate (σ), $\tau = \mu\sigma$. Also interpreted as the thickness of fluid, which is dependent on temperature, pressure, and molecular characteristics for Newtonian fluids. For non-Newtonian fluids it also depends on shear rate and potentially on time.
2. What force drives the fluid? Gravitational force. The principles of the interpretation of this device measurements are Hagen-Poiseuille's law (valid for laminar flow) Newtonian fluid, constant temperature, no acceleration.
3. Use Equations 4.8 and 4.9
   Fluid 1; $t_1 = 455$ s; $t_2 = 482$ s; no Hagenbach correction needed (flow time > 220s for capillary 75)

$$v_1 = 455 \cdot 0.008 = 3.64 \text{ mm}^2/\text{s};$$
$$v_2 = 482 \cdot 0.008 = 3.86 \text{ mm}^2/\text{s};$$
$$v_{avg} = 3.75 \pm 0.11 \text{ mm}^2/\text{s}$$

$\mu = 3.75 \ 10^{-2} \text{ cm}^2/\text{s} \cdot 0.867 \text{ g/cm}^3 = 3.25 \cdot 10^{-2} \text{ g/cm s} = 3.25 \cdot 10^{-3} \text{ Pas} = 3.25 \text{ cP}.$

   Fluid 2; $t_1 = 150$ s; $t_1 = 130$ s; Hagenbach correction needed both 0.03s (for capillary 100)

$$v_1 = 0.015 \cdot (150 - 0.03) = 2.25 \text{ mm}^2/\text{s}$$
$$v_2 = 0.015 \cdot (130 - 0.03) = 1.95 \text{ mm}^2/\text{s}$$
$$v_{avg} = 2.10 \pm 0.15 \text{ mm}^2/\text{s}; 2.10 \cdot 10^{-2} \text{ cm}^2/\text{s}$$
$$\mu = 2.64 \cdot 10^{-2} \pm 0.189 \text{ g/cm s}; 2.64 \text{ cP}$$

   Fluid 1 has a higher dynamic viscosity.
4. Possible errors can be timing delay at start and finish when the interface passes the marking, wrong observation of interface, turbulence, cleanliness of the glass tube, temperature variations, capillary not held straight, or other assumptions not being met.
5. See Section 4.6.1 for the derivation. According to Equation 4.6, K is dependent on radius of the capillary, volume of the fluid chamber, and gravitational acceleration. It is obtained from performing an experiment with a known viscosity. The theoretical value can be determined, but there might be e.g. slight variations in the capillary radius, which make the actual K deviation from the theoretical value inaccurate.

REFERENCES

API, American Petroleum Institute. *Recommended Practices for Core Analysis, RP40*. API Publishing Services, 1998.

ASTM. *Compass.astm.org*, n.d. https://compass.astm.org/home/0 (accessed April 2024).

Carr, N.L., Kobayashi, R., Burrows, D.B. "Viscosity of hydrocarbon gases under pressure." *Transactions of the AIME 201*, 1954: 264–272.

Dias, A., de Graaf, G., Wolffenbuttel, R.F., Rocha, L.A. "Gas viscosity sensing based on the electrostatic pull-in time of microactuators." *Sensors and Actuators A: Physical 216*, 2014: 376–385.

Enoksson, P., Stemme, G., Stemme, E. "Fluid density sensor based on resonance vibration." *Sensors and Actuators A 46–47*, 1995: 327–331.

ISO-a. *Oscillation-Type Density Meters*, International Organisation for Standardisation. ISO15212-1:1998. https://www.iso.org/standard/28482.html

ISO-b. *Petroleum Products. Transparent and Opaque Liquids. Determination of Kinematic Viscosity and Calculation of Dynamic Viscosity.* International Organisation for Standardisation. ISO3104:2023 https://www.iso.org/standard/81853.html

Jarrell, P.M., Fox, C.E., Stein, M.H., Webb, S.L. *Practical Aspects of CO2 Flooding.* Society of Petroleum Engineers, 2002.

Kratky, O., Leopold, H., Stabinger, H. "Dichtemessungen an Fluessigkeiten und Gasen auf 10^-6 g/cm3 bei 0.6 cm3 Praeparatvolumen." *Zeitschrift für Angewandte Physik 27*, 1969: 273–277.

Kuramoto, N., Fujii, K., Waseda, A. "Accurate density measurement of reference liquids by magnetic suspension balance." *Metrologia 41*, no. 2, 2004: 84.

Laesecke, A., Muzny, C.D. "Reference correlation for the viscosity of carbon dioxide." *Journal of Physical and Chemical Reference Data 46*, 2017.

Lake, L., Johns, R.T., Rossen, W.R., Pope, G.A. *Fundamentals of Enhanced Oil Recovery.* Society of Petroleum Engineers, 2014.

Langdon, R.M. "Resonator sensors, a review." *Journal of Physics E: Scientific Instruments 18*, 1985.

Lemmon, E.W., Bell, I.H., Marcia L., Huber, M.L., McLinden, M.O. "Thermophysical properties of fluid systems." In *NIST Chemistry WebBook*, Eds. P.J. Linstrom and W.G. Mallard. Vol. NIST Standard Reference Database Number 69. National Institute of Standards and Technology (NIST), 2024.

Muzny, C.D., Huber, M.L., Kazakov, A.F. "Correlation for the viscosity of normal hydrogen obtained from symbolic regression," *Journal of Chemical & Engineering Data 58*, 2013: 969–979.

NIST. *NIST Chemistry Web Book, SRD 69.* NIST, 2024.

Paar, A. *U-Tube Technology in Digital Laboratory Density Meters*, 2024. U-Tube Technology in Digital Laboratory Density Meters | Anton Paar Wiki (anton-paar.com) visited 11.04.2024 (accessed April 2024).

Sheng, J.J. *Modern Chemical Enhanced Oil Recovery: Theory and Practice.* Gulf Professional Publishing, 2010.

Speight, J.G. *Handbook of Petroleum Analysis.* Wiley Interscience, 2001.

Stemme, E., Ekelof, J., Nordin, L. "Measuring liquid density with a tuning-fork transducer." *IEEE Transactions on Instrumentation and Measurement IM-32*, no. 3, 1983.

Suzuki, J. -I. "GD Series vibratory gas density meters." *Yokogawa Technical Report*, English Edition, 2000. https://www.yokogawa.com/eu/library/resources/yokogawa-technical-reports/gd-series-vibratory-gas-density-meters/ visited 19.09.2024.

Viswanath, D.S., Ghosh, T.K., Prasad, D.H.L., Dutt, N.V.K., Rani, K.Y. *Viscosity of Liquids: Theory, Estimation, Experiment, and Data.* Springer Science & Business Media, 2007.

Wilke, J., Kryk, H., Hartmann, J., Wagner, D. *Theory and Praxis of Capillary Viscometry—An Introduction*. SCHOTT-GERÄTE GmbH, 2000.

Yang, P., Lin, B.Y. "Measurement of viscosity in a vertical falling ball viscometer." *American Laboratory 10*, no. 27, 2008. www.americanlaboratory.com/913-Technical-Articles/778-Measurement-of-Viscosity-in-a-Vertical-Falling-Ball-Viscometer/ (accessed April 2024).

Yang, X., Kleinrahm, R., McLinden, M.O. "The magnetic suspension balance: 40 years of advancing densimetry and sorption science." *International Journal of Thermophysics 44*, no. 169, 2023: 1–52.

5 Porosity

5.1 INTRODUCTION

Many materials around us are not always massive in form. The material contains pores. The pore volume in comparison to the total volume is called the porosity. The pores can be connected or not, dependent on the character of function of the material. Figure 5.1 shows some examples.

Soil can take up rainwater via its pores such that plants are supplied with water. Ground water flows via pores through the subsurface to be filtered. Plants like trees also contain transport paths, which create hollow spaces when dried, affecting the material density and mechanical properties. Materials like metal foams are designed with pores to reduce material use but still keep the mechanical strength. With pores trapping gas, material can function as insulation for heat or sounds. On the other hand, the optimal packing of beads or any kind of products in a confined space is important to be able to transport as much as possible without transporting empty space. So, there are many applications where pores in the material play an important role.

From the viewpoint of reservoir engineers, a reservoir rock is good when it can store much fluid. Porosity is a measure of storage capacity of a hydrocarbon reservoir and the available storage capacity in the case of CO_2 and H_2 injection in underground porous rock.

FIGURE 5.1 Examples of porous materials in daily life.

DOI: 10.1201/9781003382584-5

5.2   DEFINITION

Theoretically, porosity is defined as the ratio of the pore volume to bulk volume and is expressed as either a percent or a fraction. In equation form:

$$\varphi = \frac{pore\ volume}{bulk\ volume} = \frac{V_p}{V_b} = \frac{bulk\ volume - grain\ volume}{bulk\ volume} = \frac{V_b - V_g}{V_b} = 1 - \frac{V_g}{V_b} \quad (5.1)$$

Two types of porosity may be measured: total or absolute porosity and effective porosity. Total porosity is the ratio of all the pore spaces in a rock to the bulk volume of the rock. Effective porosity is the ratio of interconnected void spaces to bulk volume. Thus, only the effective porosity contains fluids that can be produced from wells. For granular materials such as sandstone, the effective porosity may approach the total porosity; however, for shales and highly cemented or vugular rocks such as some limestones, large variations may exist between effective and total porosity.

Porosity may be classified according to its origin as either primary or secondary. Primary or original porosity is developed during deposition of the sediment. Secondary porosity is caused by geologic processes after formation of the deposit. These changes in the original pore spaces may be created by overburden or tectonic stresses, water movement, or various types of geological activities after the original sediments were deposited. Fracturing or formation of solution cavities often will increase the original porosity of the rock, where compaction, cementation, and recrystallization can reduce the porosity. On a core level, the porosity measured can be considered the matrix porosity. The core sample shall be large enough to be representative.

For a uniform rock grain size, porosity is independent of the size of the grains. A maximum theoretical porosity of 48% is achieved with cubic packing of spherical grains, as shown in Figure 5.2a. Rhombohedral packing, which is more representative of reservoir conditions, is shown in Figure 5.2b; the porosity for this packing is 26%. If a second, smaller size of spherical grains is introduced into cubic packing (Figure 5.2c), the porosity decreases from 48% to 14%. Thus, porosity is dependent on the grain

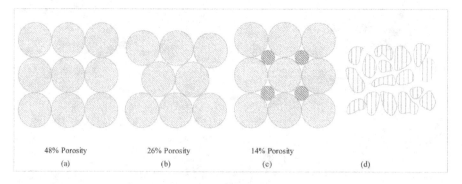

48% Porosity   26% Porosity   14% Porosity

(a)   (b)   (c)   (d)

FIGURE 5.2   (a) cubic packing, (b) rhombohedral, (c) cubic packing with two grain sizes, and (d) typical sand with irregular grain shape.

size distribution and the arrangement of the grains, as well as the amount of cementing materials. Not all grains are spherical, and grain shape also influences porosity. A typical reservoir sand is illustrated in Figure 5.2d. Typical porosities for a conventional reservoir rock range from 15–35%.

Compaction is the process of volume reduction due to an externally applied pressure. For extreme compaction pressures, all materials show some irreversible change in porosity. This is due to distortion and crushing of the grain or matrix elements of the materials, and, in some cases, recrystallization. The reversible effect of pressure on porosity, leading to compressibility, is discussed in Section 5.5.

EXERCISE 5.1

Calculate the porosity of the same spheres of diameter D arranged as shown in Figure 5.3a (cubic packing) and in Figure 5.3b (2D hexagonal packing).

SOLUTION

Consider r (grain radius) = 1

Cubic packing porosity: The square has a volume of $2r \cdot 2r \cdot 2r$, the volume of the 8 grains of which 1/8 is positioned in the corner in the unit cell is $8 \cdot 1/8 \cdot 4/3\pi r^3$.

$$\phi = \frac{V_p}{V_b} = \frac{8r^3 - \frac{4}{3}\pi r^3}{8r^3} = 1 - \frac{\pi}{6} = 0.476$$

A variant of the hexagonal packing porosity: the grain volume remains 1 grain volume but now the box volume changed to a rectangle with an area $2r \cdot sin(60°) \cdot 2r$ and a height of $2r$ giving:

$$\phi = \frac{V_p}{V_b} = 1 - \frac{V_g}{V_b} = 1 - \frac{\frac{4}{3}\pi r^3}{(2r)^3 \sin(60°)} = 0.395$$

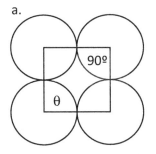

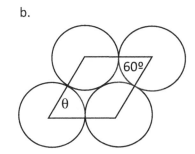

FIGURE 5.3 Porosity calculation for different types of packing, using the unit cell, the repetitive element.

In case the layers fall into each other, forming a face cubic centered packing, the unit cell sides have, based on Pythagoras knowing in the diagonal fit $4r$, a length of $2\sqrt[2]{2}\ r$. Fitting at the corners $8 \times 1/8$ and in the faces $6 \times 1/2$ grain volumes gives a total porosity of:

$$\phi = \frac{V_p}{V_b} = \frac{(2\sqrt[2]{2})^3 r^3 - (1+3)\frac{4}{3}\pi r^3}{(2\sqrt[2]{2})^3 r^3} = 1 - \frac{\pi}{3\sqrt[2]{2}} = 0.26$$

5.3 POROSITY MEASUREMENTS ON CORE PLUGS

From the definition of porosity, it is evident that the porosity of a sample of porous material can be determined by measuring any two of the three quantities: bulk volume, pore volume, or grain volume. The porosity of reservoir rock may be determined by core analysis, well logging technique, and well testing. The question of which source of porosity data is most reliable cannot be answered without reference to a specific interpretation problem. These techniques can all give correct porosity values under favorable conditions. The core analysis porosity determination has the advantage that no assumption needs to be made as to mineral composition, borehole effects, etc. However, considering representation of the horizon/formation of interest, the volume of the core is less than the rock volume, which is investigated by a logging device, therefore porosity values derived from logs are frequently more representative in heterogeneous reservoirs. In the following sections we will discuss how to estimate pore, grain, and bulk volumes from core plugs.

5.3.1 BULK VOLUME MEASUREMENT

5.3.1.1 Geometrically
The bulk volume may be computed from measurements of the dimensions of a uniformly shaped sample. In case of uniform cylindrical sample, the bulk volume can be calculated from:

$$\text{Bulk volume} = \text{cross sectional area} \cdot \text{Length} = \pi \cdot \frac{d^2}{4} \cdot L \qquad (5.2)$$

Where d and L are diameter and length of the sample respectively. Note that the accuracy of π can slightly change the outcome. Use minimum $\pi = 3.1416$.

5.3.1.2 Submersion
In the case of non-uniform irregular sample shape, the bulk volume can be measured through the usual procedure that utilizes the observation of the volume of fluid displaced by the sample. The fluid displaced by a sample can be observed either volumetrically or gravimetrically. In either procedure it is necessary to prevent the fluid penetration into the pore space of the rock. This can be accomplished a) by coating

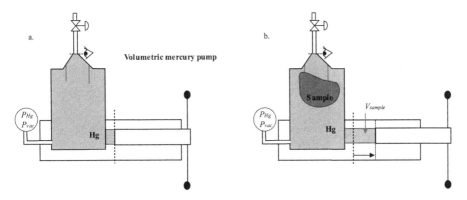

FIGURE 5.4 Schematic of mercury injection apparatus, showing in a. the volume without sample and in b. the volume change V_{sample} to fit in the sample, but keeping the same fluid level.

the sample with paraffin or a similar substance, b) by saturating the core with the same fluid into which it is to be immersed and measuring the displaced volume of the fluid, or c) by using mercury.

The bulk volume measurement procedure using a mercury porosimeter (Figure 5.4) is given here:

1. Rotate the handle forward until the mercury appears at the Lucite window and set the ruler to zero. This will be the base set point for measurement of bulk volume Figure 5.4a.
2. Rotate the handle backward to withdraw the mercury, open the cap and put the sample inside the chamber which is partly filled with mercury.
3. Rotate the handle forward to allow the mercury to be pumped into the chamber until the mercury appears again at the Lucite window. Now the sample is totally immersed in mercury, and you may read the number on the ruler which gives the total bulk volume of the sample (Figure 5.4b).

Mercury is toxic and avoided in most laboratories, so the saturation method where the sample is saturated with the same fluid as it is submerged (usually water) is most used.

Gravimetric determinations of bulk volume can be accomplished by observing the loss in weight of the sealed sample when immersed in a fluid or by change in weight of a pycnometer with and without the sealed core sample.

5.3.2 HELIUM POROSITY

Helium porosimetry is one of the most-used methods, which employs Boyle's law. The helium gas in the reference cell isothermally expands into a sample cell.

After expansion, the resultant equilibrium pressure is measured and can be correlated to the sample cell volume using Boyle's law (Bear 1972; Monicard 1980). The helium porosimeter apparatus is shown schematically in Figure 5.5.

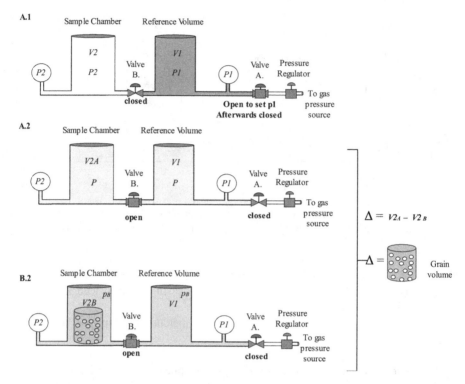

FIGURE 5.5 Schematic illustration of helium porosimeter, with experimental steps A and B to obtain the volume of the sample chamber, empty (A) and with the core (B). The difference in volume corresponds to the grain volume of the core sample. A.1 and A.2 show the application of Boyle's law before and after pressure equilibrium between V1 and V2.

Helium has advantages over other gases because: a) its small molecules rapidly penetrate small pores, b) it is inert and does not adsorb on rock surfaces as air may do, c) helium can be considered as an ideal gas (i.e., $z = 1.0$) for pressures and temperatures usually employed in the test, and d) helium has a high diffusivity and therefore affords a useful means for determining porosity of low-permeability rocks.

The schematics of the helium porosimeter shown in Figure 5.5 has a reference volume V_1, at pressure p_1 and a sample chamber with unknown volume V_2, and initial pressure p_2 (A.1). The reference volume and the sample chamber are connected by tubing. The system can be brought to equilibrium when the core holder valve B is opened, allowing determination of the unknown volume V_2 by measuring the resultant equilibrium pressure p. Pressure p_1 and p_2 are controlled by the operator; usually $p_1 = 100$ psig and $p_2 = 0$ psig, where the g in psig means gauge and indicates that the reading is done on a manometer with 0 at atmospheric pressure. When the core holder valve B is opened, the volume of the system will be the equilibrium volume V, which is the sum of the volumes V_1 and V_2 (A.2 in Figure 5.5). Boyle's law is applicable if the expansion takes place isothermally, and there is no leakage.

Thus, the pressure-volume products are equal before and after opening the core holder valve:

$$p_1 V_1 + p_2 V_2 = p(V_1 + V_2) \tag{5.3}$$

Solving the equation for the unknown volume, V_2:

$$V_2 = \frac{(p - p_1) V_1}{p_2 - p} \tag{5.4}$$

Since all pressures in Equation 5.4 must be absolute and it is customary to set $p_1 = 100$ $psig$ and $p_2 = 0$ $psig$, Eq. 5.4 may be simplified as follows:

$$V_2 = \frac{V_1 (100 - p)}{p} \tag{5.5}$$

where V_2 in cm^3 is the unknown volume in the sample cup, and V_1 in cm^3 is the known volume of the reference cell. p in $psig$ is pressure read directly from the gauge.

To determine the porosity, measurement of the sample chamber needs to be repeated with and without the core in place, respectively experiment A and B in Figure 5.5. The difference in volume between the two measurements A and B is the grain volume, assuming all accessible pores will be filled with helium.

Small volume changes occur in the system, including the changes in tubing and fittings caused by pressure changes during equalization. A correction factor, G, may be introduced to correct for the composite system expansion. The correction factor G is determined for porosimeters before they leave the manufacturer, and this correction is built into the gauge calibration in such a way that it is possible to read the volumes directly from the gauge.

The grain volume of pore samples is sometimes calculated from sample weight and knowledge of average density. Formations of varying lithology and, hence, grain density limit applicability of this method. This can be used to roughly verify the results of the helium porosity measurement.

PROCEDURE

In the helium porosity experiment, the following personal protective equipment is needed: Lab coat, safety glasses, safety shoes, and gloves (in case cores surface needs protection). The detailed procedure is given here:

1. Measure the dimensions of the core (length and diameter) using a caliper.

A. *Sample chamber with the core plug*

2. Put the clean, dry core inside the sample chamber, fill it up with some steel plates, and mount the matrix cup in the cup holder.

Pressurizing the cell

3. Make sure the regulator on the gas bottle is closed (upper position).
4. Turn/open the main valve on the gas bottle. Turn the regulator on the gas bottle, until the reference pressure (given by manufacturer) is reached on the pressure gauge.
5. Open the valve towards the reference cell.
6. Regulate the needle at the calibration mark on the helium-porosimeter.
7. Close the access to the gas bottle to isolate the system.
8. Open the valve B towards the sample holder.
9. Take the reading for volume V2 (cm³), for the matrix cup with the core inside. In case the reading is beyond the scale, the measured volume is comparably larger. Then add an extra volume to the reference cell if needed.

Pressure release from the system.

10. Make sure the valves to the gas bottle are closed.
11. Open exhaust valve to release the pressure from the closed system.
12. Take the core out of the matrix cup.
13. Close all the black valves.

B. Sample chamber without the core plug,

Repeat all the steps without the core in the sample holder. Leave the same steel plates in the cell that were used in the experiment with the core present.

The grain volume can be derived from the volume difference between step A and B.

EXERCISE 5.2

The porosity is determined by using a helium porosimeter.

1. Derive the equation you need to calculate the sample's grain volume, when using the following notation

   V_1 = the volume of chamber 1, P_1 = the initial pressure in chamber 1
   V_2 = the volume of chamber 2, P_2 = the initial pressure in chamber 2
   V_g = grain volume, P_3 = system final pressure

2. The following data were measured with Boyle's law porosimeter:

   $V_1 = V_2 = 50$ cm³
   $P_1 = 1$ bar $P_2 = 10$ bar $P_3 = 7$ bar

   Calculate the grain volume and porosity of the core sample when the sample diameter and length are 3.0 cm and length 4.5 cm respectively.

3. The sample is a limestone with 10 wt% clay. The dry sample weight is 68 g. The density for clay is 2.65 g/cm³ and for limestone is 2.73 g/cm³. Verify with this data the porosity as was determined with the Boyle's law porosimeter.

SOLUTION

1. $$P_1\left(V_1 - V_g\right) + P_2 \cdot V_2 = P_3\left(V_1 - V_g + V_2\right)$$

$$V_g = \frac{P_3\left(V_1 + V_2\right) - P_1 V_1 - P_2 V_2}{P_3 - P_1}$$

2. $$V_g = \frac{7(50 + 50) - 1 \cdot 50 - 10 \cdot 50}{7 - 1} = 25.0 \text{ cm}^3$$

$$V_{bulk} = \pi r^2 h = \pi \left(\frac{3}{2}\right)^2 \cdot 4.5 = 31.8 \text{ cm}^3$$

$$\phi = \frac{31.8 - 25}{31.8} = 0.21$$

3. $$V_g = \frac{W_{dry} \cdot 0.1}{\rho_{clay}} + \frac{W_{dry} \cdot 0.9}{\rho_{limestone}} = \frac{68 \cdot 0.1}{2.65} + \frac{68 \cdot 0.9}{2.73} = 25 \text{ cm}^3$$

$$\phi = \frac{31.8 - 25}{31.8} = 0.21$$

5.3.3 LIQUID POROSITY; GRAVIMETRIC METHOD

The pore volume can also be determined when filling the pore space of a dry core with a fluid with a known density and measuring the weight change before and after. Core saturation is in general a necessary step when performing liquid permeability experiments or multiphase flooding experiments. Therefore, weighting of the core before and after the saturation shall be done, so that other porosity measurements can be verified. The chosen liquid needs to have a known density $\rho(T)$ and shall not interact with the sample. Brine is used here as an example fluid. In case of the presence of clay the water needs to contain salt to avoid clay swelling. The special case of mercury use will be discussed in a separate paragraph later (5.3.4). The liquid needs to reach all the pores. Therefore, air needs to be removed by using a vacuum before introducing the liquid.

The following steps are recommended considering a set-up as shown in Figure 5.6.

1. Weigh the dry sample and measure dimensions.
2. Place the sample inside the desiccator in a fitting beaker to reduce the amount of liquid to be used, connect to vacuum pump, and wait for about 1 hour to remove all gas inside the sample. Vacuum pump should reach 10^{-2} mbar.
3. Saturate the core sample with brine with known density by opening the valve to the liquid supply to enable flow of liquid into the beaker.
4. Degas the liquid in advance.
5. Leave the core submerged in the vacuum for a few minutes. Liquid vapor is created so the vacuum pump needs to stand or needs protection from liquid vapor.

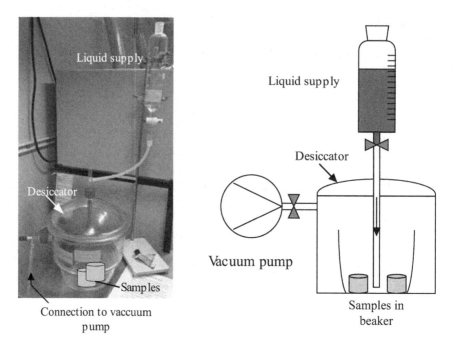

FIGURE 5.6   Set-up to saturate porous media with liquid.

6. Remove the vacuum and take the cores or beaker with cores out.

Weighting

7. One by one, remove the excess water according to a repeatable routine.
8. Weigh the saturated sample and determine the pore volume.

$$V_p = \frac{W_a - W_b}{\rho} \left[\frac{\text{g}}{\frac{\text{g}}{\text{cm}^3}} \right] \tag{5.6}$$

With W_a = core weight after saturation, W_b = dry core weight, and $\rho(T)$ = the density of fluid used. Alternatively, air can be displaced by CO_2 before this experiment so that any possible leftover gas can dissolve in the liquid, possibly pressurizing the system to stimulate dissolution.

By measuring the dry core weight and saturated core weight, the volume of the pore space can be determined. Full saturation of all connected pores is critical; if not the pore space will be underestimated. Another significant source of error is the possible presence of excess water on the surface of the core. Only the pore space shall be filled and a water film at the surface of the core will overestimate the pore space. A standardized routine is advised to remove the excess water e.g. to shake off water or rolling the sample once over a tissue paper around the circumference of the core and then pressing the top and bottom of the core on the paper once before measuring the saturated weight.

An alternative to avoid this error is weighting of the saturated core submerged in the liquid that fills the pore space. The difference in the saturated core weight measured in air m_{dry} and submerged, $m_{submerged}$ is the buoyancy force on the displaced fluid, which is the grain volume.

$$m_{submerged} - m_{dry} = V_g \cdot \rho$$

Note that by knowing the grain volume and its dry weight the grain density can be determined.

Grain volume may be measured by crushing a dry and clean core sample. The volume of crushed sample is then determined by (either pycnometer or) submersion in a suitable liquid. The change in weight Δm of the liquid filled pycnometer with and without rock powder is the difference in mass between the grains and liquid,

$$\Delta m = V_g \left(\rho_g - \rho_l \right) = m_{powder,dry} - V_g \cdot \rho_l \qquad (5.7)$$

Due to the crushing, the V_g does not contain any inaccessible pores, which normally would contribute to the grain volume, so using this value of V_g for the porosity calculation will give the total porosity.

EXERCISE 5.3

Calculate effective porosity for a core sample with the following data:

Weight of dry core sample: 259.2 g
Weight of 100% saturated core sample with water: 297.0 g
Water density: 1.0 g/cm^3
Weight of the core sample submersed in water: 161.4 g

SOLUTION

$$\phi_{eff} = \frac{Interconnected\ pore\ space}{Bulk\ volume} \quad \phi_{total} = \frac{Total\ pore\ volume}{Bulk\ volume}$$

$$V_w = V_p = \frac{m_{saturated\ core} - m_{dry\ core}}{\rho_w} = \frac{297 - 259.2}{1.0} = 37.8\ cm^3$$

$$V_{grain} = \frac{m_{dry\ core} - m_{immersed\ core}}{\rho_w} = \frac{259.2 - 161.4}{1.0} = 97.8\ cm^3$$

$$\phi_{eff} = \frac{37.8}{37.8 + 97.8} = 0.28$$

EXERCISE 5.4

A core sample has the following data:

• Diameter: 2.20 cm	• Weight of a dry core sample: 28.5 g
• Length: 3.87 cm	• Weight of the plug saturated with water: 31.5 g

The density of the water is 1.0 g/cm³. Calculate the effective porosity of the rock.
The sample was then crushed, and the weight of crushed sample submersed in water was 16.9g. Calculate total porosity of the sample.

SOLUTION

Effective porosity:

$$V_{bulk} = \pi r^2 h = \pi \left(\frac{2.2}{2}\right)^2 \cdot 3.87 = 14.7 \text{ cm}^3$$

$$V_{pore} = \frac{W_{saturated\ core} - W_{dry\ core}}{\rho_{water}} = \frac{31.5 - 28.5}{1.0} = 3.0 \text{ cm}^3$$

$$\phi_{eff} = \frac{3}{14.7} = 0.20$$

Total porosity:

$$\text{Buoyancy: } \rho_w V_{matrix} = W_{dry} - W_{immersed}$$

$$V_{matrix} = \frac{28.5 - 16.9}{1.0} = 11.6 \text{ cm}^3$$

$$V_{pore} = V_{bulk} - V_{matrix} = 14.7 - 11.6 = 3.1 \text{ cm}^3$$

$$\phi_{total} = \frac{3.1}{14.7} = 0.21$$

EXERCISE 5.5

A core got saturated with brine. See the following experimental data.

$\mu_{gas} = 0.0179$ cP
$\mu_{brine} = 0.79$ cP
$L = 13.668$ cm
$r = 1.90$ cm
weight of dry core in air = 301.1 g
weight of core with brine = 340.6 g
brine density = 1.02 g/cm³

rock density = 2.59 g/cm³
air density =1.27 mg/cm³

Derive the equation to obtain the effective porosity of the core, using the core mass before and after saturation. Calculate with that equation the effective porosity.

SOLUTION

$$m_{dry} = \pi r^2 L \left(1-\phi\right) \rho_{rock} + \pi r^2 L \phi \rho_{air}$$

$$m_{wet} = \pi r^2 L \left(1-\phi\right) \rho_{rock} + \pi r^2 L \phi \rho_{brine}$$

$$\Delta m_{w-d} = \pi r^2 L \phi \left(\rho_{brine} - \rho_{air}\right)$$

$$\phi = \frac{\Delta m_{w-d}}{\pi r^2 L \left(\rho_{brine} - \rho_{air}\right)} = \frac{340.6 - 301.1}{\pi 1.9^2 13,668 \cdot 1.02} = 0.25$$

5.3.4 MERCURY POROSITY

A variation to the liquid porosity is the mercury injection. When a rock has a small fraction of void space, it is difficult to measure porosity by the mentioned methods. In this case, mercury injection is used. The principle consists of forcing mercury under relatively high pressure into the rock pores (Cosse 1993). A typical apparatus is illustrated in Figure 5.7. A pressure gauge is attached to the cylinder for reading pressure under which measuring fluid is forced into the pores. Figure 5.7 shows a typical curve from the mercury injection method. The volume of mercury entering the core sample is obtained from the device with accuracy up to 0.01 cm³.

The mercury saturation versus pressure curve is useful in obtaining information about the pore size distribution of the rock (Jerry Lucia 1999). This will be further discussed in Chapter 10 as part of the topic of capillary pressure.

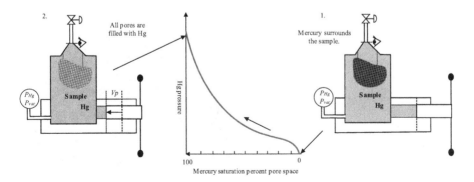

FIGURE 5.7 With increased mercury pressure from position 1 to 2, more pore space can be invaded by mercury. Measuring the total volume change of the pump, the mercury porosity can be determined.

5.3.5 PorOSITY BY IMAGE ANALYSIS

With the development of pore scale imaging, see Chapter 12, from the 2D and 3D images also the porosity can be derived. Figure 5.8 shows a micro-CT scan image of a bead pack and a cross-section through the bead pack.

The 3D image consists of voxels assigned a grey value. In case the contrast in greyscale between the pore space and solid material is sufficient, the greyscale values can be binarized (0–1) based on a threshold grey value, separating the pore space from the rock matrix. In Figure 5.8 the pore space is identified and colored. Counting the amount of pore voxels in comparison to the total amount of voxels forming the core sample will give an estimation of the pore space. Note that the porosity calculated is the total porosity. The error in this approach depends on the image contrast affecting the identification of the pores and the image resolution. The same can be done for 2D on thin sections; see Figure 5.9 for an example of sandstone. This

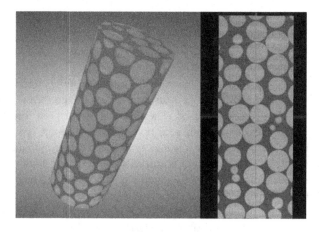

FIGURE 5.8   Micro-CT image of a 3D bead pack where the pore space has been identified (diameter 1 cm).

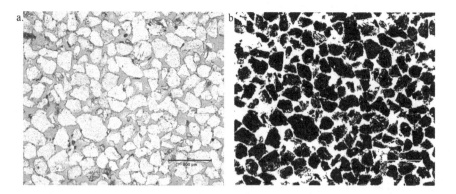

FIGURE 5.9   The determination of porosity in thin section a. using the binarized image b. The white pore area is 36.6% (courtesy of M.B.E. Mørk).

approach assumes the cross-section is representative and can be considered as capillary tube model. But pores are not like tubes and porosity is generally overestimated using 2D images.

5.4 EXPERIMENT 5.1: POROSITY MEASUREMENT METHODS

Experiments 5.1A and 5.1B: Effective porosity determination by helium and liquid porosimeter method

OVERALL OBJECTIVE

Determine porosity of a provided core by use of the helium porosimeter (5.1A) and liquid porosity (5.1B) and compare them. Consider the different sources of error and decide which value will be used in further experiments to determine the pore volume.

5.4.1 EXPERIMENT 5.1 A: HELIUM POROSITY

Objective: Determine the helium porosity of a dry, cleaned core based on Boyle's law. Follow the procedure.

As safety precaution for using helium from a gas bottle, adapt the choke of the gas bottle to the maximum pressure the set-up can handle, ideally a safety margin.

Materials needed:

- Core
- Helium

REPORTING

Parameters to measure in the helium porosity method:

Core # = _____
Length=_____Diameter=_____Bulk volume (V_b)=_____
$V_{chamber}$ = _____
$V_{chamber\ and\ core}$ = _____
Grain volume $(V_g) = V_{chamber} - V_{chamber\ and\ core}$ = _____ Φ_e =_____

5.4.2 EXPERIMENT 5.1 B: LIQUID POROSITY

Objective: Saturate the core used in 5.1A with liquid to determine liquid permeability. Saturate the core with a model brine, 36 g/l NaCl brine, ρ_{brine} = 1.02g/cm³, to avoid clay swelling.

Calculations and report:

1. Calculate the saturated brine weight, $W_{brine} = W_{sat} - W_{dry}$.
2. Calculate the pore volume (saturated brine volume), $V_p = W_{sat} / \rho_{brine}$.
3. Calculate effective porosity, $\Phi_e = V_p/V_b$.

Core No.: 1   D: 3.822 cm,   L: 4.597 cm.

See Exercise 5.6 for an example of comparing the data.

EXERCISE 5.6

In Table 5.2 data is presented from a helium and liquid porosity measurements for core 3.

1. Verify the porosity of the helium method with the liquid porosity (neglect the error range).
2. Name a third method to determine the porosity.
3. What is the largest and smallest porosity to be expected within the error margin for the liquid porosity?
4. Discuss what systematic error and random error can occur in a liquid porosity measurement.

TABLE 5.1
Porosity Determination by Liquid Saturating Method

Calculation Table with Example Values

W_{dry} (g)	W_{sat} (g)	W_{brine} (g)	V_p (cm³)	Φ_e
113.97	123.334	9.364	9.180	17.41

TABLE 5.2
Helium and Liquid Porosity Measurements for Exercise 5.6

	Core 3
Core length (cm)	4.57±0.04
Core diameter (cm)	3.82±0.02
Dry core weight (g)	124.7±0.2
Wet core weight (g)	138.9±0.2
He porosity data. Empty sample cell volume (cm³)	146.0±0.1
He porosity data. Sample cell with sample (cm³)	109.4±0.1
Brine density (g/cm³)	1.019

SOLUTION

1. $V_{bulk} = 4.57 \cdot (\frac{3.82}{2})^2 \cdot \pi = 52.376$ cm^3, with π as 3.1416

   $V_{pore} = \dfrac{(138.90 - 124.75)}{1.019} = 13.89$ cm^3

   Liquid porosity: $13.89/52.376 = 0.27$

   Helium porosity: $\dfrac{(52.376 - (146 - 109.4))}{52.376} = 0.30$

   The helium porosity is 0.03 higher compared to the liquid porosity. This is generally expected based on the expectation that liquid might not reach all the pores and the vacuum was not strong enough. Excess water can, however, counterbalance that. In general, therefore the liquid porosity is taken for further calculations.

2. Using 3D imaging like CT scanning, mercury porosity

3. $V_b = 4.53 \cdot (\frac{3.80}{2})^2 \cdot \pi = 51.26$ cm$^3 \Rightarrow \dfrac{(139.1 - 124.5) \cdot 1.019}{51.26} = 0.28$

   $V_b = 4.61 \cdot (\frac{3.84}{2})^2 \cdot \pi = 53.39$ cm$^3 \Rightarrow \dfrac{(138.7 - 124.9) \cdot 1.019}{52.83} = 0.26$

   This error margin 0.26–0.28 does only include the uncertainty on the weighting and core dimensions. Random errors are not quantified.

4. Systematic error examples: calibration error, tare error, scale accuracy. Random error examples: Draft, dirty balance, different excess liquid to the outside of the core, not full saturation, core dimensions.

5.5 COMPRESSIBILITY

So far, we discussed the porosity without acknowledgment of the fact that the rock used has experienced large stresses on the subsurface. This pressure can have a reversible effect on the rock volume, the bulk volume, and grain and pore volume, the parameters needed to describe porosity. The effect of pressure on rock is described as rock compressibility. The concept of rock compressibility can be like the action of squeezing a sponge; since a sponge is a porous material, and the rock compressibility will be similar. When sponges are pressed during cleaning, some water is expelled due to the applied pressure. A greater force exerted on the sponge leads to a more significant reduction in pore space, resulting in the release of more water. In a parallel manner, rocks experiencing compaction due to the weight of the material above, leading to a decrease in the volume of pores within the rock structure. Rock compressibility can be defined as the relative change in pore volume in response to variations in the compaction pressure experienced within the reservoir.

In any reservoir, as fluids are extracted, the pressure of the fluids in the reservoir decreases, while the burden of the material above is constant. The decompression leads to an expansion of the grain volume, while the bulk volume can additionally reduce and the fluid in the pores will not counteract this. So, the decrease in fluid pressure contributes to a reduction in pore volume and porosity, which is of primary concern to reservoir engineers. It can contribute to the recovery of the field, but it can also lead to negative effects visible at the surface as subsidence.

5.5.1 TYPES OF ROCK COMPRESSIBILITY

The general form of the compressibility can be written as:

$$c = -\frac{1}{V} \left(\frac{dV}{dP} \right)_T \tag{5.8}$$

Here, c denotes rock compressibility (measured in 1/Pa), V represents volume (measured in m³), P stands for pressure (measured in Pa), and dV/dP signifies the rate of volume change per unit pressure change (measured in m³/Pa). The subscript T signifies that the calculation of compressibility is performed under constant temperature conditions, similar to the unchanging temperature within the reservoir. The inclusion of a negative sign serves to render the rock compressibility a positive value. This adjustment is necessary because dV/dP is negative, because when pressure increases the corresponding volume decreases.

Rock compressibility is defined into three major types as follows (Geertsma 1957):

5.5.1.1 Rock Matrix Compressibility

Matrix compressibility is the fractional change in the volume of the solid rock material with a unit change in pressure and is normally expressed as:

$$c_m = -\frac{1}{V_m} \left(\frac{dV_m}{dP} \right)_T \tag{5.9}$$

Where c_m is the matrix compressibility, V_m is the matrix volume, and dV_m/dp is the change in matrix volume over change in pressure.

5.5.1.2 Bulk Compressibility

Bulk compressibility is defined as the fractional change in volume of the bulk of the rock with a unit change in pressure and is expressed with the following equation:

$$c_b = -\frac{1}{V_b} \left(\frac{dV_b}{dP} \right)_T \tag{5.10}$$

Where c_b is the matrix compressibility, V_b is the matrix volume, and dV_b/dp is the change in matrix volume over change in pressure.

5.5.1.3 Pore Compressibility

Pore compressibility is the fractional change in the pore volume of the rock with a unit change in pressure and its equation is given here:

$$c_p = -\frac{1}{V_p}\left(\frac{dV_p}{dP}\right)_T \tag{5.11}$$

An alternative expression for pore compressibility as a function of porosity is also given by noting that porosity increases with increase in the pore volume:

$$c_p = \frac{1}{\phi}\left(\frac{d\phi}{dP}\right)_T \tag{5.12}$$

Formation compressibility is considered to be equal to the pore compressibility since rock and bulk compressibilities are small compared to the pore compressibility, so:

$$c_f = c_p = \frac{1}{\phi}\left(\frac{d\phi}{dP}\right)_T \tag{5.13}$$

Typical values of formation compressibility range from $1 \cdot 10^{-10}$ to $40 \cdot 10^{-10}$ Pa^{-1} (Ahmed 2010).

Equation 5.8 can be rewritten as:

$$c_f = -\frac{1}{V_p}\frac{\Delta V_p}{\Delta P} \tag{5.14}$$

or

$$\Delta V_p = -c_f V_p \Delta P \tag{5.15}$$

Where ΔV_p and ΔP are the changes in the pore volume and pore pressure, $(V_{p,final} - V_{p,initial})$ $(P_{final} - P_{initial})$ respectively. Considering equations 5.8 and 5.9, Geertsma (1957) suggested that bulk compressibility is related to the pore compressibility by the following expression:

$$c_b = \phi c_p \tag{5.16}$$

Equation 5.13 can be rearranged as:

$$c_f \partial p = \left(\frac{1}{\phi}\right)\partial\phi \tag{5.17}$$

Integrating the previous equation gives:

$$c_f \int_{P_0}^{P} \partial P = \int_{\phi_0}^{\phi} \frac{\partial \phi}{\phi} \tag{5.18}$$

$$c_f (p - p_0) = \ln(\frac{\phi}{\phi_0}) \tag{5.19}$$

or:

$$\ln \frac{\phi}{\phi_0} = e^{c_f (p - p_0)} \tag{5.20}$$

$$\phi = \phi_0 \, e^{c_f (p - p_0)} \tag{5.21}$$

Considering Taylor expansion as:

$$e^x = 1 + x + \frac{x^2}{2!} + \frac{x^3}{3!} + \dots \tag{5.22}$$

It gives:

$$\phi = \phi_0 \left(1 + c_f (p - p_0)\right) \tag{5.23}$$

The total reservoir compressibility denoted as c_t, can be defined as:

$$c_t = S_g c_g + S_o c_o + S_w c_w + c_f \tag{5.24}$$

Empirical correlations between pore compressibility and porosity are available in the literature (Hall 1953; Newman 1973).

5.5.2 Laboratory Measurement of Compressibility

Laboratory measurement of compressibility is like squeezing a saturated sponge to remove the water inside. The reservoir process can be likened to placing a saturated sponge within a sealed plastic bag, subsequently placing multiple weighty bricks on top of it, and eventually puncturing a hole in the bag.

In petroleum reservoirs, the overburden pressure remains constant, while the pressure of the fluid within pores varies, leading to alterations in pore volume. In the lab setting, we manipulate the confining pressure applied to the core plug (overburden) while maintaining a consistent pore pressure. Knowing that the effective compaction pressure acting on the matrix is determined by the contrast between overburden and pore pressures, this principle enables us to achieve valuable outcomes during laboratory investigations (Amyx et al. 1960; Hall 1953; Geertsm 1957).

Figure 5.10 illustrates the experimental arrangement used to quantify compressibility in a laboratory setting. It involves a pump utilized to introduce oil or

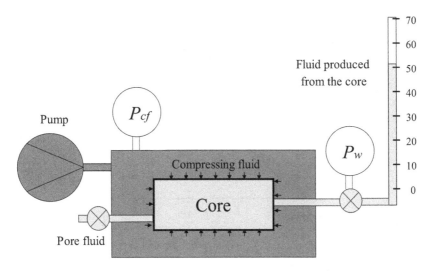

FIGURE 5.10 Schematic of the apparatus for measuring compressibility.

water as the overburden liquid, a pressure gauge, a core accompanied by a core holder, and a measuring burette designed to record the volume of the displaced fluid within the core.

PROCEDURE

The step-by-step laboratory procedure for measuring compressibility is outlined here:

1. Begin by measuring the porosity and pore volume of the core sample without compression.
2. Completely saturate the core with a liquid such as brine.
3. Place the saturated core within a core holder.
4. Initiate the procedure with atmospheric pressure as the initial confining pressure.
5. Gradually elevate the pressure to reach a level denoted as P_1. This process leads to a reduction in pore volume, expelling water that is collected in a graduated pipette connected to the system. Allow sufficient time for the system to reach equilibrium with no further fluid production.
6. Progressively increase the pressure to gather additional data points.

Once multiple data points are collected, a plot can be generated. This graph illustrates the change in pore volume at each increment (ΔV_p) compared to the product

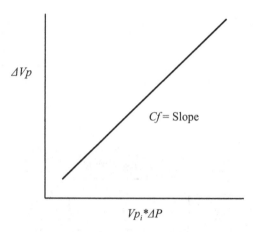

FIGURE 5.11 Fractional volume as a function of the product of fractional volume and pressure difference.

of the pore volume and the change in pressure $(V_p \cdot \Delta P)$ for each corresponding increment. As indicated by Equation 5.15, the slope of the graph gives the formation compressibility, as shown in Figure 5.11.

Note that the compressibility experiment can be extended measuring resistance over the core, to obtain the cementation factor, according Equation 6.5 (see Chapter 6, Section 6.5.3).

EXERCISE 5.5

Table 5.3 presents the outcome of a laboratory compressibility test conducted on a carbonate core. The sample's weight when dry is 149.3 g, and when fully saturated with water (with a density of 1.0 g/cm³), it weighs 161.5 g. Determine the average compressibility of the formation for this sample.

SOLUTION

Pore volume of the sample can be calculated as:

$$V_p = \frac{\text{Saturated weight} - \text{dry weight}}{\text{Fluid density}} = \frac{161.5 - 149.3}{1.0} = 12.2 \text{ cm}^3$$

Based on Equation 5.15 a plot of the initial pore volume times the pressure difference; $V_p \cdot \Delta P$, versus the cumulative volume of expelled fluid, will give the compressibility as gradient, see Figure 5.11.

From Figure 5.12 the compressibility factor is found to be $1.49 \cdot 10^{-5}$ psi^{-1}.

TABLE 5.3
Results of Compressibility Experiments on a Carbonate Core

Pressure [psia]	Production per Pressure sSep [cm³]	Cumulative Production [cm³]
0.00	0.00	0.00
1,500	0.27	0.27
2,000	0.12	0.39
3,000	0.16	0.55
5,000	0.36	0.91
7,000	0.35	1.26

TABLE 5.4
Calculated Values Using Data of Table 5.3

ΔP [psi]	$V_p \cdot \Delta P$ [cm³ psi]	ΔV_p [cm³]
0	0.0	0
1,500	18,300.0	0.27
2,000	24,400.0	0.39
3,000	36,600.0	0.55
5,000	61,000.0	0.91
7,000	85,400.0	1.26

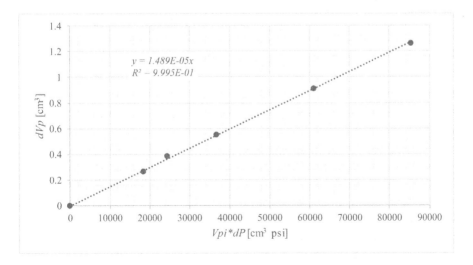

FIGURE 5.12   Plot of the $V_p \cdot \Delta P$ versus ΔV_p of the data from Table 5.4 with the compressibility as gradient.

REFERENCES

Ahmed, T. *Reservoir engineering handbook.* Gulf Publishing, 2000.

Amyx, J.W., Bass Jr., D.M., Whiting, R.L. *Petroleum reservoir engineering.* McGraw-Hill, 1960.

Bear, J.C. *Dynamic of fluids in porous media.* American Elsevier, 1972.

Cosse, R. *Basic of reservoir engineering.* Editions Technip, 1993.

Geertsma, J. "The effect of fluid pressure decline on volumetric changes of porous rocks." *Transactions of the AIME 210*, 1957: 331–340.

Hall, H.N. "Compressibility of reservoir rocks." *Transactions of the AIME 198*, 1953: 309.

Jerry Lucia, F. *Carbonate reservoir characterization.* Springer, 1999.

Monicard, R.F. *Properties of reservoir rocks; core analysis.* Edition Tecnip, 1980.

Newman, G.H. "Pore volume compressibility of consolidated, friable, and unconsolidated reservoir rocks under hydrostatic loading." *Journal of Petroleum Technology*, 1973: 129–134.

6 Electrical Properties of Porous Media

6.1 INTRODUCTION

Porous rocks are comprised of solid grains and void space. The solids, with the exception of certain clay minerals, are nonconductors of electric charge. On the other hand, among the fluids present in petroleum reservoirs (oil, gas, and brine), the oil and gas are nonconductors and brine is a conductor as it contains dissolved salts, such as NaCl, $MgCl_2$, and KCl, normally found in formation water. Current is conducted in water by movement of ions and can therefore be termed electrolytic conduction. So, the electrical properties of a rock depend on the geometry of the voids and the fluid with which those voids are filled. In the context of electrical properties of reservoir rocks, we will focus on resistivity, which is the inverse of conductivity and quantifies the material's resistance to the flow of electricity.

This chapter is devoted to important aspects related to electrical properties of reservoir rocks, which can give information on formation type (clay versus sandstone), porosity, and pore content (hydrocarbon and brine content) and wettability changes. In well logging, specifically resistivity logging and spontaneous potential logging, this is applied to obtain information of the various formations along the well bore and impact of the drilling activity in the near well-bore area. The laboratory data is used as input for the interpretation.

6.2 RESISTIVITY OF POROUS MEDIA

Concept of resistivity of porous media operates on the principles of electrical circuits. Resistivity is an intensive property of rock and fluid, which can be defined as a measure of (the inverse of) the electrical flow capacity of the rock. The resistivity of a porous material is defined by:

$$R = \frac{rA}{L} \tag{6.1}$$

where
A = cross-sectional area, $[m^2]$
L = length, $[m]$
r: resistance [Ohm; Ω]; total resistance to flow of electrical current over volume of interest. R: resistivity $[\Omega m]$; the ability to transmit a flow of electric current through the material, normalized to length.

Note: Conductance = 1/resistance	Unit: [Siemens, $1/\Omega$]
Conductivity = 1/resistivity	Unit: $[1/\Omega m]$, [Siemens/m] or [mho/m],

DOI: 10.1201/9781003382584-6

TABLE 6.1

Some Examples of Resistivity Values for Pure Materials

Medium	Resistivity [Ω m]	Medium	Resistivity [Ω m]
Quartz	10^{12}–10^{14}	Pyrite	10^{-4}
Clay	1–200	Oil-gas	∞
Brine NaCl 20°C 100–200 000 ppm	0.043-10	Pure water	∞
Brine NaCl 80°C 100–200 000 ppm	0.013-10	Pure water	∞

Source: Modified from Liu, 2017.

Table 6.1 gives some examples of resistivity values. However, for a complex material like rock as non-conductor containing non-conducting oil and conducting brine, resistivity can not be generalized. The resistivity to an electric current in porous rock is due primarily to the movement of dissolved ions in the brine that fills the pore of the rock. Therefore the resistivity of the rock depends on: Porosity and change in porosity dependent on formation stress, pore geometry, salinity of water, composition of rock, and temperature. The resistivity varies with temperature due to the increased activity of the ions in solution as temperature increases.

Due to the conductivity properties of reservoir formation water, the electrical well-log technique is an important tool in determination of water saturation versus depth measured in a well and thereby a reliable resource for in-situ hydrocarbon evaluation.

The theory of the electrical resistivity log technique generally applied in petroleum engineering was developed by Archie (Archie 1942), the so-called Archie's equation. This empirical equation was derived for clean water-wet sandstones over a reasonable range of water saturation and porosities. In practice, Archie's equation should be modified according to the rock properties such as clay contents, wettability, and pore distribution (Ellis and Singer 2007). The following is a brief presentation of the main electrical properties of reservoir rocks and related parameters.

Formation Factor: The most fundamental concept considering electrical properties of rocks is the formation factor F. Archie showed that the resistivity of a clean formation is proportional to the resistivity of the brine saturated rock. The constant of proportionality is called as resistivity factor or formation factor (F or FF or FRF). Figure 6.1a shows a. a box with length L and cross sectional area of A filled 100% with brine with resistivity of R_w and b. a box that resembles a porous media filled 100% with the same brine ($S_w = 1$) having resistivity of R_o and c. a box that resembles a porous media filled with brine and hydrocarbon ($S_w < 1$) having resistivity of R_t, depending on the water saturation.

Mathematically the formation factor can be expressed as:

$$F = \frac{R_0}{R_w} \tag{6.2}$$

where
R_o = the resistivity of the rock when saturated 100% with water, Ωm.
R_w = the water resistivity, Ωm.

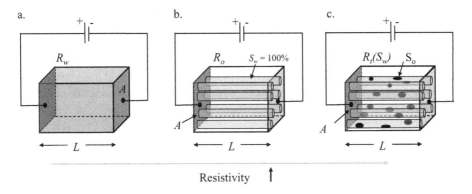

FIGURE 6.1 Schematic showing resistivity in porous media dependent on the pore content.

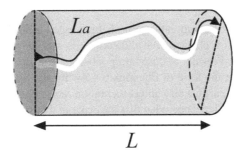

FIGURE 6.2 Definition of tortuosity, actual path length La versus core length L.

The formation factor can be easily determined by measuring R_o and R_w in laboratory. The formation factor is always larger than 1 as the rock will have a larger resistivity. With a change in R_w, F remains constant. A larger F indicates more rock resistivity.

Assuming the rock is a bundle of capillaries and the pores or capillaries are filled with brine, it is to be expected that the formation factor scales with porosity. If porosity is 20% we could expect F = 1/0.2 = 5. However, this is generally not the case. Wyllie (Wyllie 1953) developed the relation between the formation factor and other properties of rocks, like porosity F and tortuosity τ, which cause this deviation. Tortuosity can be defined as $(L_a/L)^2$, where L is the length of the core and L_a represents the effective path length through the pores as it is shown in Figure 6.2. Note that, based on the simple pore models, the following relationship can also be derived:

$$F = \frac{\tau}{\phi} \qquad (6.3)$$

where
F = formation factor
t = tortuosity of the rock with the definition $(L_a / L)^2$,
F = porosity of the rock.

So with a more tortous path, the F increases, implying more resistance to flow, due to a relative longer path way through the core. Using the aforementioned definition tortuosity > 1.

Cementation factor: Archie (Archie 1942), however, suggested a slightly different relation between the formation factor and porosity by introducing the cementation factor m:

Archies first equation:

$$F = \phi^{-m}$$
(6.4)

where:
F = porosity of the rock
m = Archie's cementation factor.

and reported that the cementation factor probably ranged from 1.8 to 2.0 for consolidated sandstones and for clean unconsolidated sands was about 1.3. Archie, as mentioned before, reported results of correlating laboratory measurements of formation factor with porosity in the form of Equation 6.4 Wyllie and Spangler (1952) investigated the influence of particle size and cementation factor on the formation factor of a variety of materials. They concluded that the cemented aggregates exhibit a greater change in formation factor with a change in porosity than the unconsolidated aggregates. Based on this study, the general form of the relation between formation factor and porosity should be:

$$F = a\phi^{-m}$$
(6.5)

where m is a constant depending on cementation and a is a constant controlled by the porosity of the unconsolidated matrix prior to cementation and usually equals 1. A comparison of some suggested relationships between porosity and formation factor is shown in Figure 6.3.

6.2   EFFECT OF CONDUCTIVE SOLIDS

The clay minerals present in a natural rock act as a separate conductor and are sometimes referred to as "conductive solids". The bound water in the clay and the ions in the water act as conducting materials. More water and more ions will increase the conductivity and respectively reduce the resistivity. The effect of the clay on the resistivity of the rock is dependent upon the amount, type, and manner of distribution of the clay in the rock (e.g. layers or coating on the grains). The presence of conducting clay or shale in a sand bed lowers the true formation resistivity R, and if not corrected, this will result in overestimating porosity and S_w, or interpreting zones as water-bearing zones that are actually oil-bearing.

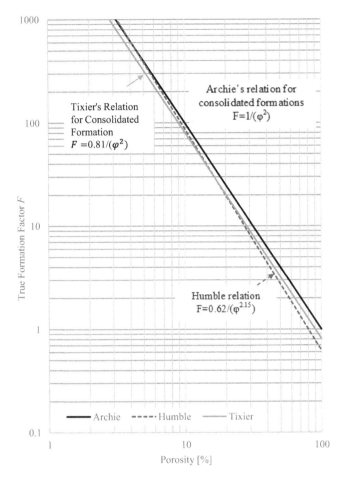

FIGURE 6.3 Formation factor as a function of porosity.

The formation factor for a clay-free sand is constant. The formation factor for clayey sand increases with decreasing water resistivity and approaches a constant value at a water resistivity of about 0.1 Ωm, as illustrated in Figure 6.4. The apparent formation factor F_a was calculated from the definition of the formation factor and observed values of R_{oa} and R_w with $F_a = R_{oa}/R_w$. Wyllie and Spangler (1952) proposed that the observed effect of clay minerals was like having two electrical circuits in parallel: The conducting clay minerals and the water-filled pores. Thus:

$$\frac{1}{R_{oa}} = \frac{1}{R_c} + \frac{1}{FR_w}$$

(6.6)

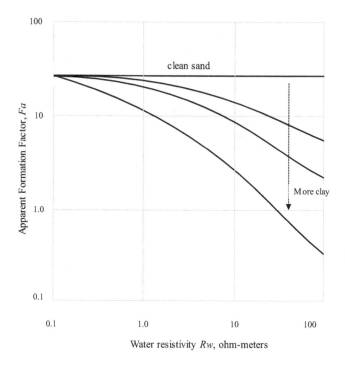

FIGURE 6.4 Apparent formation factor versus water resistivity for clay and clean sands (modified from Amyx et al. (1960)).

where Roa is the resistivity of a shaly sand when 100% saturated with water of resistivity R_w. R_c is the resistivity due to clay minerals. FR_w is the resistivity due to the distributed water, and F is the true formation factor of the rock (the constant value when the rock contains low-resistivity water).

The data presented in Figure 6.5 graphically represent the confirmation of the relationship expressed in Eq. 6.6. The plots are linear and are of the general form:

$$\frac{1}{R_{oa}} = C\frac{1}{R_w} + b \tag{6.7}$$

where C is the slope of the line and b is the intercept. Comparing Eq. 6.6 with Eq. 6.7, it may be noted that $C = 1/F$ and $b = 1/R_c$. The line in which $b = 0$ indicates a clean sand, then as $R_w \rightarrow 0$, $\lim\limits_{R_w \to 0} F_a = \dfrac{R_c}{\frac{R_c}{F}} = F$

Therefore, F_a approaches F as a limit as R_w become small. This was observed in Figure 6.4

$$\frac{1}{R_{oa}} = C\frac{1}{R_w} = \frac{1}{FR_w} \quad \text{or} \quad R_o = FR_w \tag{6.8}$$

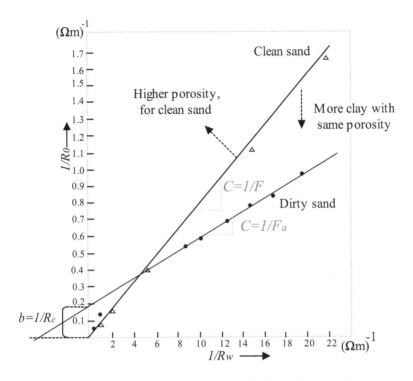

FIGURE 6.5   Water-saturated rock conductivity as a function of water conductivity.

Eq. 6.6 can be rearranged to express the apparent formation factor in term of R_c and FR_w:

$$R_{oa} = \frac{R_c R_w}{R_w + \dfrac{R_c}{F}} \quad \text{and} \quad F_a = \frac{R_c}{R_w + \dfrac{R_c}{F}} \tag{6.9}$$

Since Wyllie and Spangler in 1952 (Wyllie and Spangler 1952), several other models have been developed to describe the effect of conductive clays in rocks (Ellis and Singer 2007; Qi and Wu 2022).

6.3   EFFECT OF OVERBURDEN PRESSURE ON RESISTIVITY

Confinement or overburden pressure may cause a significant increase in resistivity. According to Equation 6.5 the porosity changes, and tortuosity (Equation 6.3) increases the formation factor (Jing 1992). This usually occurs in rocks that are not well cemented and in lower porosity rocks. F will increase linearly in the case that rocks behaves elastically, where the original state is restored when pressure is removed. When this is not the case anymore, (plastic, ductile behavior) irreversible damages are created at pore level and the change in F can deviate from the linear trend (Nourani et al. 2022).

6.4 RESISTIVITY OF PARTIALLY WATER-SATURATED ROCKS

Another fundamental notion of electrical properties of porous rocks containing both water and hydrocarbons is the resistivity index I, which is defined as the ratio of resistivity of porous media that contains brine and hydrocarbons at different saturation ratios to the saturated porous media with 100% brine. Mathematically, the resistivity index can be expressed as:

$$I = \frac{R_t}{R_o} \tag{6.10}$$

where
I = resistivity index (dimensionless)
R_t = resistivity of the rock when saturated partially with water, Ωm.
R_o = resistivity of the same rock when saturated with 100% water, Ωm.

As hydrocarbons are considered less conductive, I will always be larger than 1. When oil and gas are present within a porous rock together with a certain amount of formation water, its resistivity is larger than R_o since there is less available volume for the flow of electric current. This volume is a function of the water saturation S_w. Equation 6.10 indicates that the resistivity index is a function of water saturation and the path length. From the theoretical development, the following generalization can be drawn:

$$I = C' S_w^{-n} \tag{6.11}$$

where $I = R_t/R_o$ is the resistivity index, C' is a constant including the tortuosity, and n is the saturation exponent and S_w = water saturation as a fraction [-].

The generally accepted formation that relates water saturations and true resistivity R_t is that of Archie, which may be written in the following different form:

$$S_w = \sqrt[n]{\frac{R_o}{R_t}} = \sqrt[n]{\frac{FR_w}{R_t}} = \sqrt[n]{\frac{R_w a}{R_t \phi^m}} \tag{6.12}$$

where a is a unique property of the rock. All the equations fitted to the experimental data have assumed that both C' and n of Eq. 6.11 are constants, with $C' = 1$.

The saturation exponent n is affected by wettability, overburden pressure, and the nature and microscopic distribution of the reservoir fluids. In Archie's equation n is 2.0, but models with other n-values are also available (Figure 6.6). Generally n ranges from 1.4 to 2.2 ($n = 2.0$ if no data are given). N increases with the core being more oil wet for uniform oil wet cores > 10 (Anderson 1986; Morgan and Pirson 1964). N, m, a, and $R_t(S_w)$ can be obtained from laboratory experiments. Once the

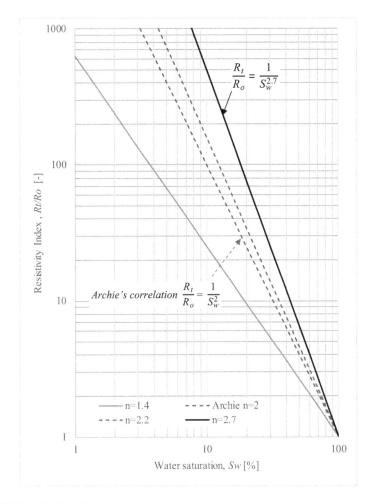

FIGURE 6.6   Resistivity index versus water saturation.

constants of Equation 6.11 are known, R_t and R_o can be obtained from well logging data, and with the Archie saturation equation S_w can be determined. Based on the material balance equation for the formation, $S_w + S_o + S_g = 1.0$, the hydrocarbon reserve in place may be calculated.

EXERCISE 6.1

The following data for the core plug and the resistivity measurement of this sample are given:

Core plug diameter: 3.71 cm
Core plug length: 6.20 cm
Porosity: 0.21

Reference resistance: 1,000 Ω
Resistivity of water: 0.188 Ωm
Measured voltage over the core plug: 0.67 V
Measured voltage over the reference resistor: 2.29 V

The sample is 100% saturated with water. Calculate the resistivity, formation factor, and cementation factor of the core sample.

$$\text{Equations: } R = r\frac{A}{L} \quad F = \frac{R}{R_w} \quad F = \frac{1}{\phi^m}$$

SOLUTION

$$\frac{r_x}{r_r} = \frac{V_x}{V_r} \Rightarrow r_x = r_r\frac{V_x}{V_r} = 1000\frac{0.67}{2.29} = 292.6 \ \Omega$$

$$R_0 = r_x\frac{A}{L} = 292.6\frac{\pi\left(3.71^2\right)}{4\cdot6.20} = 510 \ \Omega\text{cm} = 5.10 \ \Omega\text{m}$$

$$F = \frac{R_0}{R_w} = \frac{5.10}{0.188} = 27.14 \quad F = \frac{1}{\phi^m} \Rightarrow m = \frac{-logF}{log\phi} = \frac{-\log 27.14}{\log 0.19} = 2.00$$

6.5 MEASUREMENTS OF ELECTRICAL PROPERTIES

6.5.1 RESISTANCE MEASUREMENTS OF ROCKS

Figure 6.7 shows an example of a set-up to measure core resistivity under ambient pressure and temperature conditions. It consists of a generator to set a voltage difference over a reference resistor and the core sample placed in series. The core is in this case clamped between two electrodes. There are other configurations possible (McPhee et al. 2015). The current, I, can be obtained measuring the voltage difference, V_r, over a known resistance r, using Ohm's law: $V = r\cdot I$. By measuring the voltage over the sample, V_c, and knowing I as V_r/r_r, the resistance of the core sample can be derived using:

$$r_c = \frac{V_c}{V_r/r_r} \tag{6.13}$$

The resistivity of the sample can be derived when the size of the sample is known. For safety reasons the generator shall be used with AC current. Here experiments will be considered at room temperature. However because ratios of resistivity are used in the calculations, the temperature dependence is assumed to cancel out, such that room temperature data can also be used for reservoir conditions.

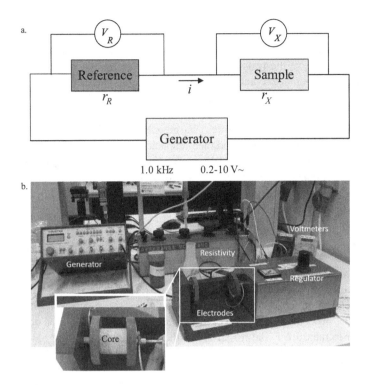

FIGURE 6.7 The electrical circuit of resistance measurements.

PROCEDURE

- Switch on the equipment.
- Wet the two electrodes using a paper wetted with saltwater, for a good connection.
- Take a 100% water-saturated core and assemble it between the electrodes. Adjust the clamping pressure if needed.
- Place the plastic cover over the core-electrode system to avoid evaporation.
- Regulate the resistance on the "reference resistance" to 1,000 Ω.
- Turn on the generator and set its output frequency to 1,000 Hz. (The amplitude setting shall be < 7 V).
- Read the values from the display of the two voltage meters, one for the reference resistance and one for the core.

Shutting down

- Turn off the generator to remove the power from the circuit.
- Take the core out of the clamps
- Turn off and unplug all devices.
- Clean the electrodes and area underneath with distilled water after the measurement.

ERRORS

Errors in the measurement can occur because of a bad connection between electrodes and core. In this case the resistance measured will be higher. If the core resistance is measured without sleeve and sleeve pressure, excess water forming a water film around the core will create a conduit for current flow so that the measured resistance (and resistivity) will be too low. Finally, when using alternative current (AC), the core can act as a capacitor that stores charge, which artificially increases the resistance measured, then called impedance (capacity plus resistance). The impedance is dependent on the frequency. High enough frequency > 10kHz would avoid the capacitor effect completely (McPhee et al. 2015). Or when using lower frequencies, the sinusoidal waves of the set V and measured I shall be less than 0.5° out of phase, then the measurement corresponds to resistance (McPhee et al. 2015).

6.5.2 DETERMINATION OF FORMATION FACTOR, F AND LIQUID RESISTIVITY, R_w

In order to calculate F according to Equation 6.2, the resistivity of 100% brine-saturated core is to be measured, R_0. The core shall be saturated according to the procedure described in Chapter 5, as part of the liquid porosity experiment (Section 5.2.3). Applying the procedure described earlier (6.5.1) the resistivity can be derived.

In order to measure F, the resistivity of the formation brine must also be known, R_w. It can be obtained from logs (Peveraro 1992) but with a clean sample from the reservoir it can be measured in the laboratory under the right conditions. Note that drilling mud invasion can contaminate the brine sample and needs to be accounted for. Then the use of a composed brine (See Section 4.2.2) is advised.

Liquid resistivity is measured in a solution between two platinum electrodes where a potential difference is set and the distance between the plates is known. Polarization of the electrode shall be limited, where an unwanted charge accumulation at the electrode causes resistance of the electrode dependent on the current. Electrode material choice is important. Also capacitive coupling of electrodes shall be avoided (Worthington et al. 1990). Calibration of the cell with a solution with a known resistance is required. Note that temperature has an effect on the liquid resistivity and needs to be reported with the data (Arps 1953). Further guidance can be obtained from literature (Worthington et al. 1990; McCoy et al. 1997).

When knowing the composition of the brine salinity the resistivity can also be derived (Tiab and Donaldson 1996). There are literature sources available, like for example Schlumberger, gen 4 chartbook_0017 (Schlumberger 2009), where the resistivity for NaCl solutions can be obtained, dependent on the temperature. However, brines most likely do not only contain NaCl salt. Therefore, the equivalent NaCl concentration needs to be calculated such that resistivity data for NaCl solutions can still be applied. The equivalent NaCl concentration is defined as:

$$C_{sm} = \sum_{i=1}^{n} M_i C_{sii} \tag{6.14}$$

where C_{sii} = ion concentration and M_i = weighting multiplier.

The weighting multiplier is used to convert the contribution of ions other than Na^+ and Cl^- to an equivalent NaCl concentration dependent on the concentration of ions present in the water. Literature tables can be found to obtain the multiplier for various ions dependent on concentration, for example Schlumberger, gen 4 chartbook_0017 (Schlumberger 2009). Ion interactions between different species is not taken in account.

6.5.3 DETERMINATION OF CEMENTATION FACTOR, m AND a

The cementation factor depends on the level of cementation of the grains or tortuosity. Experiments to measure m shall be performed on a group of rock samples with different porosities within a formation but similar pore size distribution on which resistivity tests are performed. Equation 6.5 can be rewritten as:

$$\text{Log } F = \log a - m \log \phi \tag{6.15}$$

A log-log plot of porosity versus F will give the m factor as gradient and $\log a = \log F$ for $\varphi = 1$.

When m and a can be measured in the lab, the formation factors measured in the field can also be converted to porosities, to verify the porosities obtained from the neutron density log.

6.5.4 DETERMINATION OF SATURATION EXPONENT, n

Equation 6.11 can be rewritten as

$$n \log S_w = \log \frac{R_w a}{R_t \phi^m} = \log \frac{R_o}{R_t} \tag{6.16}$$

A plot of $\log S_w$ *versus* $\log \dfrac{R_o}{R_t}$ gives as slope the saturation exponent n.

- m, a can be measured in laboratory as described in Section 6.5.3.
- R_w can come from lab test with the reservoir brine or from SP logging (see section 6.5.2).
- R_0 can be obtained from a 100% water-bearing zone or from laboratory experiment.
- R_t needs to be determined in controlled experiments, where the brine saturation S_w in the core is known (section 6.5.5).

6.5.5 EXPERIMENTAL SET-UP TO OBTAIN R_T

R_t can be measured on a core, which in steps contains different water saturations. This can be combined with various multiphase flow experiments, later described in this book.

- It can be measured during steady-state (SS) relative permeability experiments (Chapter 11) where two or more electrodes are incorporated in the set-up to measure the resistance over the core once a constant saturation is reached. Alternatively, it can be obtained during low-flow rate unsteady-state flooding experiments where a porous plate or semi-permeable membrane is incorporated for controlled core drainage (pressure controlled) changing S_w. Here the saturation might not be uniform.

- It can be run at the same time as capillary pressure measurements; like centrifuge or porous plate drainage experiment (Chapter 10). The sample is to be removed from porous plate apparatus or centrifuge after each saturation step, and resistivity can be measured in ambient conditions.

Note that measurements shall be done when the saturation is uniform, so time shall be allowed for the equilibrium to be established. Monitoring the resistance over time shall give this information (Dunlap et al. 1949).

6.5.6 MONITORING OF WETTABILITY CHANGE USING RESISTIVITY MEASUREMENTS

The surface affinity or wettability of the core for oil or water affects the fluid distribution in the core. At initial water saturation, S_{wi}, in a water wet core, brine creates a continuous, conducting film over the grains. This will give a relative low resistance. In an oil wet core brine S_{wi} is located discontinuously in individual small pores. Due to the discontinuity of the conducting phase the resistance in an oil wet core will be higher in comparison to the water wet case.

This can be used in the lab to monitor the aging process, where the core changes from water to oil wet. This occurs in time by adsorption of oil components to the rock surface, stimulating a change in fluid distribution in the core. This process can be monitored by resistivity measurements. The aging process is completed once the increase of resistance over the core has ceased.

6.6 EXPERIMENT 6.1: RESISTIVITY MEASUREMENTS OF FLUID-SATURATED ROCKS

DESCRIPTION

The objective of this experiment is to measure the main electrical properties of porous rock like water resistivity, formation factor F, a, cementation factor m, resistivity index I, and saturation exponent n.

PROCEDURE

Measure the resistivity of a core after different experiments of the SCAL analysis:

- A dry core
- A core with a saturation of 3 and 6 wt% NaCl to obtain R_0. Use the liquid porosity experiment to saturate the core with the initial salt solution. Use the single-phase permeability experiment to exchange the salt solutions (Chapter 7).

- A core at residual oil saturation to obtain R_t (S_{wi}). This can be obtained after imbibition (water injection) e.g. the Amott test (Chapter 9), the capillary pressure test using porous plate or the centrifuge (Chapter 10), or after the (un)steady-state relative permeability test (Chapter 11). Note that the R_0 shall be measured with the same brine used in the multiphase flow experiments.

Data needed from other experiments; porosity, core dimensions.

CALCULATIONS AND REPORT

Here some example tables are presented for reporting the data and calculations.

TABLE 6.2
Measurements of Fluid Resistivity

		Measured			Calculated
	Concen- tration	Cell Diameter D	Cell Length L	Cell Resistance, r_x	Water Resistivity R_w [W.m]
Sample	[mg/l]	[m]	[m]	[W]	at $T =$____°C

TABLE 6.3
Formation Factor and Cementation Factor; Calculation Table

				Measured			Calculated	
Core No.	Core D [m]	Core L [m]	r_0 [Ω] T=__°C	Porosity ϕ	Water Resistivity R_w [Ω m]	Formation Factor, F	Cementation Factor, m	a

TABLE 6.4
Resistivity Index and Saturation Exponent; Calculation Table

			Measured					Calculated	
Core No.	Core D [m]	Core L [m]	Sw [-]	rt [Ω]	r_0 [Ω]	R_0 [Ω m]	Rt [Ω m]	Resistivity Index, I	Saturation exponent, n

EXERCISE 6.2

In the following, some example data and measurements are given.

1. Calculate water resistivity, R_w for room temperature at 20°C

Equation:

$$R_w = \frac{r_x A}{L} = \frac{r_x \pi D^2}{4L}$$

2. Calculate formation factor, F, a, and cementation factor, m

R_w is equal to the value of R_w in Table 6.5.
From Figure 6.8, the m can be found as the negative gradient and a as the negative constant. This gives a cementation factor of 1.96 and an a of 0.62.

TABLE 6.5
Measurements of Fluid-Saturated Rocks. Calculation Table. Water Resistivity

	Measured			Calculated
	Cell Diameter	**Cell Length**	**Cell Resistance,**	**Water Resistivity**
Sample	**D (m)**	**L (m)**	**r_x (Ω)**	**R_w (Ω.m)**
Brine-36 g NaCl/l	0.0378	0.0418	6.33	0.17

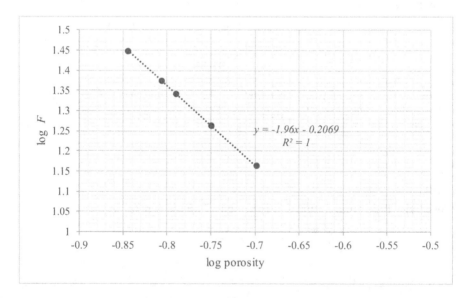

FIGURE 6.8   The log of porosity versus the logarithm of F to obtain the cementation factor m and constant a, using Equation 6.5.

TABLE 6.6

Formation Factor, Tortuosity, and Cementation Factor Calculation Table

			Measured		Calculated
Core	Core D	Core L	Porosity, ϕ	R_o	Formation
No.	[m]	[m]	[−]	[Ω. m]	Factor, F
1	0.0378	0.0418	0.14	4.77	28.06
2	0.0361	0.0416	0.16	4.03	23.69
3	0.0375	0.0420	0.18	3.11	18.29
4	0.0360	0.039	0.16	3.74	22.00

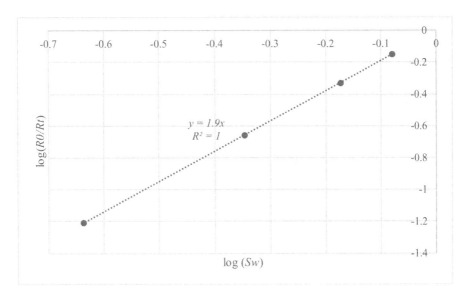

FIGURE 6.9 The plot the log of the brine saturation versus the inverse resistivity index to obtain the saturation gradient.

3. Calculate resistivity index, I, saturation exponent, n, using core 1 from Table 6.6.

Plot the log of the brine saturation versus the inverse resistivity index $I = \dfrac{R_t}{R_o}$, such that the gradient of the curve provides the saturation factor, n (see Figure 6.9).

TABLE 6.7

Resistivity Index and Saturation Exponent; Calculation Table

	Measured					Calculated	
Core No.	Core D [m]	Core L [m]	Sw [-]	R_t (Ω m)	*R_o (Ω m)	Resistivity Index, I	Saturation exponent, n
1	0.0378	0.0418	0.23	77.84	4.77	16.32	1.9
			0.45	21.75		4.56	
			0.67	10.21		2.14	
			0.83	6.80		1.42	

*R_o is equal to the value of core 1 in Table 6.6.

EXERCISE 6.3

In a resistivity experiment (see Figure 6.7 for the set-up), the following values are measured:

Temperature (°C)	30
Resistance r_R (Ohm)	1,000
Voltage, V_R (V)	2.1
Voltage V_x (V)	0.9
Mass dry core (g)	106.87
Mass wet core (g)	107.75
Core diameter (cm)	3.776
Core length (cm)	4.196
brine salinity (ppm)	20,000
Porosity [-]	0.123
brine density (g/cm³)	1.009
1 g/cm³ = 1,000,000 ppm	

1. How is current conducted in a rock saturated with oil and brine at initial water saturation?
2. Calculate the core resistivity R.
3. According to the capillary bundle model the core resistivity would be linearly inversely proportional to the porosity, according $R = \dfrac{R_w L}{\varphi A}$. The formation factor $F = \dfrac{R_o}{R_w}$ would be inversely proportional to the porosity. It is not. Why not; how is F influenced?
4. What is tortuosity? Calculate the tortuosity using; F formation factor $F = \dfrac{\tau}{\varphi}$, with φ = porosity and T the tortuosity.

5. What would the porosity be based on Humble's formula?
6. Describe a potential source of error in the resistance measurement that would increase the obtained resistivity and one that would decrease the obtained resistivity.

SOLUTION

1. Via the brine and possible clay-bound water.
2. Calculate the core resistivity R using Ohms law; $V = I \cdot r$ for a serial circuit, with I as the current and $R = \dfrac{r_x A}{L}$, with R as specific core resistivity, $r_x =$ core resistance, A as the cross sectional area, [m^2] and L as the core length [m].

$$r = V/I = \frac{0.9}{2.1/1000} = 428.6 \text{ Ohm}$$

$$R = \frac{428.6 \cdot \pi \cdot (\dfrac{0.03776}{2})^2}{0.04196} = 11.44 \text{ Ohm m}$$

3. Tortuosity of the pore structure increases the pathway of the current, so there is more resistivity than expected based on the capillary bundle model. When Ro becomes higher, F will be higher.

4. Formation factor $F = \dfrac{R_0}{R_w} = \dfrac{\tau}{\varphi}$, with $\varphi =$ porosity and τ the tortuosity.

For $T = 30°C$ and brine salinity of 20,000ppm = 0.002g/l, R_w will be 0.29 Ohm m.

$$F = 11.44 / 0.29 = 39.443$$
$$\tau = F \cdot 0.123 = 4.85$$

5. Humbles equation $a = 0.62$, $m = 2$: $F = 0.62/\phi^2 \rightarrow \phi = \sqrt{\dfrac{0.62}{F}} = 0.125$

6. An increase can be caused by bad connected clamps. A resistance decrease can occur due to excess water at the surface of the core.

6.7 SPONTANEOUS POTENTIAL AND ELECTROKINETICS

So far it was assumed in this chapter that there is no interaction of the grains with the brine, except for clay. This is sufficient for considering electric circuits; it is, however, not reality. A solid submerged in liquid acquires a surface charge due to dissociation, partial charges, or atom exchange (Gin et al. 2021). To balance this charge towards the brine, which is neutral, a so-called electronic double layer (EDL) is formed, counterbalancing the surface charge. Figure 6.10 shows how this is built up according to the Stern model, having two bound layers – the Stern layer

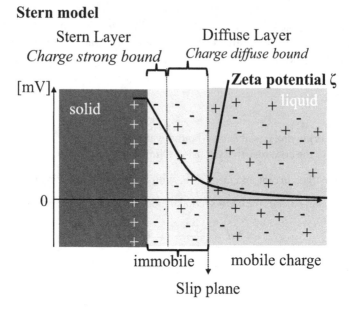

FIGURE 6.10 The electronic double layer according to the Stern model, consisting of a bound Stern and Diffuse layer. The voltage at the slip plane between the mobile and immobile charge defines the zeta potential ζ, a measure for the surface charge.

and diffuse layer – and a mobile phase in the bulk fluid. This created an electric field around the particle, which creates a voltage depending on the distance from the solid surface.

This can exist both for clays as well as silica grains. Part of that electronic double layer where charge is accumulated is mobile and can be moved with flow. When Darcy's flow occurs, the mobile charge is separated and as a consequence a macroscopic electric field is created and a voltage difference builds up along the flow direction; in case of a core, it would be along the core length (~mV). A small conductive countercurrent will be established, such that the voltage difference (dV) becomes constant. This voltage difference can be measured and is called the streaming potential. It is dependent on the surface charge, so material properties and composition but also wettability, the fluid conductivity, temperature, pH, and flow rate. This is one type of electrokinetic phenomenon. Streaming potential can be used to derive zeta-potential, a measure for surface charge, important e.g. for suspension stability, but it has also been studied for use in wettability characterization (Sadeqi-Moqadam et al. 2016; Jackson et al. 2016; Gomari et al. 2020).

A similar effect of charge separation occurs mainly in clays around the well bore, when a diffusive flow can be created through the mudcake due to salt concentration differences between the formation brine and the drilling fluid, where different ions tend to balance out the concentration difference. As not all ions are equally mobile, a charge separation arises, which creates a potential difference, which is measured during spontaneous potential (SP) logging (Ellis and Singer 2007). Note that to measure this, electrodes are used, between which the dV is measured, but such that no electric current is flowing and no electrical circuit shall be created. It is like measuring the voltage difference of a battery. This is different from resistivity logging or measurements where the battery/generator creates a dV, which forms an external circuit to reduce the dV.

There are more electrokinetic phenomena, with the basis of use of flowing particles or fluid to create an electric field measured as dV or to set a dV and measure charge movement in fluids or during particle movement. There are different ways charge can be separated by flow; it can be chemical potential driven or viscous flow driven, where the fluid can flow along the particles like the streaming potential but also particles can move, e.g. during sedimentation, creating a sedimentation potential. The inverse can be done where charged particles are placed in an electric field, such that the particles move towards the electrode with opposite charge. This electrokinetic phenomenon is called electrophoresis. This is applied in equipment to measure the zeta potential of particles in suspension; see standard ISO 13099-1, (ISO 2012). This is often combined with particle size measurements, tracing particle movement due to Brownian motion, both using lasers to track particle movement. See for further information the standard ASTM E2834-12 (ASTM n.d.). The opposite of electrophoresis can occur where a liquid moves relative to a stationary charged surface under the influence of an electric field, electroosmosis. See Table 6.8 for a summarizing overview.

Considering electrical properties of solid porous media, the streaming potential and electroosmosis are topics of interest. Surface charge characterization and the effect on fines migration (Yang et al. 2022; Tang and Morrow 1999), wettability, low salinity flooding (Katende and Sagala 2019; Morrow and Buckley 2011) and soil remediation (Reddy and Cameselle 2009) are active research topics, where we refer to literature for more details (Chang 2016) and the latest developments (Virkutyte et al. 2002; Jouniaux and Ishido 2012).

TABLE 6.8
Overview of the Main Electrokinetic Phenomena

Electrokinetic Effect	Moving Phase	Static Phase	Set Parameter	Measurement
Streaming potential	Liquid	Porous medium (core)	Flow rate	dV
Sedimentation potential	Particles	Liquid	Sedimentation rate (g)	dV
Electrophoresis	Particles (powders, in stable suspension)	Liquid	dV	Particle velocity
Electroosmosis	Liquid	Surface	dV	Liquid flow velocity

REFERENCES

Amyx, J.W., Bass Jr., D.M., Whiting, R.L. *Petroleum Reservoir Engineering*. McGraw-Hill, 1960.

Anderson, W.G. "Wettability literature survey-part 3: The effects of wettability on the electrical properties of porous media." *Journal of Pertoleum Technology* 38 (1986): 1371–1378.

Archie, G.E. "The electric resistivity log as an aid in determining some reservoir characteristics." *AIME* (1942): 54.

Arps, J.J. "The effect of temperature on the density and electrical resistivity of sodium chloride solutions." *Petroleum Transactions, AIME* 198 (1953): 327–330.

ASTM. *Compass.astm.org*, n.d. https://compass.astm.org/home/0 (accessed April 2024).

Chang, Q. "Electrical properties." In *Colloid and Interface Chemistry for Water Quality Control*, edited by Q. Chang, 79–136. Academic Press, 2016.

Dunlap, H.F., Bilhartz, H.I., Shuler, E., Bailey, C.R. "The relation between electrical resistivity and brine saturation in reservoir rocks." *Journal of Petroleum Technology* 1, no. 10 (1949): 259–264.

Ellis, D.V., Singer, J.M. *Well Logging for Earth Scientists*. Springer, 2007.

Gin, S., Delaye, J.-M., Angeli, F., Schuller, S. "Aqueous alteration of silicate glass: State of knowledge and perspectives." *NJP Materials Degradation* 42 (2021): 1–20.

Gomari, S.R., Amrouche, F., Santos, R.G., Greenwell, H.C., Cubillas, P. "A new framework to quantify the wetting behaviour of carbonate rock surfaces based on the relationship between zeta potential and contact angle." *Energies* 13, no. 993 (2020): 1–14.

ISO. "Colloidal systems methods for zeta-potential determination. Part 1: Electroacoustic and electrokinetic phenomena." 2012. International Organisation for Standardisation. ISO13099-1:2012. https://www.iso.org/standard/52807.html

Jackson, M.D., Al-Mahrouqi, D., Vinogradov, J. "Zeta potential in oil-water-carbonate systems and its impact on oil recovery during controlled salinity water-flooding." *Scientific Reports* 6, no. 1 (2016): 1–13.

Jing, X.D., Archer, J.S., Daltaban, T.S. "Laboratory study of the electrical and hydraulic properties of rocks under simulated reservoir conditions." *Marine and Petroleum Geology* 9, no. 2 (1992): 115–127.

Jouniaux, L., Ishido, T. "Electrokinetics in earth sciences: A tutorial." *International Journal of Geophysics* 2012 (2012): 1–16.

Katende, A., Sagala, F. "A critical review of low salinity water flooding: Mechanism, laboratory and field application." *Journal of Molecular Liquids* 278 (2019): 627–649.

Liu, H. "Electrical logging." In *Principles and Applications of Well Logging*, edited by H. Liu, 9–58. Petroleum Industry Press and Springer-Verlag Berlin Heidelberg, 2017.

McCoy, D.D., Warner, H.R. Jr., Fisher, T.E. "Water-salinity variations in the Ivishkak and Sag River reservoirs in Prudhoe Bay." *SPE Reservoir Engineering*, no. SPE 28577 (1997): 37–44.

McPhee, C., Reed, J., Zubizaretta, I. *Core Analysis. A Best Practice Guide in "Developments in Petroleum Science, Volume 64"*. Elsevier, 2015.

Morgan, W.B., Pirson, S.J . "The effect of fractional wettability on the archie saturation exponent." In *SPWLA 5th Annual Logging Symposium*. Society of Petrophysicists and Well Log Analysts, Midland, TX, 1964.

Morrow, N.R., Buckley, J. "Improved oil recovery by low-salinity waterflooding." *Journal of Petroleum Technology* 63, no. 5 (2011): 106–112.

Nourani, M., Pruno, S., Ghasemi, M., Fazlija, M.M., Gonzalez, B., Rodvelt, H.-E. "Analytical models for predicting the formation resistivity factor and resistivity index at overburden conditions." In *The 35th International Symposium of the Society of Core Analysts*. SCA 06, 2022.

Peveraro, R. "Determination of water resistivity: Part 4. Wireline methods." In *AAPG Special Volumes, ME 10: Development Geology Reference Manual*, 170–173. AAGO Datapages, Inc. Archives, 1992.

Qi, Y., Wu, Y. "Electrical conductivity of clayey rocks and soils: A non-linear model." *Geophysical Research Letters* 49 (2022): e2021GL097408. https://doi.org/10.1029/2021GL097408

Reddy, K.R., Cameselle, C. *Electrochemical Remediation Technologies for Polluted Soils, Sediments and Groundwater*. John Wiley & Sons, 2009.

Sadeqi-Moqadam, M., Riahi, S., Bahramian, A. "Monitoring wettability alteration of porous media by streaming potential measurements: Experimental and modeling investigation." *Colloids and Surfaces A: Physicochemical and Engineering Aspects* 497 (2016): 182–193.

Schlumberger. *Log Interpretation Charts, 2009 Edition*. Sugar Land, TX, 2009.

Tang, G., Morrow, N.R. "Influence of brine composition and fines migration on crude oil/brine/rock interactions and oil recovery." *Journal of Petroleum Science and Engineering* 24, no. 2–4 (1999) : 99–111.

Tiab, D., Donaldson, E.C. *Petrophysics, Theory and Practice of Measuring Reservoir Rock and Fluid Transport Properties*, 3. Elsevier, 1996.

Virkutyte, J., Sillanpää, J., Latostenmaa, P. "Electrokinetic soil remediation—critical overview." *Science of the Total Environment* 289, no. 1–3 (2002): 97–121.

Worthington, A.E., Hedges, J.H., Pallatt, N. "SCA guidelines for sample preparation and porosity measurement of electrical resistivity samples part I -Guidelines for preparation of brine and determination of brine resistivity for use in electrical resistivity measurements." *The Log Analyst* (1990): 20–28.

Wyllie, M.R.J. "Formation factor of unconsolidated porous media: Influence of particle shape and effect of cementation." *Transactions of the AIME* 198 (1953): 103–110.

Wyllie, M.R.J., Spangler, M.B . "Application of electrical resistivity measurements to problem of fluid flow in porous media." *Bulletin AAPG* 36, no. 2 (1952): 359–403.

Yang, Y., Yuan, W., Hou, J., You, Z. "Review on physical and chemical factors affecting fines migration in porous media." *Water Research* 214, no. 118172 (2022): 1–16.

7 Absolute Permeability

Permeability is an intrinsic property of the porous medium and a measure of the capacity of the medium to transmit fluids. Permeability describes the ease of flowing through a porous medium. Sponges shall be permeable to take water up, where brick in the wall or roof tiles shall be impermeable against weather influences. Wood is often made impermeable by painting or impregnation such that its pores are filled, and permeability of liquid and gases is limited.

A reservoir rock shall have a high permeability, linking all the pores to the injection or production well, where a cap rock shall have limited permeability avoiding leakage. A higher permeability enables more reservoir production with a given pressure reduction in the field or it enables injection at higher flow rates without risking too-high injection pressures reaching the fracture pressure of the rock. Very often the permeability varies by several magnitudes, and such heterogeneity will of course influence any oil recovery. Opposite to porosity, which is a static property of the porous media, permeability is a dynamic property of the porous media; it is direction dependent and can be measured in the laboratory only by fluid flow experiment. This chapter is devoted to absolute permeability of the reservoir rock, which is when the rock contains one single phase (gas or liquid).

7.1 DARCY'S LAW

Henry Darcy (Darcy 1856), a hydraulic engineer, performed a series of experiments on the relationship effecting the downward flow of water through sands while he was in charge of purification of the water in Dijon, France. The experiments were designed to understand water flow through sand layers and the relationship between flow rate and pressure drop. His experiment and methodology were simple, and, with some modifications, the method is used in core laboratories today. In Darcy's experiment, which was mainly for purification of the water, the water flow rate was linear proportional to the pressure gradient across the core and cross-sectional area of the sample area and inversely proportional to the length of the sand pack, the parameters shown in Figure 7.1. He used only water, but also for other fluid similar proportionalities could be found, where the viscosity of the passing fluid was inversely proportional to the flow rate. The constant is defined as the absolute permeability of the porous media.

The generalized equation called Darcy's law may be written in the form:

$$\bar{\nu} = -\frac{\bar{k}}{\mu}\left(\nabla P + \rho \bar{g}\right) \tag{7.1}$$

Where $\bar{\nu}$ is superficial velocity, $\bar{k}$ is permeability tensor, μ is dynamic fluid viscosity, ∇P is pressure gradient, ρ is fluid density, and $\bar{g}$ is gravitational vector.

                    DOI: 10.1201/9781003382584-7

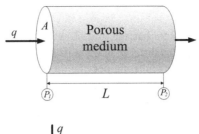

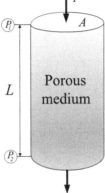

FIGURE 7.1 Parameters of Darcy's equation for horizontal or vertical positioned porous media.

Writing the superficial $\bar{v}$ as the ratio of volumetric rate to cross-sectional area perpendicular to flow as q/A distance L, Darcy's law can be expressed as:

$$q = \frac{kA}{\mu}\frac{\Delta P}{L} \tag{7.2}$$

And permeability can be calculated by rewriting Equation 7.2 as

$$k = \frac{q}{\Delta P} \cdot \mu \cdot \frac{L}{A} \tag{7.3}$$

ASSUMPTIONS

Darcy's law in the form of Equations 7.1 and 7.2 can be applied under the following conditions:

- Incompressible flow.
- Steady state (no acceleration) flow.
- Laminar flow.
- Newtonian fluid.
- Isothermal conditions.
- No reactivity of fluid with porous medium.

UNITS

Permeability in the SI system has the unit of m², which can be derived from a unit analysis of Equation 7.3. The order of magnitude of common rock permeabilities lies then below 10^{-10} m². In petroleum engineering therefore the unit Darcy is common, a *cgs* system unit. A porous medium is supposed to have permeability of 1 Darcy when a single-phase fluid with a known viscosity of 1 centipoise (cP) completely saturates the media and flows through it at a rate of 1 cm³/s under a laminar flow regime and a pressure gradient of 1 atm/cm through a cross sectional area of 1 cm². Therefore, from Equation 7.3:

$$1\,\text{Darcy} = \frac{\left(\frac{cm^3}{s}\right)(cP)(cm)}{(atm)\left(cm^2\right)}$$

With 1 cP = 1mPas and 1 atm = 10.1325 N/cm², so

$$1\,\text{Darcy} = \frac{\left(\frac{cm^3}{s}\right)\left(10^{-7}\frac{N\cdot s}{cm^2}\right)(cm)}{\left(10.1325\frac{N}{cm^2}\right)\left(cm^2\right)} = 9.869 \cdot 10^{-9}\ cm^2,$$

$$\text{or}\quad 9.869 \cdot 10^{-13}\ m^2 \approx 1\cdot 10^{-12}\ m^2 = 1\mu m^2$$

EXERCISE 7.1

Flow experiments are performed on a core with two different fluids, respectively with an oil with a viscosity of 0.89 cP and water with 1.0 cP viscosity. Use Equation 7.2 and Figure 7.1 as guide.

1. For which fluid is the ΔP measured higher at constant q?
2. For a set-up with a core twice as long, using the same q; what happens to the measured ΔP?
3. If a similar ΔP is measured for a core of similar dimensions but the core has a higher permeability, what happens to q?

SOLUTIONS

1. According to Equation 7.2, when k, L, and A are equal for both cases then q is proportional to $\Delta P/\mu$. If q is constant, then the ΔP must become higher for a higher viscosity.
2. According to Equation 7.2, when k, μ, and A are equal for both cases, q is then proportional to $\Delta P/L$. So, if q is constant, for a length of L twice as long, ΔP must become twice as high. Note that the pressure gradient $\Delta P/L$ remains constant.

3. According to Equation 7.3, for both fluids it is valid that ΔP, μ, A, and L are equal and constant, such that k scales with q, so, a higher k with a constant ΔP results in higher flow rates. This is beneficial for any kind of production from natural reservoirs, where ΔP is set by bottom hole pressure (outlet) and reservoir pressure (inlet) is constant. With a higher rock permeability, the resulting flow rates will be higher. Similarly, low-viscosity liquid will produce more quickly as well.

7.2   KOZENY-CARMAN MODEL

Formation permeability may be determined or estimated based on core analysis, well tests, production data, well log interpretations, or correlations based on pore structure. One of these often-used pore models is Kozeny-Carman model. A detailed presentation of this model is given in Feder et al. (2022) which is summarized later.

The model considers the rock to be a bundle of capillary tubes. The volumetric flow rate q in one horizontal capillary of radius R and length L_t is given by Hagen-Poiseuille's equation:

$$q = \frac{\pi R^4 \Delta P}{8\mu L_t} \tag{7.4}$$

and the average velocity in the tube is:

$$\bar{v} = \frac{q}{\pi R^2} = \frac{R^2 \Delta P}{8\mu L_t} \tag{7.5}$$

We must transform this equation to the scale of a representative element volume (REV) of the rock. The REV is defined as a volume below which local fluctuations in permeability is large, when taking averages over that volume. Above the REV the average represents the total average. If we make the travel time in the capillary tube equal to that in a REV, then:

$$\left(\frac{L_t}{\bar{v}} \right)_t = \left(\frac{L}{v_{pore}} \right)_{REV} \tag{7.6}$$

With L_t the path length of the fluid flow through the rock (capillary tube length) and L the length of the porous media, measured externally, $\bar{v}$ the average velocity through the capillary tube or the interstitial velocity (as this is not constant based on the presence of pore throats, which work as constrictions in the capillary). v_{pore} is the pore velocity of the fluid in the core and can be derived from Darcy's equation as $v_{Darcy} = \frac{q}{A}$. The actual surface in which the flow is flowing is $A \cdot \phi$, see Figure 7.2, so the pore velocity or interstitial velocity is defined as $v_{pore} = \frac{q}{A\phi} = \frac{k\Delta P}{\mu L\phi}$. The relation between pore and Darcy velocity can then be written as $v_{pore} = v_{Darcy}/\phi$.

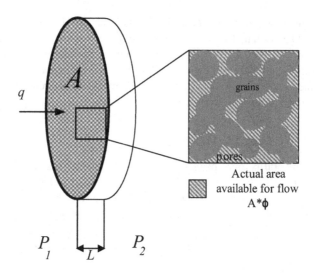

FIGURE 7.2   The actual area of flow through the porous media $A \cdot \phi$.

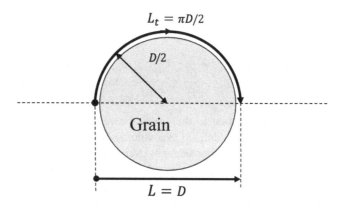

FIGURE 7.3   Flow length L and L_t in a system with flow along spherical grains. The tortuosity is $\pi/2$.

Applying this to equation 7.6 and Darcy's law to eliminate v_{Darcy} and substituting Hagen Poiseuille for $\bar{v}$ we obtain the permeability component k as:

$$k = \frac{R^2 \phi}{8\tau^2} \qquad (7.7)$$

where τ is the tortuosity, which is defined as L_t/L, being larger than one. Note that sometimes tortuosity is defined inversely, as L/L_t. See Figure 7.3 for a sketch of the difference between L and L_t, which is defined as tortuosity.

Tortuosity is an important media property and is usually estimated from electrical resistivity measurements. The tortuosity is in the range of 2 to 5 for most reservoir rocks.

Alternative derivation: considering all tubes in the porous media

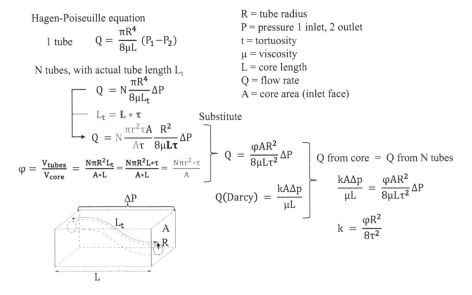

Hagen-Poiseuille equation

1 tube $Q = \dfrac{\pi R^4}{8\mu L}(P_1 - P_2)$

N tubes, with actual tube length L_t

$Q = N\dfrac{\pi R^4}{8\mu L_t}\Delta P$

$L_t = L * \tau$

$Q = N\dfrac{\pi r^2 \tau A}{A\tau}\dfrac{R^2}{8\mu L\tau}\Delta P$

$\varphi = \dfrac{V_{tubes}}{V_{core}} = \dfrac{N\pi R^2 L_t}{A*L} = \dfrac{N\pi R^2 L*\tau}{A*L} = \dfrac{N\pi r^2 *\tau}{A}$

R = tube radius
P = pressure 1 inlet, 2 outlet
t = tortuosity
μ = viscosity
L = core length
Q = flow rate
A = core area (inlet face)

Substitute

$Q = \dfrac{\varphi A R^2}{8\mu L\tau^2}\Delta P$ Q from core = Q from N tubes

$Q(\text{Darcy}) = \dfrac{kA\Delta p}{\mu L}$

$\dfrac{kA\Delta p}{\mu L} = \dfrac{\varphi A R^2}{8\mu L\tau^2}\Delta P$

$k = \dfrac{\varphi R^2}{8\tau^2}$

FIGURE 7.4   An alternative derivation for the Kozeny-Carman equation, using pore volume.

The capillary radius R in Eq.7.7 is difficult to define for a porous medium but may be approximated by the hydraulic radius R_h that expresses the ratio between volume open to flow and the wetted surface area. For a porous and permeable medium, it is derived as:

$$R_h = \frac{\pi R^2 L}{2\pi R L} = \frac{R}{2} \tag{7.8}$$

$$R_h = \frac{\phi}{a(1-\phi)} \tag{7.9}$$

where a is internal surface area per volume, an important intrinsic property of permeable media. By substituting these expressions for R_h in Eq. 7.7 and solving the equation for an assemblage of uniform spheres of diameter D, with $a = 6/D$, one can get the Kozeny-Carman equation:

$$k = \frac{1}{72\tau}\frac{\phi^3 D^2}{(1-\phi)^2} \tag{7.10}$$

where D is sphere or particle diameter, showing that permeability is a strong function of particle size and packing through Φ and τ.

EXERCISE 7.2

The Kozeny-Carman equation for calculating the permeability of a rock sample is given as:

$$k = \frac{\phi^3}{2a^2 \left(1-\phi\right)^2 \tau^2}$$

1. Define the parameters that influence the value of permeability and name the equations used to derive the Kozeny-Carman equation.
2. Calculate the permeability of a spherical grain rock. The average grain diameter is 0.1 mm; porosity is 26% and τ is 1.25.

SOLUTION

1. ϕ is porosity, a is $\dfrac{Lateral\, surface\, area}{Volume}$, and tortuosity is $\tau = L_e/L$

   Hagen Poiseuille equation, Darcy's equation, correlation between Darcy velocity and pore velocity, description of the hydraulic radius.

2. $k = \dfrac{0.26^3}{2\left(\dfrac{6}{0.1\cdot10^{-3}}\right)^2 \left(1-0.26\right)^2 1.25^2} = 2.85 \cdot 10^{-12}\ \text{m}^2 = 2{,}89\ \text{D}$

See Figure 7.5 for an example of a sandstone, where the porosity, mean grain diameter, and tortuosity are estimated from the thin section. Note that there is an error in these due to the 2D character of the slice. Grains might not be cut through the center, so the diameter might not be representative. The grains are not spherical, so what

Example calculation of the permeability from a sandstone thin section:

$$k = \frac{\varphi^3 D^2}{72\tau(1-\varphi)^2}$$

$$\varphi = \frac{V_p}{V_b} = \frac{Dark\, gray\, area}{Total\, area} = 0.19 \text{ (from a.)}$$

$$D = Mean\, grain\, diameter = 0.35 \text{ mm (from b.)}$$

$$\tau = \left(\frac{Length\, of\, line}{Total\, length}\right)^2 = 3 \text{ (from c.)}$$

Then we get:

$$k = \frac{(0.35 * 10^{-3})^2 * 0.19^3}{72 * 3 * (1-0.19)^2} = 5.92 * 10^{-12}\text{m}^2$$
$$= 6.01\text{D}$$

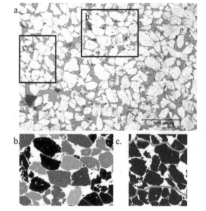

FIGURE 7.5 Example calculation of the permeability from a sandstone thin section (Courtesy of M.B.E. Mørk).

diameter shall be used? Pores are also 3D irregular structures, so the cross-section is likely questionable. Statistics on the data are needed to get representative values.

The Kozeny-Carman equation is often used to make order-of-magnitude estimates of pore size from knowledge of permeability or estimating of permeability using thin sections. However, the capillary tube model is of limited value since it does not provide alternate pathways for fluid flow within each REV. The consequence is that we cannot predict relative permeability or trapped phase saturations, parameters of major importance in the oil recovery processes. With the increased possibility to image rocks and obtain 3D models, attempts are made to estimate core properties applying the physical law for fluid flow, using Navies-Stokes or Lattice Boltzmann approaches. This is referred to as digital rocks or digital rock analysis (Fredrich et al. 2014). Absolute permeability estimations are also a part of this analysis.

Based on Kozeny-Carman equation it becomes clear that geological factors affecting the permeability are the shape, grain size, and grain size distribution. The smaller the particles, the larger the tortuosity, the greater the resistance to flow, and permeability generally decreases. Orientation of the particles can affect permeability in different directions. Commonly a horizontal or vertical (absolute) permeability is defined, or permeability is oriented along layer orientation. Sample selection can play an important role here.

PERMEABILITY ALTERATION

In geological time, deposited sediments can have been altered by different processes. Cementation, precipitation, and compaction are, for example detrimental for permeability, reducing the hydraulic radius. Oppositely processes such as (micro) fracturing at the core scale or dissolution can increase permeability. On a core scale in general the permeability of the matrix is measured rather than fracture permeability. Awareness of these factors is beneficial for the sample selection and interpretation of the measured data.

Besides alteration of permeability in geological time, alteration of permeability can also occur due to human activities, for example close to the well; both during production as well as injection. Fracturing due to high injection pressures can occur, but this is generally on a scale beyond core scale. On a core scale, scaling effects can be studied occurring around the well bore. Scaling is the precipitation of salts around the wellbore while injecting brines that have a different salinity than the original formation water. Comparably, when CO_2 is injected, it will dry the formation, leaving the salt precipitated, which will affect the permeability of the formation. This will occur especially around the well bore with a high throughput of pore volumes. This is a topic of interest for the development of the injection strategy for CO_2 sequestration (Bacci et al. 2011; Ringrose 2020). On the other hand, injection of CO_2 reduces the pH, which might stimulate dissolution of specific minerals.

Also, effects can occur further away from the well, for example fines migration during (low salinity) water flooding (Tang and Morrow 1999) or retention of polymer during polymer flooding can lead to plugging and reduction of permeability (Mishra et al. 2014). Conformance control is a measure to intentionally reduce the permeability of high permeable zone to divert flow to lower permeable zone (Seright and

Brattekås 2021). So, various activities can have effect on permeability. Absolute permeability tests before and after different activities can be performed to understand the impact of the activity on the permeability and study how to optimize it.

7.3 LABORATORY MEASUREMENT OF ABSOLUTE PERMEABILITY

Laboratory measurement of absolute permeability usually is based on application of the Darcy equation (Equation 7.2) in which the sample's physical properties such as length and area as well as fluid viscosity are known. Independent variables such as flow rate and pressure difference between inlet and outlet of the core should be either set or measured during the experiment. When the flow rate is regulated, pressure is to be measured and reversed. Measurement of the flow rate at the outlet is still advised.

Most laboratory measurements for rock samples are conducted on samples with well-defined geometry, such as cylindrical core plugs with 3.81 cm (1.5 inch) diameter and length varying from 5 to 10 cm. Permeability tests are performed on samples that have been cleaned and dried and a fluid (either gas or liquid) is used as flowing phase in the test; see Chapters 3 and 4 for the details.

Various measurement conditions can be used for permeability measurements of cores, depending on sample dimensions and shape, degree of consolidation, type of Newtonian fluid used, ranges of confining pressure and fluid pressure applied, and range of permeability of the core. Note that confining pressure and possible compressibility of the porous media will have a direct influence on the permeability and representation of the in-situ reservoir permeability. See Chapter 5 for more details on compressibility linked to porosity. Here compressibility control will not be part of the experimental equipment discussion. A distinction will be made between measurement using a liquid – further discussed in this section – and a gas in Section 7.4.

7.3.1 Liquid Permeability

The procedure for permeability measurement is based on the application of Darcy's law in porous media. In the following a procedure for single phase liquid permeability is described. A simple schematic of the equipment is given in Figure 7.6, where flow rate is controlled and pressure difference measured. Essential is that the fluid used is not reactive and Newtonian. HSE can also play a role to choose no harmful liquids to work with. The flooding fluid is introduced in the system using a displacement pump (1) that ideally functions with one pump fluid that depletes a fluid reservoir (2) filled with the fluid to perform the experiment with. Fluid flow can be led through a core holder (5) with the porous medium mounted in it, here a consolidated core (6). To avoid fluid bypassing the core, a confinement is used by pressurizing a sleeve surrounding the core using a hydraulic pump (4). To purposely bypass the core, e.g. to remove a second unwanted phase, like gas bubbles when using a liquid, a bypass line can direct the fluid flow around the core. With a flow over the core set with the pump, the pressure can be measured using a differential pressure transducer (4). In case the outlet pressure is to be higher than atmospheric pressure a back

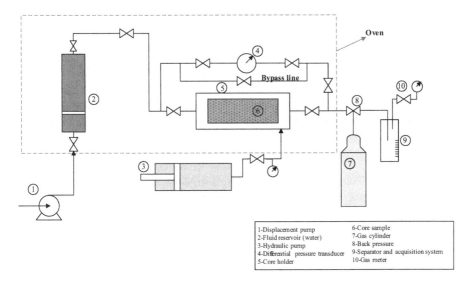

1-Displacement pump	6-Core sample
2-Fluid reservoir (water)	7-Gas cylinder
3-Hydraulic pump	8-Back pressure
4-Differential pressure transducer	9-Separator and acquisition system
5-Core holder	10-Gas meter

FIGURE 7.6   Schematic drawing of a liquid permeability apparatus.

pressure regulator can be used (8) with gas supply from a bottle (7). Flow rates can be monitored at the outlet either measuring liquid quantities (9) or gas flow rates (10).

7.3.2   LIQUID PERMEABILITY PROCEDURE: PRE-FLOODING PREPARATION

PLUG SATURATION

Plug saturation is a necessary step for liquid permeability and can be combined with the determination of the liquid porosity; see for more details Chapter 5.

Alternatively, the core can be saturated inside the core holder and flooded with CO_2 to remove air, pulling the vacuum afterward and then introducing the liquid to the system. The pressure can be increased afterward to enable possible remaining CO_2 molecules to dissolve in the fluid.

SYSTEM MOUNTING AND FLUID FILLING

Presuming the set-up is mounted as in Figure 7.6, the pump fluid supply tank needs to be filled and the fluid reservoir (2) needs to be full of the injection fluid. Before connecting the lines to the core holder, lines need to be purged.

SAMPLE LOADING

The core is loaded in a core holder. There are different designs of core holders depending on the measurements, porous medium used, and the conditions of the measurement. Figure 7.7 shows an example where a consolidated core can be mounted with a flexible length.

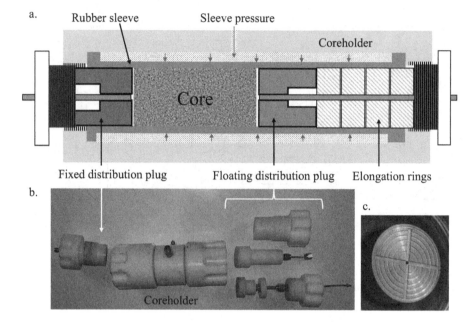

FIGURE 7.7 An example of a core holder (a, b) with a fixed sleeve and two end pieces, where one can be placed flexibly such that samples with different core length can be used. c. Shows the liquid distribution pattern on a metal end piece (2-inch diameter).

A critical point of loading a saturated core is that gas needs to be avoided in the inlet flowlines. If that gets into the core, an unwanted additional phase is added that cannot be removed unless the core is cleaned, dried, and saturated again. The following procedure is suggested.

1. Fill the flow line with the fluid the core is saturated with till the inlet of the core holder.
2. Load the sample in the core-holder by sliding a properly prepared core into the rubber sleeve in the core holder. Load the *inlet* endcap so that it contacts the core sample. No gas shall be trapped.
3. Load the endcap *outlet* distribution plug and slide it until the core is firmly installed between the fixed inlet and floating distribution plug.
4. Initiate the confining pressure by injecting confining fluid through the pump and increase the pressure to a pre-calculated value and keep it constant by using "constant pressure" mode of the pump.
5. Initiate the back pressure by injecting gas under pressure and adjust the pressure with the mode "constant pressure" of the pump. Wait at least 30 minutes for pressure stabilization due to thermal effect. Remember that at any time the following relation must be valid:

Back Pressure < Outlet Pressure < Inlet Pressure < Confining Pressure

7.3.3 EXPERIMENTAL PROCEDURE FOR LIQUID PERMEABILITY MEASUREMENT

To conduct liquid permeability measurement, the following steps are recommended. The procedure is based on injecting liquid at different flow rates and measuring pressure differences across the core. Permeability will then be derived from Darcy's equation. Prior to the experiment, check that pore pressure is lower than the confining pressure, and perform a leakage test for the confinement and flooding system.

- Inject liquid at a given rate (e.g. 0.1 cm³/min) and wait for stabilization of the pressure difference over the core plug (ΔP). The first injection rate should not be more than 0.5 or 1 cm³/minute to prevent fines migration inside the core.
- When ΔP is stabilized after injection of at least two pore volumes, increase fluid injection rate and repeat previous step (see Figure 7.8a). To judge whether the data is realistic, compare ΔP's for cases where the flowrate is doubled. The ΔP shall be doubled then as well.

The Darcy equation can be written as:

$$\Delta P = \frac{q\mu L}{kA} \qquad (7.11)$$

So, a plot of ΔP vs. q for different flow rates should fall on a straight line $y = ax + b$ with a slope equal to $(\mu L)/(kA)$ as shown in Figure 7.8b. Variations can be made on which variables to plot versus each other, e.g. $\Delta PA/\mu L$ versus q will result in k as the gradient. Note that averaging of the k calculated for each ΔP or flow rate gives an increased error in the data. The judgment of reaching a steady state is crucial and can lead to errors in case it was not.

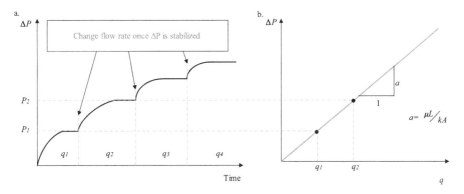

FIGURE 7.8   a) Pressure difference versus time dependent on flow rate $q_1 < q_2 < q_3 < q_4$. The plateau values for pressure are plotted versus flow rate b) to derive the permeability k from the gradient.

EXERCISE 7.3

A permeability experiment is performed with the experimental set-up shown in Figure 7.6, where the differential pressure transducer is replaced by the set of differential pressure transducers, as shown in Figure 7.9. Brine is used, flow rates are set, and pressure differences over the core are measured in psi [psi].

The following data is available:

TABLE 7.1
Core Data to Exercise 7.3

	Standard Core Data
Core length (cm)	4.57
Core diameter (cm)	3.82
Total core volume (cm³)	52.26
Dry weight (g)	106.24
Wet weight (g)	117.76
Brine viscosity (cP)	1.16
Brine density g/cm³	1.019
Pressure transducer accuracy	1% of range

1. Which parameters can the permeability of a rock be dependent on? Name three.
2. What is the unit of permeability in SI units and field units?
3. What is the objective of having more than one differential pressure transducer?
4. Flow rates shall represent field rates; (Darcy velocity) from 0.5 to 1.5 ft/day. What is the permeability range this set-up can cover using brine as fluid? Use the core data from Table 7.1.

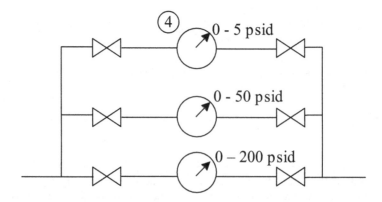

FIGURE 7.9   Experimental set-up for liquid permeability experiment used in Exercise 7.3.

The following data (see Table 7.2) was generated for a single-phase brine flooding experiment. The error estimation is obtained from the pressure fluctuations compared to the average pressure once it is stabilized.

5. Comment on the data; would you use all for curve fitting? Why or why not?
   a) Which permeability would you report and b) why? c) Would there be arguments for using the other curve instead? Discuss.

SOLUTION

1. According to the Kozeny-Carman equation, permeability depends on pore radius, pore structure/tortuosity, and porosity.
2. The units for permeability in SI is m^2 and in cgs Darcy.
3. More accurate pressure measurements at low pressures are then possible. The differential pressure meters have each their own range. The wider the pressure range that can be measured, the less accuracy the meters can deliver. Using differential pressure transducers with smaller ranges enables accurate measurement of P for lower pressures.

TABLE 7.2
Experimental Data for Exercise 7.3

Q [ml/min]	ΔP [bar]
3.7	0.18 ± 0.02
4.1	0.18 ± 0.01
7.9	0.30 ± 0.02
10.0	0.38 ± 0.03
12.0	0.42 ± 0.02

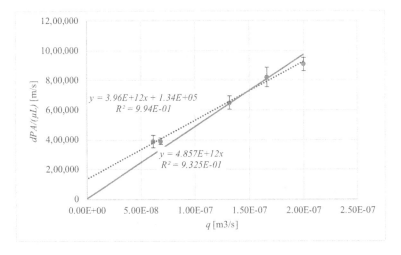

FIGURE 7.10 Plotted experimental data of Exercise 7.3 for the determination of liquid permeability.

4. $Q = \dfrac{kA\Delta p}{\mu L}$; $k = \dfrac{Q\mu L}{A\Delta p}$

The velocity in SI units is $v = Q/A = \dfrac{0.5\,\text{ft}\,/\,\text{d} \cdot 0.30\,\text{m/ft}}{24\text{h} \cdot 3600\,\text{sec/d}} = 1.7 \cdot 10^{-6}\ \text{m/s}$

Or for 1.5 ft/d, v is $5.2 \cdot 10^{-6}$ *m/s*.

From Darcy's equation (Equation 7.2) we see that the lowest permeability is respectively found by combining the lowest flow rate q and lowest viscosity μ with the highest ΔP and the reverse for the higher permeability.

	q [ft/d]	μ [cP]	dP [psi]
k lowest	lowest: 0.5 ft/d	brine: 1.16 (lowest)	highest: 200 psi
k highest	highest: 1.5 ft/d	brine: 1.16 (highest)	lowest: 1% of 5 psi = 0.05 psi*

* $\Delta P = 0$ is not correct; within error range of P transducer of 1%, one should switch to device with lower
  ΔP range.

$$k_{\text{lowest}} = [\dfrac{0.5 \cdot 0.30}{84600} \cdot \dfrac{\dfrac{1.16}{1000} \cdot 0.0457}{\left(200 \cdot 10^5\,/\,14.5\right)}]/\,9.869 \cdot 10^{-13} = 0.067 \cdot 10^{-3}\ \text{D} => 0.067\text{mD}$$

$$k_{\text{highest}} = [\dfrac{1.5 \cdot 0.30}{84600} \cdot \dfrac{\dfrac{1.16}{1000} \cdot 0.0457}{\left(0.05 \cdot 10^5\,/\,14.5\right)}]/\,9.869 \cdot 10^{-13} = 0.811\text{D}$$

5. a) In the table for ΔP is 0.18 bar for the flow rates 3.7 and 4.1 cm³/min. This does not follow Darcy's law. No error bars are given for q, so scattering is unclear.
   b) Darcy's law is valid when $q = 0$, $\Delta P = 0$. So, the trendline shall go through origin; $y = ax$; $k = 1/4.86 \cdot 10^{12} = 2.06 \cdot 10^{-13}\ \text{m}^2$
   (c) When there are systematic errors in calibration of the pressure transducers or flow devices, the other trendline $y = ax + b$ can be used, as this shows a better linear fit. $k = 1/3.96 \cdot 10^{12} = 2.53 \cdot 10^{-13}\ \text{m}^2$

The estimated error bars are smaller than the variation possible in the trendline choice. This can indicate either that the estimated random errors are underestimated or there is a systematic error not included. In this case it was judged that the latter is the case.

7.4   GAS PERMEABILITY

Darcy's equation defining permeability as given in Equation 7.2 is valid for laminar flow in porous media. Any phase can be used that is non-reactive and incompressible. Besides using a liquid, gas also can be used. The advantage is that there can be assumed limited interaction of the gas with the clays compared to liquid, and

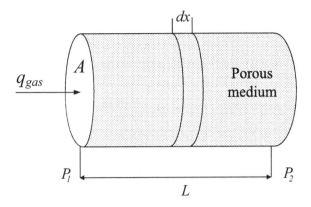

FIGURE 7.11   Linear flow of an ideal gas in a porous medium.

the sample can be easily used afterwards for other experiments without redoing the cleaning procedure. However, gas is compressible and laminar flow is not always achieved. Darcy's equation can be adapted for compressible flow. Let us suppose that there is a porous medium (Figure 7.11). Upstream and downstream pressure P_1 and P_2 are constants.

The mass flow $\dot{m}$ is the same in all slices and each slice is related to the volume flow rate q by:

$$\dot{m} = q \cdot \rho \tag{7.12}$$

where ρ is gas density in the slice when pressure is P. Using Boyle's law, $PV = nRTz$, and $n = m/M$, the relationship $\rho = m/V = bP$ can be applied where b is the constant $\dfrac{M}{RTz}$. Then:

$$q = \frac{1}{bP}\dot{m} \tag{7.13}$$

By substituting q in Darcy's equation, a differential relation between P and distance x may be obtained.

$$bPdP = \frac{\mu}{A}\frac{\dot{m}}{k_g}dx \tag{7.14}$$

After integration for distance L:

$$\frac{b}{2}\left(P_1^2 - P_2^2\right) = \frac{\mu}{A}\frac{\dot{m}}{k_g}L \tag{7.15}$$

An alternative derivation of Darcy's law for compressibility of gas

Consider the mean flow rate q_m and mean pressure P_m in the core.

$$q_m = -\frac{k_g A \Delta p}{\mu L}$$

From Boyle's law we have $pV = c_1$

$q * P_{\text{any where in the core}} = q_{atm} * P_{atm}$

$\rightarrow q_m * P_m = q_{atm} * P_{atm}$

$$q_m = \frac{q_{atm} P_{atm}}{P_m}$$

$$\frac{q_{atm} P_{atm}}{P_m} = -\frac{k_g A \Delta p}{\mu L}$$

$$q_{atm} = -\frac{k_g A P_m \Delta p}{\mu P_{atm} L}$$

$$P_m = \frac{P_{out} + P_{in}}{2}$$

$$\Delta P = P_{outlet} - P_{inlet}$$

$$q_{atm} = -\frac{k_g A (P_{out} + P_{in})(P_{out} - P_{in})}{2\mu P_{atm} L}$$

$$q_{atm} = -\frac{k_g A}{\mu L}\frac{\left(P_{out}^2 - P_{in}^2\right)}{2P_{atm}}$$

FIGURE 7.12   An alternative derivation of the Darcy's law for compressible fluids.

and considering that $\dot{m} = qbP$, we get:

$$q_{atm} = \frac{A k_g}{\mu L}\frac{\left(P_1^2 - P_2^2\right)}{2P_{atm}} \tag{7.16}$$

with P_1 as the gas pressure at the inlet and P_2 at the outlet, with P_{atm} the pressure at which the flowrate q_{atm} is measured. The gas flow is measured at standard conditions, thus $P_{atm} = 1$ atm.

7.4.1 KLINKENBERG CORRECTION

Applying Equation 7.16 for different pressure differences results in a varying pressure dependent k_g, where the single-phase permeability shall be rock dependent. Klinkenberg (Klinkenberg 1941) has reported variations in permeability measured by using gases as the flowing fluid compared to those obtained when using (non-reactive) liquids. These variations occur due to slippage, a phenomenon well known with respect to gas flow in capillary tubes. The phenomenon of gas slippage occurs when the diameter of the capillary openings approaches the mean free path of the gas. The mean free path of a gas is a function of molecular size and the kinetic energy of the gas. So, flow rates will be larger due to slippage at low gas pressures compared to no slippage. Therefore, permeability measured with a gas depends on factors that influence the mean free path, such as temperature, pressure, and the molecular size of the gas.

Figure 7.13 is a plot of the permeability of a porous medium as determined at various mean pressures using three different gases. Note that for each gas a straight line is obtained for the observed permeability as a function of the reciprocal of the mean pressure of the test. At low pressures the calculated permeability is overestimated. All the lines when extrapolated to infinite mean pressure ($1 / P_m = 0$) intercept the permeability axis at a common point. This point is designated as absolute

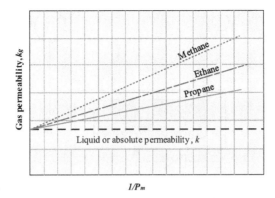

FIGURE 7.13 Variation in gas permeability versus mean pressure and type of gas (modified from Monicard (Monicard 1980)).

permeability k or the equivalent liquid permeability. At this point P_m is infinitely large, where gas can be assumed to behave like a liquid, such that slippage can be neglected.

Klinkenberg has related apparent permeability k_g measured for gas for an average pressure P_m to the true absolute permeability k by:

$$k_g = k\left(1 + \frac{b}{P_m}\right) \qquad (7.17)$$

where b is constant depending upon the average free movement λ of the molecule at P_m:

$$b = \frac{4C'\lambda P_m}{r} \qquad (7.18)$$

where r is channel radius and $C' \approx 1$. The approach to derive the absolute permeability from gas data is called the Klinkenberg correction.

7.4.2 PROCEDURE FOR MEASURING ABSOLUTE PERMEABILITY USING AIR

PPE: Lab coat, safety glasses, safety shoes, gloves (in case cores surface needs protection). Familiarize yourself with risk assessment and make adaptions depending on the activities you intend.

Some notes before you start: Make sure the internal core holder volume has similar dimensions as the core. The in- and outlet need to have a good connection with the two sides of the core, as mentioned also in the liquid permeability description in Section 7.3.2.

Familiarize yourself carefully with the valves and flowline system. The set-up presented in Figure 7.14 will be used as example for the experiment. In this

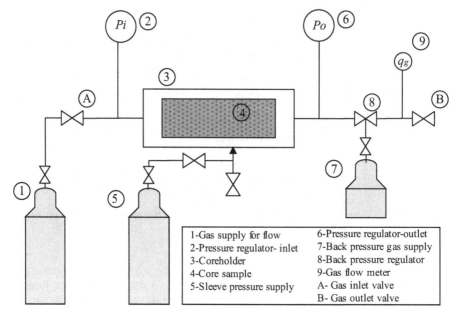

FIGURE 7.14 Experimental set-up for determination of absolute permeability using nitrogen, regulating the pressure difference over the core sample.

set-up the gas is supplied via a gas bottle. In the set-up the gas pressure in front of the core holder (2) and at the back (6) are regulated, while the gas flow rate is measured (9) after the backpressure regulator (8). The core is confined applying a sleeve, which is pressurized via an independent system using a gas bottle with a choke restricting the available pressure below the pressure limit of the sleeve system. This experiment can also be executed using a similar set-up presented for the liquid permeability Figure 7.6. where then the gas flow is regulated with a pump.

PROCEDURE

1. Make sure all valves are closed, except the outlet valve B.
2. Place the core in the sleeve inside the core holder.
3. Setting sleeve pressure. Install a choke on the gas bottle for safety such that the maximum pressure of the sleeve/system cannot be reached with the gas supply from the bottle.
    a. Open the main valve on the nitrogen gas bottle.
    b. Turn on the regulator from the gas bottle, until the desired gas pressure is reached on the pressure gauge of the bottle.
    c. Open valve A to the core holder SLOWLY. This valve is used later to disconnect from the gas supply.

4. Perform a sleeve leakage test:
   a. Close the valve to the core holder and monitor the pressure on the pressure gauge, which shows the pressure in the sleeve. Minimum five minutes. Record the initial pressure.
      i. In case sleeve pressure goes down without pressure supply: Leakage.
      ii. In case pressure remains constant: Open the valve to the core holder and continue with the next step.
5. Set the initial air flow.
6. Perform a pressure leakage test in the core holder:
   a. Close the outlet valve B and set the inlet pressure regulator to 1 bar.
   b. Close the gas inlet valve A and monitor the pressure of the two pressure gauges.
      i. In case pressure goes down without pressure supply: A leakage in the core holder is to be checked. Release the pressure and re-mount the endpiece. Use foam to check for leakage.
      ii. In case pressure remains constant: Set both pressure regulators to zero and open first SLOWLY the outlet valve. The pressure gauges at in and outlet shall show zero. Open then the valve to the gas supply and continue with the next step.
7. Measurement: You are now set to get started when the two pressure gauges, P1 and P2, show zero.
   a. Increase the inlet and outlet pressure by using the two regulators.
   b. The pressure difference should be constant 0.2 bar.
   c. Read the flowrate on the flowmeter q_g at each step.
   d. You can increase the pressure in steps but with a constant 0.2 bar pressure difference for all measuring steps.

When you are completely done:

- Turn inlet and outlet pressure down to zero, P1 and P2.
- Close the main valves on the nitrogen gas bottle for flow.
- Remove the gas pressure from the core.
  - Open gas bottle regulator valve and the pressure gauge on the gas bottle will show zero.
  - Close the gas bottle regulator.
- In case of back pressure, reduce the back pressure to zero.
- Reduce the gas pressure in the sleeve by SLOWLY opening the exhaust valve.
  - Close exhaust valve afterwards.
- Check that the pressures in the sleeve and core are zero.
- Open the core holder and take the core out.
- Unplug the two-pressure gauge and the flowmeter.

EXERCISE 7.4

A linear permeability measurement is conducted with air on a core sample with a diameter of 2.54 cm and length of 2.54 cm. The core sample is cleaned, dried, and placed in a standard permeability apparatus and the following data is recorded (Table 7.3):

TABLE 7.3
Recorded Data for Exercise 7.3

$P_1 \left(10^5 \, Pa \right)$	q_2 (cm³ / min)
1.133	6.4
1.679	35.6
3	132.8

The rate at the outlet, q_2, is measured at atmospheric conditions ($1.013 \cdot 10^5 \, Pa$). Air viscosity is $0.018 \cdot 10^{-3} \, Pa \cdot s$.

a. Explain the Klinkenberg effect.
b. Calculate Klinkenberg-corrected absolute permeability.

SOLUTION

a. When the pore sizes in the porous medium approach the mean free path length of the gas, then the gas will have a non-zero velocity at the grain boundaries. This leads to a higher measured permeability than the liquid permeability for the same porous medium.
b. Apply Equation 7.16 where P_{atm} and P_2 (as outlet pressure) are equal so the

equation can be written as $\dfrac{q_2}{A} = \dfrac{k_g \left(P_1^2 - P_2^2 \right)}{2 \mu L p_2}$

Answer: At $1 / P_m = 0$: $k_g = k_{abs}$, so the absolute permeability = 1.8 mD

EXERCISE 7.5

A cylindrical core sample with diameter 5 cm and length 30 cm has a porosity of 0.24 and an irreducible water saturation (S_{wi}) of 0.28, which is assumed to be constant during the experiment. The effective permeability of the gas is $0.25 \cdot 10^{-12} \, m^2$. The viscosity of the gas is $0.015 \cdot 10^{-3} \, Pa \cdot s$. Calculate the Darcy velocity and the real gas velocity at the mean pressure in the pore system for a pressure difference over the core plug of $6.5 \cdot 10^5 \, Pa$.
    Equations:

$$q^* = \frac{kA \left(P_1^2 - P_2^2 \right)}{2 \mu L P^*} \text{, with * the conditions q was measured in.}$$

$$V_{real} = \frac{V_{Darcy}}{\phi \left(1 - S_{wi} \right)}$$

$$P_m = \frac{P_1 + P_2}{2}$$

TABLE 7.4

Table of Calculation for Exercise 7.3

$P_2(10^5\,Pa)$	$P_1\,(10^5\,Pa)$	q_2(cm³/min)	$P_m\,(10^5\,Pa)$	$\dfrac{1}{P_m}$ [1/bar]	$k_g\,(mD)$
1.013	1.133	6.4	1.073	0.93	7.80
1.013	1.679	35.6	1.346	0.74	6.06
1.013	3.001	132.8	2.0067	0.50	5.08

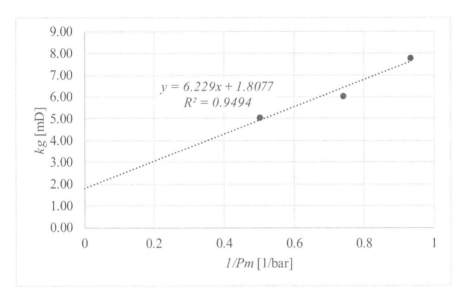

FIGURE 7.15 Plot of k_g versus $1/P_m$.

SOLUTION

q at mean pressure $P^* = P_{m:}$ $q = \dfrac{kA\left(P_1^2 - P_2^2\right)}{2\mu LP} = \dfrac{k}{\mu L}\dfrac{\left(P_1 - P_2\right)\left(P_1 + P_2\right)}{2\dfrac{\left(P_1 + P_2\right)}{2}}$

$$V_{Darcy} = \dfrac{q}{A} = \dfrac{0.25\cdot 10^{-12}\left(6.5\cdot 10^5\right)}{0.015\cdot 10^{-3}\cdot 0.3} = 0.036\,\dfrac{m}{s}$$

$$V_{real} = \dfrac{V_{Darcy}}{\phi\left(1 - S_{wi}\right)} = \dfrac{0.036}{0.24\left(1 - 0.28\right)} = 0.209\,\dfrac{m}{s}$$

7.4.3 HIGH VELOCITY FLOW

For high flow rates, Darcy's law is not valid. This is, for example, of importance to predict gas flow rates of wells in gas fields. The range of flow rate at which laminar flow exists is dependent on the Reynolds number, which is a dimensionless quantity. The Reynolds number for porous media is defined as:

$$Re = \frac{inertia\ forces}{viscous\ forces} = \frac{\rho \bar{v} d}{\mu} \tag{7.19}$$

where $\bar{v}$ is average Darcy velocity (q/A), ρ is fluid density, μ is fluid viscosity, and d is the average sand grain diameter (Forchheimer 1901). For example, in sand, transition from laminar to turbulent flow occurs in the range of Reynolds numbers from 1 to 10. At higher velocities, inertia of the fluid will start to have an effect.

Many models were suggested to replace or modify Darcy's law for high-velocity flow. The most accepted model for non-Darcy flow was proposed by Forchheimer (Forchheimer 1901):

$$-\nabla P = av + bv^2 \tag{7.20}$$

where a and b are constants and $v = q/A$, rewritten in an equation where a and b are expressed in terms of fluid and rock properties as:

$$-\nabla P = \frac{\mu}{k} v + \beta \rho v^2 \tag{7.21}$$

where β is a high-velocity coefficient, k is absolute permeability, and μ and ρ are viscosity and density of the gas, respectively. The additional term accounts for the inertia forces. The high velocity coefficient is a property of the formation rock that accounts for the deviation from Darcy's law.

The deviation is more pronounced in gas flow than in oil flow, due to higher velocities in comparison to the difference in density. Many correlations for β exist in the literature. Usually, β is taken as a property of the reservoir rock, which may be estimated from:

$$\beta = \frac{const.}{k^{\alpha}} \tag{7.22}$$

where a is constant which can be determined experimentally from a known permeable formation (Amyx and Whiting 1960). To identify whether the velocities or flow rates used can lead to the non-validity of the Darcy equation, the Re number can be estimated before the experiment. This can help to determine the appropriate flow rates for the experiment. Based on Eq. 7.21 it becomes clear that the measured pressure difference is higher compared to when viscous forces are dominant. If then the Darcy equation is used without correction, the permeability is underestimated.

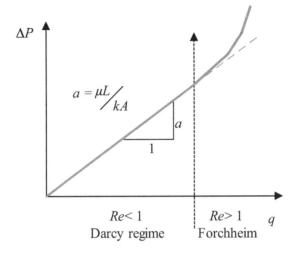

FIGURE 7.16 The effect of high Reynolds numbers on the q-dP curve, deviating from linearity.

7.5 EXPERIMENT: ABSOLUTE PERMEABILITY MEASUREMENT

DESCRIPTION

The objective of this experiment is to measure the absolute permeability of a core using air and brine based on the Darcy theory. Make a comparison of the air and brine permeability experiment. What is the absolute permeability to be reported?

7.5.1 EXPERIMENT 7.1: ABSOLUTE PERMEABILITY WITH BRINE

PROCEDURE

1. Weigh a dry Berea plug W_{dry} and measure its diameter D and length L, with calipers. Saturate the core with 36 g/l NaCl brine and weigh the plug, $W_{sat.}$
2. After core mounting and leakage tests, measure a minimum of five combinations of pressure difference and flow rate, between 2 and 5 cm³/min. Collect each measurement the water production V_w, over a time interval $\Delta t = 60$ seconds.
3. Plot a line through the three ΔP versus V_w data. Calculate the absolute permeability k_{abs}.

CALCULATIONS AND REPORT

Core No.: x D: 3.78cm, L: 4.18 cm, W_{dry}: 106.9 g, W_{sat}: 112.946 g

Liquid data:

$$\rho_w = 1.020 \text{ g/cm}^{3,} \quad \mu_w = 1.04 \text{ cp}$$

TABLE 7.5

Absolute Permeability Measurement of Water Calculation Table

Flow rate [cm³/min]	ΔP [bar]
2.0	0.059
2.5	0.082
3.0	0.100
3.5	0.131
4.0	0.152

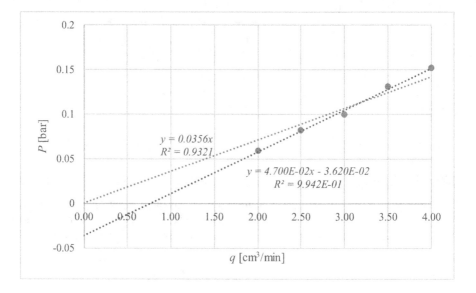

FIGURE 7.17 Results of Table 7.5 plotted.

Calculation of k_{abs}:

$$k_{abs} = \frac{(\mu_w L)}{(qA\Delta P)}$$

$$\frac{(\mu_w L)}{(k_{abs} A)} = \text{slope of the } \Delta P \text{ versus } q \text{ plot}$$

A = Cross section area of the core, cm²

The data in Table 7.5 is plotted in Figure 7.17 and the gradient is determined in two ways, using a trendline $y = ax$ and $y = ax + b$. Based on the R² depicted in the graph (regression) the match is better for $y = ax + b$. This indicated a systematic error in the data, as based on Darcy's equation $b = 0$. Accepting the systematic error and using the gradient 0.0470 [$\dfrac{\text{bar}}{\text{cm}^3/\text{min}}$], the permeability is

$$k_{abs} = \frac{\left(\dfrac{1.04}{1000}\,\text{Pas}\cdot 0.0418\text{m}\right)}{0.0470\,\dfrac{[\text{bar}]}{\dfrac{\text{cm}^3}{\text{min}}}\cdot\dfrac{10^5[\text{Pa}/\text{bar}]}{\dfrac{10^{-6}[\text{m}^3/\text{cm}^3]}{60[\text{sec}/\text{min}]}}\cdot\left(\pi\left(\dfrac{0.0378\text{m}}{2}\right)^2\right)}$$

$$= 1.37\cdot 10^{-10}\,\text{m}^2 = 1.39\cdot 10^{-1}\,D$$

Pay attention to rounding the final result.

7.5.2 EXPERIMENT 7.2: MEASUREMENT OF PERMEABILITY WITH AIR

PROCEDURE

The constant pressure method is used to measure the air permeability. The measured air permeability is influenced by the mean pressure P_m of the core. The mean pressure is regulated by the upstream and downstream values on the sides of core holder. The rate is measured at atmospheric conditions. Air viscosity as a function of temperature at atmospheric conditions is shown in Figure 7.19.

Four measurements of air permeability will be taken at different pressures. It is important to keep the pressure difference $(P_1 - P_2)$ constant, because the air flow at

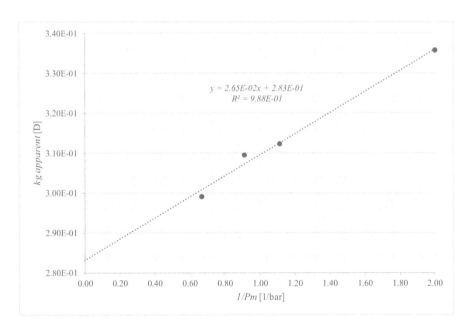

FIGURE 7.18 Data belonging to the example data of Experiment 7.2.

the core sample must be laminar. To avoid turbulent flow, use a maximum $(P_1 - P_2) =$ 0.2 bar. The procedure is as follows:

1. Calculate the air permeability k_g from Eq. 7.16 for each pair of mean pressure and gas flow rate.
2. Plot k_g versus $1/P_m$ and calculate k.
3. Calculate the Klinkenberg constant b.

RESULTS AND CALCULATION

Example results from an experiment are shown here.

From Figure 7.19 the nitrogen viscosity at 30°C can be found, being $\mu_{nitrogen} =$ 0.0179 cP.

Based on the trendline at $1/Pm = 0$, the $k = 0.283$ D.

The gradient is $k \cdot b$. The gradient is $2.65 \cdot 10^{-2}[D \, bar]$, $b = \dfrac{2.65 \cdot 10^{-2}}{2.85 \cdot 10^{-1}} = 9.29 \cdot 10^{-2} \, bar$

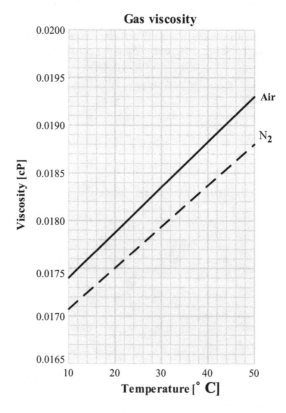

FIGURE 7.19 Dynamic gas viscosity as a function of temperature, at atmospheric P (modified from Pirson (Pirson 1958)).

TABLE 7.6

Measurement of Gas Permeability Calculation Table Using Nitrogen, at 30°C

$P\,1$ [bar]	$P\,2$ [bar]	P_m [bar]	Q [cm³/s]	Q [m³/s]	$P\,1^2 - P\,2^2$ [bar²]	$1/P_m$ [1/bar]	k_g [m²]	k_g [D]
0.60	0.40	0.50	4.94	4.96E-06	2.00E+09	2.000	3.864E-13	3.92E-01
1.00	0.80	0.90	7.80	7.79E-06	3.60E+09	1.111	3.391E-13	3.44E-01
1.20	1.00	1.10	12.8	1.26E-05	6.00E+09	0.909	3.341E-13	3.38E-01
1.60	1.40	1.50	12	1. 20E-05	6.00E+09	0.667	3.135E-13	3.18E-01

This fits quite well with estimations from literature with an order of magnitude between 0.1–1.0 (Jones 1987; Amyx and Whiting 1960).

COMPARISON EXPERIMENTS 7.1 AND 7.2

The values of the found absolute permeabilities can vary between air and liquid, here between 180 and 281 mD. For the comparison it is good to consider errors that can occur in both experiments. Gas permeability can experience errors due to use of too-high flow rates causing turbulence. Leakage can cause erroneous readings, but incorrectly calibrated gas flow rate meters or gas pressure regulators also can introduce errors. The liquid permeability is generally lower than the air permeability measured (Amyx and Whiting 1960; Swanson 1981). It might be due to the fact that the saturation is not 100%, such that trapped gas causes friction and a reduction of permeability. This effect is reduced if the gas is compressed with an increased mean pressure. If k increases with increasing ΔP, it might be an indication gas is present. Note that when the saturated core is placed in the core holder, the flow lines in front of the core need to be gas free; if not, the gas can be flooded through the core and get trapped, leading to inaccurate measurements. Interactions of the clay with the injection brine might cause clay swelling, reducing the permeability using brine. Deviations can be significant. Therefore, generally for the determination of the relative permeability, as discussed in Chapter 11, the absolute permeability of liquid is used as input.

7.6   VISCOSITY DETERMINATION IN CORES FOR NON-NEWTONIAN FLUIDS

For the Darcy equation to be applied, the viscosity needs to be known and the fluid used is to be Newtonian. Using Darcy's equation, the viscosity can be derived from flow through a porous medium with a known permeability for a Newtonian fluid using $\mu = \dfrac{kA\Delta P}{qL}$. In a porous media there is a variety of pore and pore throat sizes. Velocities vary within the pore based on changes of the areal cross flow A. For Newtonian fluids the viscosity is independent of velocity. The viscosity found using a porous medium shall therefore be equal to the viscosities found in the bulk measurements, which methods were introduced in Chapter 4 based on movement in tubes or plate/cylinder geometries.

For non-Newtonian fluid the flow behavior in a porous media is more complicated as the in situ viscosity depends on the velocity and so the local pore structure. The average velocity over a porous media cannot be used to determine an average viscosity using bulk data measured on non-Newtonian fluids. To solve this, an apparent average viscosity can be measured using the introduced flooding set-ups here for liquid, using a porous media with a known permeability. Darcy's equation can be applied to derive for each flow rate a porous medium specific viscosity for the non-Newtonian fluid, which can then be used to predict fluid flow of non-Newtonian fluids on a larger scale (Delshad et al. 2008).

7.7 TRANSIENT STATE PERMEABILITY TESTS FOR LOW-PERMEABILITY ROCKS

For Darcy's equation to be applicable the flow needs to reach a steady state, meaning constant flow rates and pressures. In cases of ultralow permeabilities, e.g. to be expected for cap rock or tight gas reservoirs, in the micro- to nano-Darcy range, this will take a significant amount of time. Then unsteady-state or transient state permeability experiments can be performed as an alternative. There are several experimental techniques developed, of which two will be briefly introduced here, the pulse decay test and pressure fall off test or the Gas Research Institute (GRI) test. For further details we refer to the literature (Sander et al. 2017).

PULSE DECAY TEST

The pulse decay test is a test first described by Brace (Brace and Frangos 1968) and various variants have been developed since (Sander et al. 2017). In Figure 7.21 a sketch is shown of the common set-up. A confined core is in contact at both sides with reservoirs of ideally equal size. The experiment is initiated with an instant drop or increase in pressure in one of the reservoirs of 1–5% of the initial pressure (Sander et al. 2017). The pressure vs. time behavior in both reservoirs, $P1$ and $P2$, is observed as the pore fluid moves through the sample from one reservoir to the other, where the pressure in the high-pressure reservoir decreases while in the other reservoir the pressure increases. Generally, high gas pressures are used to reduce gas slippage and reduce the measurement time. Several modifications of this method are developed; see Sander et al. (2017) for an overview.

There are different solutions to the transient flow regime of compressible fluid through a compressible medium. According to Brace (Brace and Frangos 1968), assuming no storage in the sample and a constant pressure gradient over the sample, the decay or increase in pressure can be correlated to the permeability of the sample according to:

$$k = \frac{\alpha \mu c_f L}{A \left(\dfrac{1}{V_u} + \dfrac{1}{V_d} \right)} \qquad (7.23)$$

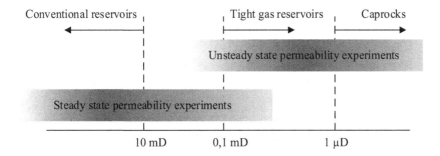

FIGURE 7.20   Range of permeabilities where the steady and unsteady-state permeability methods can be applied (Sander et al. 2017; Gensterblum et al. 2015).

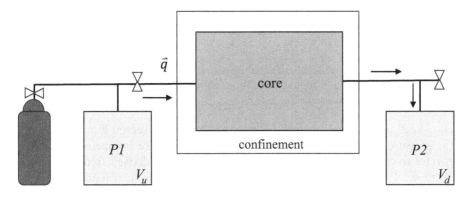

FIGURE 7.21   Pulse decay test set-up, to determine permeabilities in transient pressure state.

Where α is the slope of the pressure decay curve in the linear regime, plotting t versus the pressure difference in reference to the ultimate pressure, on a semi log plot and where μ is the liquid dynamic viscosity, C_f the fluid compressibility, L the core length, A the area of perpendicular to flow, and the up and down stream volumes V_u and V_d. Methods to compensate for the fluid storage adaptions to this solution can be found in the literature, as the discussion is beyond the scope of this chapter. A recommended reference is Sander et al. (2017).

Note that for very dense materials, other physical effects are going to play a role and the no-slip boundary condition of the Darcy's equation is not valid anymore (Wang et al. 2021).

PRESSURE FALL-OFF TECHNIQUE

In Figure 7.22 a sketch is given of the set-up. A set-up like helium porosity is used (Chapter 5), where in a closed reservoir a pressure is set ($P1$). Consequently, the

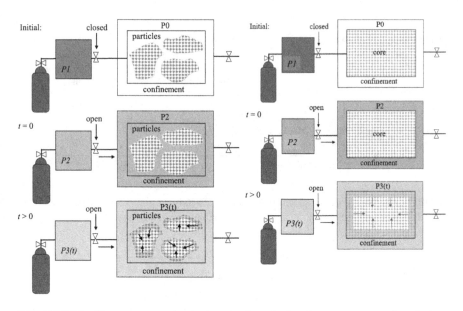

FIGURE 7.22 The gas penetration in the sample, reducing the pressure over time during a pressure fall-off test for a. particles like drilling cuttings or b. solid samples.

closed cell with the rock sample is connected with the reservoir at $t = 0$. Initially the pressure drops to $P2$, filling the accessible pore space instantaneously, comparable to the helium porosity test. The permeability is derived from the further decrease of system gas pressure over time $P3(t)$, where in the helium porosity the pressure remains constant. Sometimes this set-up is used for crushed samples, which speeds up the analysis, as long as the crushing does not affect the permeability. The method is widely used for permeability measurements of shale samples (Profice and Lenormand 2020; Tinni et al. 2012).

REFERENCES

Amyx, J.W., Bass Jr., D.M., Whiting, R.L. 1 960. *Petroleum Reservoir Engineering*. McGraw-Hill.

Bacci, G., Korre, A., Durucan, S. 2011. "Experimental investigation into salt precipitation during CO2 injection in saline aqufers." *Energy Procedia 4* 4450–4456.

Brace, W.F., Frangos, J.B. 1968. "Permeability of granite under high pressures." *Journal of Geophysical Research* 2225–2236.

Darcy, H. 1856. *Les fountaines publiques de la ville de Dijon: exposition et application. Victor Dalmont, Paris, 1856*. Victor Dalmont.

Delshad, M., Kim, D.H., Magbagbeola, O.A., Huh, C., Pope, G.A., Farhad, T. 2008. "Mechanistic interpretation and utilization of viscoelastic behavior of polymer solutions for improved polymer-flood efficiency." In *SPE Symposium on Improved Oil Recovery*. SPE. https://doi.org/10.2118/113620-MS.

Feder, J., Flekkøy, E.G., Hansen, A. 2022. *Physics of Flow in Porous Media*. Cambridge University Press.

Forchheimer, P. 1901. "Wasserbewegung durch Boden." *Zeitschrift des Vereins Deutscher Ingenieure 45* 1781–1787.

Fredrich, J.T., Lakshtanov, D.L., Lane, N.M., Liu, E.B., Natarajan, C.S., Ni, D.M., Toms, J.J. 2014. "Digital rocks: Developing an emerging technology through to a proven capability deployed in the business." In *SPE Annual Technical Conference and Exhibition*. Amsterdam. http://doi.org/10.2118/170762-MS.

Gensterblum, Y., Ghanizadeh, A., Cuss, R.J., Amann-Hildenbrand, A., Krooss, B.M., Clarkson, C.R., Harrington, F.J., Zoback, M.D. 2015. "Gas transport and storage capacity in shale gas reservoirs—A review. Part A: Transport processes." *Journal of Unconventional Oil and Gas Resources 12* 87–122.

Jones, S.C. 1987. "Using the inertial coefficient, ß, to characterize heterogeneity in reservoir rock." In *SPE Annual Technical Conference and Exhibition*. Dallas, TX. https://doi.org/10.2118/16949-MS.

Klinkenberg, L.J. 1941. "The permeability of porous media to liquids and gases." *Drilling and Production Practice, American Petroleum Institute* 200–213.

Mishra, S., Bera, A., Mandal, A. 2014. "Effect of polymer adsorption on permeability." *Journal of Petroleum Engineering 2014* (ID: 395857) 9.

Monicard, R.F. 1980. *Properties of Reservoir Rocks; Core Analysis*. Edition Technip.

Pirson, S.J. 1958. *Oil Reservoir Engineering*. McGraw-Hill.

Profice, S., Lenormand, R., 2020. "Low-permeability measurement on crushed rock: Insights." *Petrophysics 61* (2) 162–178.

Ringrose, P. 2020. *How to Store CO2 Underground: Insights from Early-Mover CCS Projects*. Springer Briefs in Earth Sciences.

Sander, R., Pan, Z., Donnell, L.D. 2017. "Laboratory measurement of low permeability unconventional gas reservoir rocks: A reveiw of experimental methods." *Journal of Natural Gas Science and Engineering* 248–279.

Seright, R., Brattekås, B. 2021. "Water shutoff and conformance improvement: An introduction." *Petroleum Science 18* 450–478.

Swanson, B.F. 1981. "A simple correlation between permeabilities." *Journal of Petroleum Technology* 2498–2504.

Tang, G., Morrow, N.R. 1999. "Influence of brine composition and fines migration on crude oil/brine/rock interactions and oil recovery." *Journal of Petroleum Science and Engineering 24* (2–4) 99–111.

Tinni, A., Fathi, E., Agarwal, R., Sondergeld, C., Akkutlu, Y., Rai, C.S. 2012. "Shale permeability measurements on plugs and crushjed samples." In *SPE Canadian Unconventional Resources Conference, Paper SPE 162235*. SPE.

Wang, R., Chai, J., Luo, B., Liu, X., Zhang, J., Wu, M., Wei, M., Ma, Z. 2021. "A review on slip boundary conditions at the nanoscale: Recent development and applications." *Beilstein Journal of Nanotechnology 12* 1237–1251.

8 Surface and Interfacial Tension

8.1 INTRODUCTION

Surface and interfacial tension of fluids result from molecular interactions happening at the boundary between different phases. Surface tension is the tendency of a liquid to minimize its free surface when exposed to gases. In fact, surface tension of a liquid arises due to imbalance between the intermolecular attractive forces at its interface. Unlike molecules within the bulk of the liquid, those at the interface do not have similar molecules on all sides, causing a net inward force, which tends to draw these molecules back into the bulk liquid; see Figure 8.1. This process minimizes the number of molecules remaining at the liquid's interface. In thermodynamics, surface tension is measured as the work needed to change the size of a liquid surface by a unit area. The SI unit for surface tension is Newton per meter [N/m], and the CGS (centimeter-gram-second) unit is dynes per centimeter [dynes/cm]. Similarly, interfacial tension reflects a tendency that exists when two *immiscible* liquids come into contact. In the following, interfacial tension will be denoted for both surface and interfacial tension.

Effects of interfacial tension and reduction of the surface energy of liquid-liquid systems or gas-liquid systems is common in daily life. The Plateau-Rayleigh instability is where liquid streams break up in droplets where the droplets tend to be spherical, like water from a tap or streams of rainwater. Small insects or leaves can remain and move over the water surface of a pond due to surface tension. For the purpose of cleaning in all kinds of daily activities, interfacial tension is reduced by using soap, for instance for doing the dishes, personal hygiene, or cleaning clothes. Foam breaks down or coarsens over time, for example dying beer, reducing the amount of interfacial area.

For flow in porous media, it will be shown in the relevant chapters that the interfacial tension is of importance for the topics of:

- Capillary pressure causing flow diversion in the porous media and affecting the fluid distribution.
- Residual oil where droplet or bubble trapping causes recovery to be never 100% and production is not optimal.
- Wettability.
- EOR method to reduce interfacial tension for increased oil production, like surfactant flooding.

Figure 8.1 illustrates a spherical cap exposed to interfacial tension (denoted as σ) around the base of the cap and two normal pressures P_1 and P_2 at each point on the

DOI: 10.1201/9781003382584-8

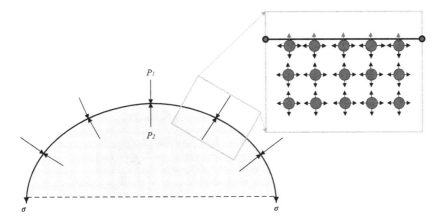

P_1

P_2

σ σ

FIGURE 8.1 Capillary equilibrium of spherical cap.

surface. The effect of the interfacial tension σ is to reduce the size of the sphere unless it is opposed by a sufficiently great difference between pressures P_1 and P_2.

This equilibrium is described by the Young-Laplace equation, where governing the mechanical equilibrium of an arbitrary surface can be expressed as:

$$P_2 - P_1 = \sigma\left(\frac{1}{r_1} + \frac{1}{r_2}\right) \tag{8.1}$$

where r_1 and r_2 represent the principal radii of curvature. Introducing the mean radius of curvature, denoted as r_m defined by:

$$\frac{1}{r_m} = \frac{1}{2}\left(\frac{1}{r_1} + \frac{1}{r_2}\right), \tag{8.2}$$

the Young-Laplace equation becomes:

$$P_2 - P_1 = \frac{2\sigma}{r_m}. \tag{8.3}$$

Note that the phase located on the concave side of the surface must exhibit a higher pressure, denoted as P_2, in comparison to the pressure P_1, on the convex side.

EXERCISE 8.1

In the set-up shown in Figure 8.2, two soap bubbles are hanging. They are connected via a tube. Before opening valve 1, bubble A with a pressure $P2$ is smaller than bubble B, with a pressure $P1$. What happens when valve A is opened? Explain why.

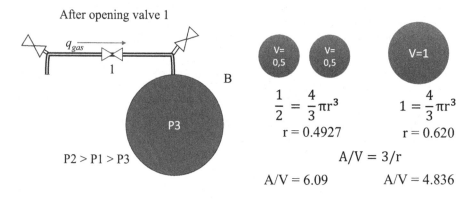

Before opening valve 1

A

P2

1

B

P1

P2 > P1

FIGURE 8.2 Two soap bubbles connected by a closed valve. What happens when valve 1 is opened?

After opening valve 1

q_{gas}

1

B

P3

P2 > P1 > P3

$V= 0,5$ $V= 0,5$ $V=1$

$$\frac{1}{2} = \frac{4}{3}\pi r^3 \qquad 1 = \frac{4}{3}\pi r^3$$

$r = 0.4927 \qquad\qquad r = 0.620$

$$A/V = 3/r$$

$A/V = 6.09 \qquad A/V = 4.836$

FIGURE 8.3 The final state of the two-bubble system as pictured in Figure 8.2 once valve 1 has been opened.

Solution: Based on Young's Laplace equation and the bubble radii, it is valid that P2 > P1. So, with a gas flow flowing from high to low pressures, bubble A empties into bubble B; see Figure 8.3. It can also be explained by surface energy, where the area/volume (A/V) ratio with one larger bubble is smaller than that with two equally sized bubbles. Note that the radius of the film of bubble A ends up having a similar radius as bubble B.

Similarly, the effect of foam coarsening can be explained in beer, where the gas flow between the bubbles with different radii takes place, from high pressure to low pressure via the gas diffusing through the liquid films between the bubbles. Gas solubility in the liquid will therefore significantly affect the coarsening process.

The resulting shape of surface minimization in different geometries and conditions can be modelled using e.g. the open-source software Surface Evolver of Ken Brakke (Brakke 1992).

8.2 PARAMETERS AFFECTING THE INTERFACIAL TENSION

The interfacial tension is determined by the balance of molecular forces between the two fluid phases: The cohesive *intra*molecular forces and adhesive forces as *inter*molecular force between the two phases. When the forces – and so the force balance – change, the resulting interfacial tension will be different. Note that the cohesive forces for gases are assumed much smaller in comparison to the liquid cohesive forces. So, also with minimum adhesive forces, liquids in a gas phase will consequently always want to form a droplet.

The interfacial tension is dependent on conditions of temperature and pressure and on the composition of the phases. The general order of magnitude of surface tensions lies between 15–75 mN/m. Table 8.1 shows typical values for known systems.

TEMPERATURE-PRESSURE

Generally, the surface tension of pure liquids decreases as temperature increases. The cohesion forces reduce due to higher temperature (Shaw 1992). Pressure can have an effect on density and on cohesive forces. With higher external pressure, $P1$ increases in Equation 8.3 and the measured surface tension decreases. It is therefore important to report with the data the temperature and pressure during the measurement.

PHASES

The interfacial tension of two contacting phases, gas-liquid or liquid/liquid, surface is found to depend on the nature of the phases. As mentioned earlier, the cohesive and adhesive force balance is important. Mixtures of additional components will affect the interfacial tension. e.g. for brine, salinity has an influence with both composition and quantity. The ions of NaCl in water can strengthen the bonds between the water molecules, increasing the surface tension close to the interface (Ozdemir et al. 2009). But also, for solvent content or dissolved gas, gas oil ratio (GOR) will

TABLE 8.1
Typical ST/IFT Values for Special Fluid Pairs at Ambient Conditions

System	ST/IFT (mN/m)
Air/mercury	480
gas/oil	24
Gas/brine	72
Oil/brine	32

play a role. To measure realistic interfacial tensions, the compositions and conditions need to be considered.

ADDITIVES

One specific component can have a large effect on the interfacial tension. These are surface active components, called surfactants. Surfactants generally prefer to be at the interface, having a hydrophobic tail and hydrophilic head e.g. of a hydrocarbon chain; see in the inset of T. The head can be ionic or non-ionic. The head resides best in the polar fluid, the tail in the non-polar fluid. The molecules in this way form a bridge between the two phases affecting the force balance at the interface. With increasing concentration, they reduce the surface tension, till the surface is saturated: the critical micelle concentration; see T. At higher concentrations the surfactant molecules can form micelles with the part of the molecule that prefers the fluid towards the outside. In the case of two fluids, the preference of the micelles depends on solubility preferences of the surfactants, type of surfactants, and generally on temperature and salinity, which is described as surfactant phase behavior. The interfacial tension development depends on the phase behavior, where at optimum temperature and salinity an ultralow interfacial tension can be found. Further details can be found in Lake et al. (Lake et al. 2014) or Sheng (Sheng 2011).

The function of the surfactant is to reduce the interfacial tension to create a surface, (a situation with increased surface energy). Surface active molecules can be

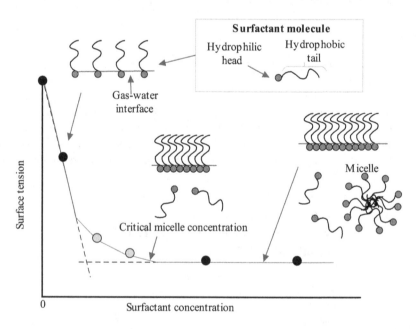

FIGURE 8.4   The development of surface tension with surface concentration and the model behind based on the surfactant arrangement.

naturally present in the oil phase or added in the water phase (surfactant flooding, see further Chapter 11) to reduce in a reservoir the interfacial tension between oil and water but also between gas and water. In an oil reservoir this can result in improved mobilization of oil, or in water reservoirs gas will form foams, which increase the gas viscosity, leading to a better displacement efficiency (mobility ratio) or volumetric sweep efficiency, increasing the storage potential (see further Chapter 11).

Note that pollutants often collect at the interface, which can seriously affect the measurement results in the lab. Cleanness of the equipment is therefore crucial as preparation of the experiment.

It might take some time till the surface-active molecules reach the surface. So, there is a time dependence in the interfacial tension measurements. In this context dynamic interfacial tension also can be studied, where the development of interfacial tension with time is measured (Dukhin et al. 1995; Miller et al. 2017).

8.3 LABORATORY MEASUREMENT OF SURFACE AND INTERFACIAL TENSION

There are several methods to measure surface and interfacial tension. Here the most common methods will be discussed. All methods are based on a force balance, where capillary force is one of the forces. An overview of the methods to be presented is given in Table 8.2. The measured quantity can be forced, but also will result in droplet shape/volume or column height.

8.3.1 DU NOÜY RING METHOD

The ring or Du Noüy ring method, also called ring tensiometer, of measuring surface and interfacial tension is commonly used. For measuring interfacial tension, a

---

TABLE 8.2
Overview of the Commonly Applied Methods to Measure Surface or Interfacial Tension

Method	Measurement Principle	Quantity Measured
Du Noüy ring	Surface force on ring	Force
Wilhelmy plate	Surface force on plate	Force
Stalagmometry method	Droplet formation based on force interaction of surface tension and hydrostatics	Droplet numbers or weight
Pendant drop	Equilibrium droplet shap. Equilibration of surface tension versus gravity	Droplet shape
Spinning drop	Equilibrium droplet shape, by equilibration of surface tension versus centrifugal forces	Droplet shape
Capillary rise	Fluid rise in a tube. Capillary forces balance gravity forces	Liquid column height

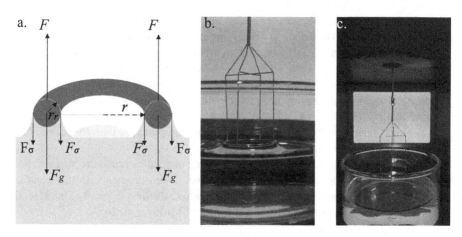

FIGURE 8.5   The Du Noüy ring method (a) the force balance on the ring when pulled out of the water (b) before detachment (c) after attachment.

platinum ring is submerged in the test liquid. The force required to withdraw it from the liquid is determined. Figure 8.5a shows the force balance. When the ring is completely wetted by the liquid ($\theta = 0$), the following equation applies:

$$F = m_r g - b + 2 \cdot (2\pi r \sigma) \tag{8.4}$$

In this equation, F represents the measured force, r stands for the radius of the ring at its center (with the radius of the platinum thread being insignificant compared to ring radius r), m_r is mass of the ring in air, g is the gravitational acceleration, and b corresponds to the buoyancy force acting on it when it's submerged in the liquid. For interfacial measurements, the ring is positioned at the interface, and the force required to break the interfacial film with the ring is determined.

The instrument can be regulated in such a way that the ring weight and buoyancy effect are taken care of with a correction factor C:

$$\sigma = C \frac{F}{2 \cdot (2\pi r)} = C \cdot \sigma_a \tag{8.5}$$

With σ_a as apparent surface or interfacial tension, mN/m. In general, this method is used for surface tension measurements at atmospheric conditions. For an accurate measurement, the surface of the liquid needs to be dust free, the ring needs to be clean and perfectly cylindrical and it needs to be pulled out parallel to the surface. It is advised to make a minimum of one or two repeating experiments to confirm the values obtained.

The correction factor C is dependent on the size of ring, the diameter of the wire used in the ring, the apparent interfacial tension, and the densities of the two phases at the given pressure and temperature (P, T). The commonly used relationship for C, introduced by Zuidema and Waters (1941), is given as:

$$C = 0.7250 + \sqrt{\frac{0.01452\sigma_a}{\pi^2 R^2 \left(\rho_l - \rho_u\right)} + 0.04534 - \frac{1.679r}{R}} \quad (8.6)$$

where
C = correction factor
R = radius of the ring, cm
r = radius of the wire of the ring, cm
σ_a = apparent surface or interfacial tension, mN/m
ρ_l = density of the lower phase, g/cm³
ρ_u = density of the upper phase, g/cm³

If surface tension was measured in the container which has been open to the air during measuring, (dry) air density may be approximated by the ideal gas law:

$$\rho_{air} = \frac{PM}{TR} \quad (8.7)$$

where
ρ_{air} = air density at P and T, kg/m³
P = pressure, Pa
T = temperature, K ($T(K)$= T(°C) + 273.15)
R = universal gas constant 8.31446261815324 m³ Pa/(K · mol)
M = molecular mass of air (kg/mol), for dry air 0.0289652 kg/mol

In the modern devices the ring motion and its calculations and corrections are automated. The presented interfacial tension are then the actual values and the input data for the correction is to be put in manually. This shall be checked for each individual device. See Section 8.4 for the advised experimental procedure.

8.3.2 WILHELMY PLATE METHOD

This methodology is based on the research conducted by Ludwig Wilhelmy, a German scientist. The principle of the Wilhelmy plate method is similar to the ring tensiometer based on measuring a force when being pushed out or in the surface, in this case using a plate made from glass or platinum, with an area in the order of a few square centimeters (Figure 8.6). It measures equilibrium surface or interfacial tension at an air-liquid or liquid-liquid interface. Platinum is chosen as the best plate material as it is chemically inert and easy to clean using solvents as toluene, acetone, and methanol. Applying this method to measure surface tension, the plate is positioned perpendicular to the interface. As the plate is moved into and out of a liquid, the change in force, F, which is the force acting on it, can be measured. Force, F, is proportional to the wetted perimeter and to the surface tension σ of the interface.

The force F on the plate can be measured using a tensiometer or microbalance and surface tension can be calculated as:

$$\sigma = \frac{F}{l \cdot \cos\theta} \tag{8.8}$$

Where l is the wetted perimeter equal to $2(w+d)$, where w is the plate width, d is the plate thickness, and θ is the contact angle between the liquid phase and the plate. In case of complete wetting, θ is assumed to be zero. The force on the plate at the time of entering or leaving the water can be described as:

$$F = W_p + 2(x+y)\sigma \tag{8.9}$$

where $2(x + y)$ signifies the contact area between the liquid and the plate, and W_p represents the weight of the plate equal to the mass of the plate, m times gravitational acceleration g. In this case, the force to consider for calculating the surface tension is when the plate is no longer submerged in the liquid, so that the weight of the plate experiences *no* buoyancy force from the fluid.

In the static method, the plate is held in the position as shown in Figure 8.6 and the corresponding equation including a correction for buoyancy becomes:

$$F = W_p - b + 2(x+y)\sigma\cos\theta \tag{8.10}$$

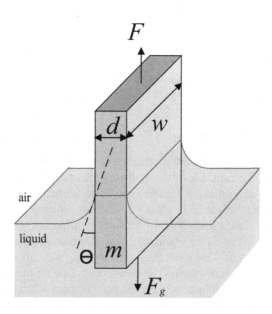

FIGURE 8.6   Wilhelmy plate method.

With b as the correction for buoyance on the part of the plate submerged, equal to the volume of the plate that is submerged and the density of the liquid the plate is submerged in:

$$b = V_{\text{plate submerged}} \cdot \rho_{\text{liquid}}$$

8.3.3   STALAGMOMETRY METHOD

The drop weight method offers a precise and straightforward approach for determining surface tension and ranks among the most widely used techniques. The principle of the method is based on more surface tension gives larger drop size. Schematic of the Stalagmometer (from Greek *stálagma*, which means drop) is shown in Figure 8.7.

It consists of a capillary tube with a liquid reservoir in the form of a bubble. The method can be carried out in two different ways, a) drop weight method and b) drop number method.

8.3.3.1   Drop Weight Method

The drop weight method of measuring the interfacial tension involves measurement of the weight of a droplet with a density (ρ) falling from a capillary tube with a radius (R).

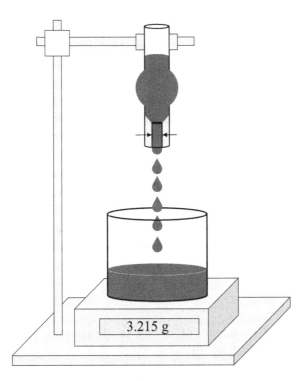

FIGURE 8.7   Schematic of the Stalagmometric method, with liquid led through a small capillary to form droplets that can be weighted.

These droplets are allowed to fall into a container until a sufficient quantity have been collected to accurately determine the weight per droplet. The underlying principle of this method is that the size of the drop falling from a capillary tube directly depends on the surface tension of the liquid. The droplet falls down when its weight just exceeds surface tension.

The maximum amount of liquid, W, which can be suspended from a capillary tube with radius r without falling, depends on the surface tension and can be expressed as $2\pi r\sigma$ and the weight acting downwards due to the mass of the drop is equal to $W = mg$. At equilibrium condition, these two forces should be equal:

$$2\pi r\sigma = mg \qquad (8.11)$$

The droplet weight ratio for two liquids that descend from the same capillary at nearly identical rates the following can be formulated:

$$\frac{w_1}{w_2} = \frac{m_1 g}{m_2 g} = \frac{2\pi r\sigma_1}{2\pi r\sigma_2} = \frac{\sigma_1}{\sigma_2} \qquad (8.12)$$

Here, m_1 and σ_1 represent the mass and surface tension of the reference fluid and m_2 and σ_2 the mass and surface tension of the fluid of interest. Normally, water can be used as a reference fluid:

$$\sigma = \sigma_{water} \cdot \frac{m}{m_{water}} \qquad (8.13)$$

Considering the surface tension value of 72.8 mN/m for water (20°C), it is possible to calculate the surface tension of the other fluid. It is extremely important that the water is pure so the theoretical values can be used and therefore knowledge of the temperature is crucial. In Table 8.3 the temperature dependent values for pure water can be found. Interpolation can be performed using the following equation:

$$\sigma_{water}\left(T\right) = -0.15 \cdot T + 75.74 \qquad (8.14)$$

T is the temperature given in °C, with T ranging between 10 and 30 °C.

The Stalagmometry method was improved by Chickanov et al. (2002), who measured weight of a fixed number of droplets rather than counting the droplets per time period.

TABLE 8.3

The Temperature Dependence of Surface Tension of Pure Water at Atmospheric Pressure (Vargaftik et al. 1983)

T [°C]	10	15	20	25	30
σ_w [mN/m]	74.2	73.5	72.8	72.0	71.2

8.3.3.2 Drop Number Method

The drop number method is based on counting the number of drops from a specific volume of the liquid. In this method, counting the number of drops for water and fluid of interest is sufficient to calculate the surface tension. It is similar to drop weight method except that the number of drops is counted when the liquid level falls from mark A to B instead of weighing. Suppose that the volume of single drop of the liquid volume v (ml) is equal to: $\dfrac{v}{n}$ where n denotes the number of drops that formed v ml of a liquid. If r is the density of the liquid, then mass of single drop of liquid is equal to: $m = \dfrac{v \cdot \rho}{n}$. Equation 8.11 can therefore be written as

$$2\pi r\sigma = \frac{v \cdot \rho}{n} \cdot g \tag{8.15}$$

For two liquids with the same volume and densities ρ_1, and ρ_2 and the number of drops n_1 and n_2, Equation 8.15 becomes:

$$\frac{\sigma_1}{\sigma_2} = \frac{n_2 \cdot \rho_1}{n_1 \cdot \rho_2} \tag{8.16}$$

Knowing the number of drops, fluid densities, and surface tension of the reference fluid, that of the other can be calculated. Water is usually taken as the standard and reference fluid as its surface tension is known at different conditions. See Section 8.4 for the advised experimental procedure.

8.3.4 PENDANT DROP

Pendant drop is an optical method for measurement of surface and interfacial tension using pendant drop shape analysis, visualizing a droplet using a camera and back light; see Figure 8.8.

Small drops naturally tend to adapt a spherical shape due to influence of surface forces, which depends on surface area; see Figure 8.9a. In principle, one can

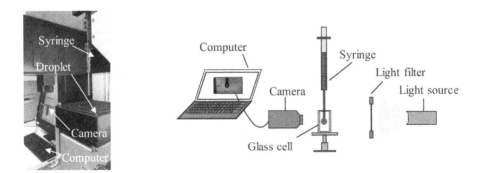

FIGURE 8.8 Pendant drop set-up.

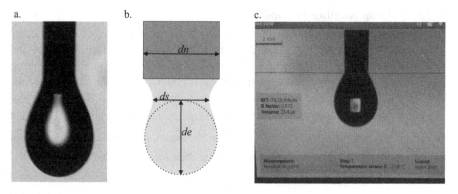

FIGURE 8.9 a. Pendant drop of heptane rich phase surrounded by methane rich phase at 85°C and 200 bar (Karimaie and Torsæter 2010). b. Relationship between dimensions of a pendant drop (modified from Adamson (Adamson 1982)). c. the automated shape analysis on a water droplet in air at 21°C.

determine the interfacial tension by examining the shape of the drop. In the context of a pendant drop, the most practical and measurable shape-dependent parameter is denoted as S and defined as: $S = d_s/d_e$ where, as indicated in Figure 8.9b, d_e represents the equatorial diameter (the maximum diameter) and d_s signifies the diameter measured at a distance d_e from the bottom of the drop. The interfacial tension can be calculated by the following equation:

$$\sigma = \frac{\Delta \rho g d_e^2}{H} \tag{8.17}$$

In this equation, H represents a shape determining variable. The relationship between the shape-dependent quantity H and the experimentally measured shape dependent quantity S is established through empirical means. The quantity of S is calculated after measuring d_e and d_s from the shape of the pendant drop, and, subsequently, $1/H$ can be determined using look-up tables, see e.g. Table 8.4 or the following empirical correlation matching the data:

$$\frac{1}{H} = 0.3156 \, S^{-2.613} \quad \text{or} \quad \frac{1}{H} = 0.3156 \left(\frac{d_s}{d_e}\right)^{-2.613} \tag{8.18}$$

So, $1/H$ becomes smaller with an increased d_s/d_e ratio, varying between 0.3 and 1. A larger surface tension enables larger droplets to form, which can be more distorted by gravity before they pinch off.

Applying this method, the droplet size shall be large enough—to the point of break off—so that the distortion is clearly visible in the shape of the droplet. In

TABLE 8.4

Tabulated Values of 1/H versus S for Pendant Drop Method Fragment between 0.300 and 0.449 (Adamson 1982)

S	0	1	2	3	4	5	6	7	8	9
0.30	7.098	7.039	6.981	6.924	6.867	6.811	6.755	6.700	6.646	6.593
0.31	6.539	6.487	6.435	6.384	6.333	6.283	6.233	6.184	6.135	6.087
0.32	6.039	5.992	5.946	5.900	5.854	5.809	5.764	5.720	5.676	5.633
0.33	5.590	5.548	5.506	5.465	5.423	5.383	5.343	5.303	5.263	5.224
0.34	5.186	5.147	5.110	5.072	5.035	4.998	4.962	4.926	4.890	4.855
0.35	4.820	4.785	4.751	4.717	4.683	4.650	4.617	4.584	4.552	4.520
0.36	4.488	4.457	4.426	4.395	4.364	4.334	4.304	4.275	4.245	4.216
0.37	4.187	4.159	4.130	4.102	4.075	4.047	4.020	3.993	3.966	3.940
0.38	3.913	3.887	3.862	3.836	3.811	3.786	3.761	3.736	3.712	3.688
0.39	3.664	3.640	3.616	3.593	3.570	3.547	3.524	3.502	3.479	3.457
0.40	3.435	3.413	3.392	3.370	3.349	3.328	3.307	3.286	3.266	3.246
0.41	3.225	3.205	3.185	3.166	3.146	3.127	3.107	3.088	3.069	3.051
0.42	3.032	3.014	2.995	2.977	2.959	2.941	2.924	2.906	2.889	2.871
0.43	2.854	2.837	2.820	2.804	2.787	2.771	2.754	2.738	2.722	2.706
0.44	2.691	2.675	2.659	2.644	2.629	2.614	2.599	2.584	2.569	2.554

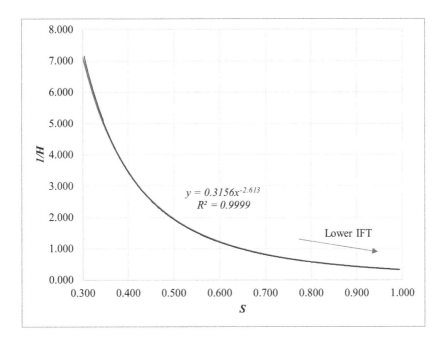

FIGURE 8.10   1/H vs. S for pendant drop method, derived from tabulated data of which a fragment is shown in Table 8.4. The trendline equation can be used to find values of 1/H for S between 0.300 and 1.000.

commercial equipment the droplet shape matching is done automatically using a droplet shape factor B; see Figure 8.9c (Berry et al. 2015).

The pendant drop method is widely used and has a high level of accuracy. See Section 8.4 for the advised experimental procedure.

8.3.5 SPINNING DROP METHOD

In this method, a droplet of a less-dense fluid is injected into a container filled with denser fluid, and the entire system is rotated as illustrated in Figure 8.11. Under the influence of the resulting centrifugal forces, the less-dense liquid tends to center around the axis of rotation, which causes the droplet to stretch along the axis of rotation. The interfacial tension acts in opposition to this elongation, as it tends to minimize the system's free energy by minimizing the surface area. This method closely

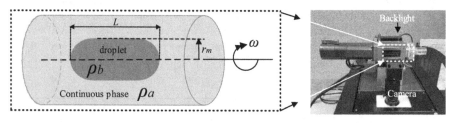

FIGURE 8.11 Schematic figure of a spinning drop in the rotating capillary.

- Rotational energy:

$$E_r = \frac{1}{2}I\omega^2 = \frac{1}{4}mr^2\omega^2$$

  - Include density difference: $m = V\Delta\rho$

$$E_r = \frac{1}{4}V\Delta\rho r^2\omega^2$$

I = momentum of inertia
ω = angular velocity
r = droplet radius
V = droplet volume
$\Delta\rho$ = density difference
between the two phases

- Surface energy based on a cylindrical shape:

$$E_s = 2\pi r l\sigma = \frac{2V_{cylinder}\sigma}{r}$$

- Minimization of total energy $E = E_r + E_s$:

$$\frac{dE}{dr} = 0 \Rightarrow \frac{1}{2}V\Delta\rho r\omega^2 - \frac{2V\sigma}{r^2} = 0$$

- Solve for interfacial tension:

$$\sigma = \frac{\Delta\rho\omega^2}{4}r^3 \qquad \text{If } l > 8r$$

FIGURE 8.12 The derivation of Equation 8.9, the equation for interfacial tension using the spinning drop method.

resembles the pendant drop technique, with the gravitational acceleration g replaced by the appropriate acceleration term for a centrifugal field.

If the fluid densities are ρ_A for the continuous phase and ρ_B, for the droplet and the angular velocity ω of rotation is known, it is possible to calculate the interfacial tension from the measured droplet profile. When droplet length is much greater than the radius r_m, $1 > 8r$, forming a cylindrical shape, the following approximate expression applies:

$$\sigma = \frac{(\rho_A - \rho_B)\omega^2 r_m^3}{4} \tag{8.19}$$

See Figure 8.11 for the simplified derivation. For a more detailed derivation see e.g. Rapp (Rapp 2017).

The spinning drop device has been widely used in recent years to measure very low interfacial tensions. Unlike the other methods, no contact between the fluid interface and a solid surface is required.

EXERCISE 8.2

Determine the interfacial tension for the following data set (Table 8.5) generated with the spinning drop method.

SOLUTION

Rewrite Equation 8.19 to

$$\frac{\sigma}{\omega^2} = \frac{(\rho_A - \rho_B)r_m^3}{4}$$

According $y = ax$; with $x = \dfrac{1}{\omega^2}$ and $y = \dfrac{(\rho_A - \rho_B)r_m^3}{4}$, the gradient in a plot of $\dfrac{1}{\omega^2}$ and $\dfrac{(\rho_A - \rho_B)r_m^3}{4}$ will give the interfacial tension σ.

TABLE 8.5
Data for Exercise 8.2

rpm	radius [mm]	T= 22°C		
1000	5.22	Droplet density	0.89	g/cm3
3000	2.63	Outer fluid density	0.98	g/cm3
5000	1.82			
8000	1.25			
1000	5.22			

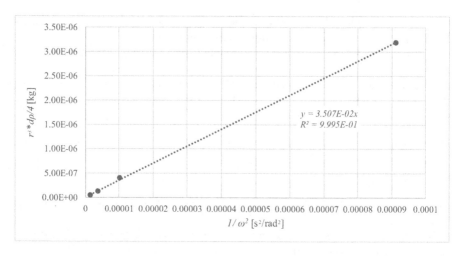

FIGURE 8.13 Spinning drop data plotted to obtain the interfacial tension, Exercise 8.2.

From Figure 8.13 the gradient is the interfacial tension: 35.07 mN/m.

8.3.6 CAPILLARY RISE METHOD

This method is based on the rising of a liquid in a capillary tube and the fact that
the height of the liquid depends on interfacial tension. When a capillary tube is
dipped in a liquid that exhibits a strong attraction to the glass surface (e.g. water),
it rises in the tube. The level of water in the tube is above the free surface in the
beaker (known as capillary rise). Conversely, when a capillary tube is dipped in the
liquid that does not exhibit affinity for the glass (e.g. mercury), it is pushed down in
the tube, resulting in the liquid level within the tube sinking below the free surface
of the liquid in the surrounding beaker (referred to as capillary fall). Therefore, the
rise of liquid in the tube is known as capillary rise while depression of the liquid
level is known as capillary fall. The affinity of the liquid to the glass in the presence of a gas phase is described by wettability, which is quantified by the angle the
liquid forms with the solid surface, the contact angle. In Chapter 9 the parameter
wettability is further explained.

Consider a capillary tube with a radius denoted as r, wetted by the liquid to
be tested. The liquid, having a density represented as ρ, immediately rises to a
height h above the free liquid level in the container (as depicted in Figure 8.14). To
sustain the column of liquid within the capillary tube against the force of gravity,
a force is required, known as capillary suction. We can express this relationship
as follows:

$$2\pi r \sigma \cos \theta \ \text{(Capillary suction)} = \rho g h \pi r^2 \ \text{(Gravity pull)} \qquad (8.20)$$

where θ is a contact angle between the liquid and glass tube and g is acceleration of
gravity.

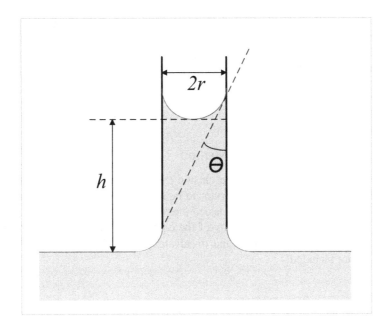

FIGURE 8.14   Capillary-rise method.

Consequently, the value of σ is calculated by:

$$\sigma = \frac{\rho g h r}{2 \cos \theta} \tag{8.21}$$

Further, for water in laboratory glass tube, θ is small, therefore $\cos \theta = 1$ and:

$$\sigma = -\rho g h r 2 \ \cos \theta$$

8.18 simplifies to:

$$\sigma = \frac{\rho g h r}{2} \tag{8.22}$$

The measurement of r (the capillary tube radius) can be avoided using relative method. In this method, h is also determined for a second liquid with unknown surface tension $\sigma 2$.

$$\sigma_1 = \frac{\rho_1 g h_1 r}{2} \text{ and } \sigma_2 = \frac{\rho_2 g h_2 r}{2} \tag{8.23}$$

$$\frac{\sigma_1}{\sigma_2} = \frac{h_1 \rho_1}{h_2 \rho_2} \tag{8.24}$$

$$\sigma_2 = \frac{h_2 \rho_2}{h_1 \rho_1} \cdot \sigma_1 \tag{8.25}$$

Where σ, h and ρ represent surface tension, capillary height, and density, respectively. The subscripts 1 and 2 denote the known fluid (water) and unknown fluid.

8.3.7 Choice of Measurement Method

Several methods to measure surface tension (ST) or interfacial tension (IFT) are presented in this chapter. It is important to be aware of the advantages and disadvantages of the methods, in order to determine for each application the most suitable method, depending on the availability of the equipment. Table 8.6 shows the main technical possibilities advantages and disadvantages. Availability, HSE, and economics are not included.

TABLE 8.6
The Advantages and Disadvantages of the Methods Presented in Section 8.3 (McPhee et al. 2015; Padday and Russell 1960)

Method	Mainly Used For:	Range [mN/m]	Advantage	Disadvantage
Du Nouy ring (Detachment)	ST	1 to 2,000	-Easy measurement, applicable at RT	-Not suitable for IFT -Ring cleaning, fragile competent
Wilhelmy plate (Detachment)	ST	1 to 2,000	-Easy measurement, applicable at RT	-Plate cleaning
Stalagmo-metric method	ST		-Very accurate with sufficient number of droplets -No resolution limitation	-Accuracy depends on size and number of the droplets
Pendant drop	ST/IFT	0.01 to 2,000	-Dynamic IFT -Relatively low IFT possible -Flexible in choice of droplet fluid based on buoyancy -Adaptable to high T, P	-Transparency of fluid surrounding the bubble/droplet -Uncertain effect of impurities
Spinning drop	ST/IFT	10^{-6} to 2,000	-Suitable for ultra-low IFT (surfactants). -Dynamic IFT -Adaptable to high T, P	-Densest fluid needs to be transparent
Capillary rise	ST/IFT	1 to 2,000	-Simple method	-Cleaning capillary -Densest fluid needs to be fully wetting or contact angle needs to be known

8.4   EXPERIMENTAL PROCEDURES

8.4.1   GENERAL EXPERIMENTAL PREPARATION

CLEANING

To do accurate measurements, independent of the method, the equipment and liquids needs to be extremely clean or respectively treated such that no pollution can affect the measurements. This can be extremely tedious. The deionized water shall be freshly tapped from a filtered source. After a prewash with toluene, heptane, methanol, and/or acetone, all glassware is advised to be cleaned in a laboratory dishwasher. In case of thorough pollution, more aggressive chemicals like chromic acid can be applied to remove any adhering greasy substances. Consider the HSE standards as advised in the material safety data sheets.

Directly before its use, rinse all glassware three times with the fluid you intend to work with. This also applies to syringes, syringe tips, and cuvettes used.

HSE CONSIDERATIONS

Consider, based on the used fluids and the potential exposure to vapors, whether the set-up needs to be in the fume hood. Perform a risk assessment to evaluate further risks that can arise while performing the experiments.

8.4.2   DU NOÜY RING METHOD

The platinum ring used is fragile and expensive. Be careful handling the ring. Hold the ring by the hook/cross wire and not the ring itself.

It is important to keep the ring clean, especially when changing from one liquid to another. Lowering the ring into acetone removes the fluid molecules left from earlier use. Finish cleaning the ring using a lighter/flame to burn off all remaining pollutants. To avoid ash on the ring, use the blue part of the flame.

If the ring is not completely covered by the fluid and/or not parallel to the fluid surface, the measured value will be incorrect.

EXPERIMENTAL PROCEDURE

The procedure of modern ring tensiometers is automated. Follow the instructions in the manual. The included steps involve:

1. Taring and calibration of tensiometer.
2. Interfacial tension measurement. Two or three readings including the given standard deviation should be taken, so an average value may be used for calculating interfacial tension. This can be done with the fluid samples three times but also by renewing the fluid and repeating the complete measurement.
3. Record the measurement conditions.

4. It needs to be checked whether the correction as discussed in Section 8.3.1 is applied or not. This needs to be applied if not included.
5. Compare the data with literature data available.

CALCULATION AND REPORT

TABLE 8.7
Measurement of IFT with the Ring Tensiometer; Calculation Table

Measurement T = _____ °C	Correction Factor, C	Apparent Value, σ_a [mN/m]	Corrected Value, σ [mN/m]	Literature Value [mN/m]
Water (3% wt NaCl)				
Zalo (1wt% solution)				
Oil (Exxsol D60)				

EXERCISE 8.3

The following data is generated using a ring tensiometer using water with different sodium chloride (NaCl) solutions at measurements from 24–26°C. Calculate the corrected surface tension values dependent on NaCl concentration. Discuss the results concerning error and comparison to the literature data.

Additional data available:

Radius of the ring = 0.955 cm
Radius of the wire of the ring = 0.02 cm

TABLE 8.8
Data for Exercise 8.3

Molarity NaCl in Water ~24–26°C	Apparent Surface Tension [mN/m]					Literature σ, 23°C [mN/m]	Literature Std σ, 23°C [mN/m]*
0	65.19	64.88	64.63	65.10	64.76	72.72	0.03
2.0	67.93	67.91	67.70	67.74	66.67	75.54	0.05
2.5	69.09	68.87	68.76	68.43	68.19	76.52	0.05
3.0	70.57	70.42	70.36	69.99	70.08	77.48	0.05
4.0	71.60	71.51	71.41	71.49	71.05	79.31	0.02

*Source: (Ozdemir et al. 2009)

TABLE 8.9
Density Difference between Air and NaCl Solutions at 25°C

M NaCl	0.0	2.0	2.5	3.0	4.0
$\Delta\rho$ [g/cm³] air-brine 25°C	1.00	1.04	1.08	1.12	1.17

SOLUTION

Use Equation 8.5 and 8.6 to correct the surface tension measured and plot the data.

With the correction, the surface tension values (SFTcor) increased. It is expected that the surface tension increases with increasing NaCl, which the data shows. The literature data found (Ozdemir et al. 2009) measured at 23°C shows higher values than the measured data. With increasing temperature, the surface tension will decrease. With an average measurement temperature of 25°C, the data fits this trend. Considering Equation 8.14, the surface tension for pure water is 72.29 mN/m, versus 72.00 mN/m at 25°C, a difference of 0.29 mN/m. Comparing the data, the difference is averagely −2.89 mN/m. It appears to be a systematic error in the measurement. The advice is to check the calibration weight for the balance, a constant temperature during the measurement, and the shape and cleanness of the ring.

TABLE 8.10
The Processed Data of Table 8.9

M NaCl	$\sigma_{av,a}$ [mN/m]	Std $\sigma_{av,a}$ [mN/m]	C (Eq. 8.6)	$\sigma_{av,c}$ [mN/m]	Std $\sigma_{av,c}$ [mN/m]	Literature 23°C [mN/m]	Std [mN/m]
0	64.91	0.21	1.075920	69.84	0.22	72.72	0.03
2	67.59	0.47	1.075953	72.72	0.50	75.54	0.05
2.5	68.67	0.32	1.071218	73.56	0.34	76.52	0.05
3	70.28	0.22	1.068092	75.07	0.23	77.48	0.05
4	71.41	0.19	1.063513	75.95	0.20	79.31	0.02

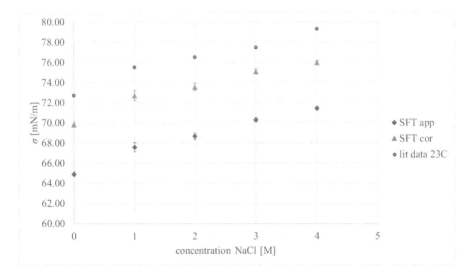

FIGURE 8.15 Data of Table 8.10 plotted, Exercise 8.3.

8.4.3 STALAGMOMETRY METHOD: DROP WEIGHT AND DROP NUMBER METHOD

First the experiment is performed with purified water and afterwards with the fluid of interest.

EXPERIMENTAL PROCEDURE FOR DROP WEIGHT METHOD

1. The equipment must undergo a cleaning process as discussed previously.
2. The apparatus should be fixed in a clamp stand. Deionized water is sucked into the Stalagmometer using a Peleus ball, or a syringe can be used to fill the equipment.
3. Record the room temperature and pressure.
4. Position an empty weighing bottle beneath the capillary outlet, allowing ten drops of water to descend into the weighing bottle and then record the total weight of the weighing bottle with the drops. The more droplets, the more accurate the recording becomes.
5. Subsequently, rinse the apparatus with acetone and allow it to dry.
6. In the next step, the fluid of interest is sucked into the Stalagmometer.
7. Following that, precisely ten liquid drops of liquid are allowed to fall into the weighing bottle and weight of weighing bottle with drops is recorded.

CALCULATIONS

Weight of empty bottle = $m_1.g$
Weight of bottle + n drops of water = $m_2.g$
Weight of bottle + n drops of fluid = $m_3.g$

Weight of n drops of water = $(m_2 - m_1).g$
Weight of n drops of fluid = $(m_3 - m_1).g$

Then the surface tension of liquid will be given by:

$$\sigma(T) = \sigma_{water}(T) \cdot \frac{m_n}{m_{nwater}} = \sigma_{water}(T) \cdot \frac{m_3 - m_1}{m_2 - m_1}$$

for 20°C this becomes $\sigma = 72.8 \cdot \dfrac{m_3 - m_1}{m_2 - m_1}$ [mN/m]

EXPERIMENTAL PROCEDURE FOR DROP NUMBER METHOD

The experimental procedure for the drop number varies only from the drop weight measurement (point 4 and 7) in measuring per given time, the number of drops for distilled water, and the experimental fluid. Measurements of the droplet counting shall be repeated a minimum of three times to obtain a minimum statistical. The advised time to count droplets is 1 to 2 minutes.

TABLE 8.11

Scheme Example for the Calculation of Surface Tension Using Drop Number Method

Exp No.	Time [sec]	Number of Droplets Reference Phase: Water n #	Average Water Droplets #	Time [sec]	Number of Droplets of Fluid of Interest (FOI) to Measure Surface Tension #	Average FOI #	Calculated Surface Tension [mN/m] $t_{av,w} = t_{av,foi}$
1			$n_{av,w}$			$n_{av,foi}$	
2			$t_{av,w}$			$t_{av,foi}$	$\dfrac{\sigma_w}{\sigma_{foi}} = \dfrac{\eta_{foi} \cdot \rho_w}{\eta_w \cdot \rho_{foi}}$
3							
$T =$ °C	$\rho_w =$ $\sigma_w =$			$\rho_{foi} =$			

The surface tension of the fluid can then be calculated from Equation 8.16.

$$\frac{\sigma_1}{\sigma_2} = \frac{n_2 \cdot \rho_1}{n_1 \cdot \rho_2}$$

8.4.4 CAPILLARY RISE METHOD

EXPERIMENTAL PROCEDURE

1. Any capillary can be used, provided that is its well cleaned; ideally the inner radius is known.
2. Record the room temperature and pressure.
3. Place the capillary tube in a beaker filled with water and measure the height to which the water rises within the tube (h_1).
4. Afterward, repeat the cleaning process for the capillary tube using acetone and ensure it is completely dry.
5. Immerse the capillary tube in a beaker containing the fluid of interest and measure the height to which the fluid ascends within the capillary tube (h_2).
6. For this method also the densities of both fluids need to be known. See Chapter 4 in case measurements are needed. These need to be performed under the same conditions.
7. Use Equation 8.25 to calculate the surface tension of the unknown fluid.

CALCULATION AND REPORT

The smaller the radius of the capillary the better, as then the capillary rise will be higher. It is required that the capillary still be well cleaned. It is also important to place the capillary straight and at the exact same level for both fluids such

TABLE 8.12

Scheme for Calculation of Surface Tension Using Capillary Rise Method

	Capillary Rise		
Conditions	Room T = °C Glass Tube Radius = [mm]		
	Water, 1	Fluid of interest, 2	Surface tension [mN/m]
Liquid data	$\rho_1 =$ [kg/m³] $\sigma_1 =$ [mN/m]	$\rho_2 =$ [kg/m³]	$\sigma_2 = \dfrac{h_{2av}\, \rho_2}{h_{1av}\, \rho_1} \cdot \sigma_1$
Exp no.	Water h_1 [cm]	Fluid of interest h_2 [cm]	
exp 1			
exp 2			
average h	h_{1av}	h_{2av}	

that $h = 0$ is equal. It is advised to make a calculation of the surface tension using the radius of the capillary as control. Especially when the rise of the fluid with the unknow surface tension is compared with the pure water results it is a requirement that the water is clean. The deviation from the control measurement is first the uncertainty in the capillary radius. This is solved by the comparison to a known value of water, assuming the liquid is pure and gives the theoretical value to be used.

8.4.5 Pendant Drop Method

EXPERIMENTAL PROCEDURE

The pendant drop method is applied to determine the interfacial tension between two liquids. This method is typically suited for liquid pairs with interfacial tensions that fall within the normal range (not excessively low or high). The pendant drop IFT measurements are often performed together with contact angle measurements. Detailed information about the contact angle apparatus, often referred to as a Goniometer, can be found in Chapter 9.

It needs to be determined which phase will be the continuous phase and which phase will be used to form a droplet and respectively a bubble. Availability of fluid, HSE limitations for the continuous phase, and compatibility with the syringe need to be considered. In case the density of the continuous phase is higher than the droplet/bubble, a hook needle can be used instead, such that larger buoyancy forces can act, forcing the bubble/droplet to direct upwards. The syringe to be used for the droplet is ideally as large as possible to form a large droplet, ~1 mm in diameter for standard fluids. For low IFT systems the syringe can be smaller.

Here is a step-by-step procedure for conducting interfacial tension measurements using the pendant drop method, here with Exxsol D60 and pure water as example where Exxcol D60 will form the droplet surrounded by pure water:

1. Fill the cell with the densest fluid, pure water.
2. Use a syringe with a diameter 1.1 mm. Measure the outer diameter of the syringe with a caliper (d_s). Be careful not to deform the syringe.
3. Create a pendant drop of the lightest fluid, Exxsol D60.
4. Ensure the drop is stable and symmetrically hanging from the syringe.
5. Ensure that the droplet image is well focused before capturing it using the camera.
6. Record the temperature.

In case of manual calculation

7. Measure d_e and d_s, and then calculate S.
8. Determine $1/H$ based on the calculated S value; use Equation 8.18.
9. Use the known syringe diameter to calibrate d_e using:

d_e (real size of d_e) = d'_e (d_s/d_{img}), mm, with d_{img} the diameter of the syringe as measured in the image and d_s as measured with the caliper and d'_e the maximum droplet diameter as measured in the image.

10. Determine with Equation 8.17 the interfacial tension.

In case of automated calculations

11. Calibrate the image using the measured syringe tip diameter.
12. Define the end of the syringe tip.
13. Check the B value (matching parameter) and note the obtained values of IFT.
14. In case of a time dependent measurement, repeat the image capturing each defined time step. Dependent on the setting, evaporation or dissolution can affect the droplet volume. This should not affect the calculations, but due to compositional changed the measured IFT can change, nevertheless.

CALCULATIONS AND REPORT

TABLE 8.13

Interfacial Tension (IFT) Measurement; Pendant Drop Method Calculation Table

				Image Picture Sizes					
System	ρ_w (g/cm³)	ρ_0 (g/cm³)	$\Delta\rho$ (g/cm³)	d_e (mm)	d_s (mm)	d_{img} (mm)	S (d_s/d_e)	$1/H$	σ (dyne/cm)

8.5 EXPERIMENT 8.1: DETERMINATION OF SURFACE AND INTERFACIAL TENSION

OBJECTIVE

Determine for distilled water, 3%wt NaCl solution, Exxsol D60, and an SDS solution with the ring tension meter the surface tension (ST). Additionally, measure the interfacial tension between oil and brine and reproduce the surface tension of distilled water and brine with pendant drop. Compare the data of the different methods with comparable literature values.

Special note: Repeat the experiments to obtain a good accuracy.

The pendant drop measurements show a larger statistical accuracy than the ring tensiometer data. According to Table 8.3, the surface tension at 25°C of water is 72.0 mN/m. The pendant drop measurement is slightly higher, where the ring tensiometer is ~5mN/m lower. The quality and cleanness of the ring is one source of error that can highly affect the quality of the data. Additionally, cleanliness of the glass container can influence the data. Since water coming from the same source gives for the pendant drop values comparable to the literature and for the ring tensiometer slightly less, the error appears not to come from a pollution of the water directly.

EXERCISE 8.4

1. Explain the difference between interfacial tension and surface tension. Which forces are of importance and on what parameters are they dependent on?
2. Which method can be used to measure ultra-low interfacial tensions $< 10^{-2}$ mN/m? Explain the principle of the method.
3. The interfacial tension between a highly vaporizing alcohol and oil is to be measured at room temperature. The alcohol has a lower density than the oil. Explain how the set-up looks when the pendant drop method is to be used.

TABLE 8.14
Example Data of Surface Tension Measurements with Pure Water Using the Pendant Drop and Ring Tensiometer

	Surface Tension – Pure Water	
Test ~25°C	Pendant Drop	Ring Tensiometer (Corrected Values)
	[mN/m]	[mN/m]
1	72.71	67.08
2	72.49	66.43
3	72.42	66.60
4	72.62	68.85
5	72.74	67.03
6	72.40	
Average	72.56 + −0.15	67.20 + −0.96

TABLE 8.15

Data for Exercise 8.4.4

Measure-ment	1			2			3			4			5		
IFT n-Decane-water [mN/m]	41.11	41.1	41.11	40.89	40.89	40.91	40.6	40.57	40.6	40.41	40.38	40.41	40.24	40.21	40.25

4. The following data in Table 8.15 for n-decane – pure water was measured at 25.5°C using the ring tensiometer. What is the interfacial tension value with standard deviation that shall be reported?

SOLUTION

1. Interfacial tension describes the net surface force at the interface between two immiscible fluids, due to a difference between adhesive and cohesive forces. Consequently, the interface behaves like a stressed film. This can occur between two liquids or a gas and a liquid. When the fluids are a gas and liquid this tension can be also called surface tension. Interfacial tension depends on temperature, pressure, and the composition of the two fluids including additives as solvents and surface active molecules.

2. The spinning drop method can be used for measurements of ultralow interfacial tensions. In a capillary a droplet or bubble is placed in a heavier fluid, after which the capillary will be spun around its axis. Based on centrifugal forces the lighter fluid shall position along the axis of rotation elongating the droplet/bubble, which tendency is counteracted by the interfacial tension. The resulting droplet/bubble shape equilibrates the two forces, which then can be used to derive the interfacial tension dependent on the rotational speed.

3. As the alcohol is quickly vaporizing, it can be wise to create an alcohol droplet surrounded by oil. A closed container can be chosen also to avoid vaporization of the oil in case applicable. As the alcohol is lighter than the oil, a hook needle is to be used.

4. The average can be calculated $\sum_{n=1}^{15} \rho_n / 15 = 40.65$ and based on Equation 2.6 the standard deviation can be std $= \sigma = \sqrt{\dfrac{\sum_{i=1}^{15}\left[40,65 - x_i\right]^2}{15}}$, which is 0.32. So, the result can be reported as the interfacial tension between n-decane and water (at 23°C) 40.65 + −0.32 mN/m.

REFERENCES

Adamson, A.W. 1982. *Physical Chemistry of Surfaces*. John Wiley and Sons.

Berry, J.D., Neeson, M.J., Dagastine, R.R., Chan, D.Y.C., Tabor, R.F. 2015. "Measurement of surface and interfacial tension using pendant drop tensiometry." *Journal of Colloid and Interface Science* 454: 226–237. https://doi.org/10.1016/j.jcis.2015.05.012.

Brakke, K.A. 1992. "The surface evolver." *Experimental Mathematics* 1 (2): 141–165.

Chickanov, S.V., Proskurina, V.E., Myagchenkov, V.A. 2002. "Estimation of micelloformation critical concentration for ionogenic and non-ionogenic surfactants on the data of modified Stalagmometric method." *Butlerov Communications* 3 (9): 33–35.

Dukhin, S.S., Kretzschmar, G., Miller, R. 1995. *Dynamics of Adsorption at Liquid Interfaces: Theory, Experiment, Application*. Elsevier.

Karimaie, H., Torsæter, O. 2010. "Low IFT gravity drainage in fractured reservoirs." *Journal of Petroleum Science and Engineering (JPSE)* 70 (1–2): 67–73.

Lake, L., Johns, R.T., Rossen, W.R., Pope, G.A. 2014. *Fundamentals of Enhanced Oil Recovery*. Society of Petroleum Engineers. http://doi.org/10.2118/9781613993286.

McPhee, C., Reed, J., Zubizarreta, I. 2015. *Core Analysis: A Best Practice Guide*. Vol. Developments in Petroleum Science: Volume 64. Elsevier.

Miller, R., Aksenenko, E.V., Fainerman, V.B. 2017. "Dynamic interfacial tension of surfactant solutions." *Dynamic Advances in Colloid and Interface Science* 247: 115–129. https://doi.org/10.1016/j.cis.2016.12.007.

Ozdemir, O., Karakashev, S.I., Nguyen, A.V., Miller, J.D. 2009. "Adsorption and surface tension analysis of concentrated alkali halide brine solutions." *Minerals Engineering*: 263–271.

Padday, J.F., Russell, D.R. 1960. "The measurement of the surface tension of pure liquids and solutions." *Journal of Colloid Science* 15: 503–511.

Rapp, B.E. 2017. *Microfluidics: Modelling, Mechanics and Mathematics*. Elsevier. https://doi.org/10.1016/C2012-0-02230-2.

Shaw, D.J. 1992. *Introduction to Colloid and Surface Chemistry*. Fourth edition. Butterworths.

Sheng, J.J. 2011. *Modern Chemical Enhanced Oil Recovery, Theory and Practice*. Gulf Professional Publishing.

Vargaftik, N.B., Volkov, B.N., Voljak, L.D. 1983. "International tables of the surface tension of water." *Journal of Physical and Chemical Reference Data* 12: 817–820.

Zuidema, H., Waters, G. 1941. "Ring method for the determination of interfacial tension." *Industrial & Engineering Chemistry Analytical Edition*: 312–313.

9 Wettability

9.1 INTRODUCTION

When a liquid is brought into contact with a solid surface, the liquid either spreads over the whole surface or forms small drops on the surface. The liquid either will wet the solid completely or form a droplet having a contact angle with the surface. The wettability of a reservoir rock-fluid system is defined as the ability of one fluid in the presence of another to spread on the surface of the rock.

Wettability plays an important role in the production of oil and gas as it not only determines initial fluid distributions, but it is also a main factor in the flow dynamics in the reservoir rock. The concept of wettability finds its manifestation in everyday scenarios. For instance, the material used in crafting umbrellas exhibits a reluctance to interact with water, explaining why raindrops roll off the umbrella's surface (hydrophobic tendency). Here, the fabric is deemed non-water wet or oil wet (see Figure 9.1a). Water droplets on leaves also show hydrophobic surface properties, where here surface structure additionally plays a role. (see Figure 9.1b). As another example, compare a freshly waxed car to one that hasn't been waxed for a long time. On a freshly waxed car, water drops prefer to stay as isolated droplets while on a car with no wax or old wax water droplets tend to spread more and wet bigger areas of the carrosserie. This is due to the different wettability behavior of the surface. Oppositely, paint shall wet the surface to obtain a homogeneous coverage. Analogously, reservoir rocks also exhibit a proclivity to interact with either oil or water.

Wettability plays a pivotal role in fluid displacement like water injection or gas injection for improved oil recovery. The impact of waterflooding can diverge significantly between a water-wet rock and an oil wet rock. In a water-wet rock, water tends to adhere to the pore surfaces, facilitating the easy displacement of oil resulting in good displacement efficiency at break through. Conversely, in an oil wet rock where oil adheres to the pore walls, water bypasses oil and the breakthrough can occur sooner, leaving more oil trapped in the reservoir at the time of water breakthrough.

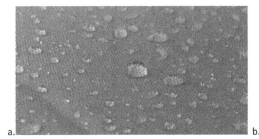 

a.   b.

FIGURE 9.1 Water droplets on an umbrella (a) and a shamrock leaf (b), showing the hydrophobic nature of the leaf.

DOI: 10.1201/9781003382584-9

9.2   CONCEPT OF WETTABILITY

To attain a deeper understanding of the wettability concept, it is imperative to revisit the distinct forces that come into play when liquids and solids come into contact, namely adhesive and cohesive forces.

The combination of adhesive and cohesive forces between the solid and liquid determine wettability. The following definition is accepted in the physics:

> Forces of attraction between molecules of the same substance is called **cohesive force**.
>
> Forces of attraction between molecules of the different substances is called **adhesive force**.

The term "cohesive forces" refers to the comprehensive array of intermolecular forces (like hydrogen bonding and van der Waals forces) that contribute to the property of liquids resisting separation. These attractive forces are prevalent among molecules of the same substance. For example, rain falls as droplets, not as a fine mist, due to the pronounced cohesion in water. This force draws water molecules together, generating cohesive clusters. Such behavior emerges from molecules' inclination to minimize interactions with their surroundings.

Similarly, the term "adhesive forces" reflects the attractive interactions between dissimilar substances. For instance, in the context of a liquid wetting a surface, adhesion prompts the liquid to adhere to the resting surface. Pouring water onto clean glass, for instance, results in its propensity to spread out, creating a thin and even film that covers the glass's surface. Figure 9.2 shows cohesive and adhesive forces acting on a solid-liquid system.

Note that cohesive forces are weak in gases and often neglected.

The extent of wettability is determined by the equilibrium of these two forces between each of the phases. In a system with one solid phase and two liquids, interaction takes place between the solid and liquid one and, respectively, liquid two, and interaction between the two liquids in the form of interfacial tension σ. The consequence of the force balance is that each phase attempts to minimize its interfacial

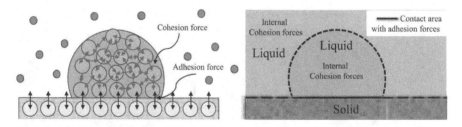

FIGURE 9.2   Adhesion and cohesion forces in a. a liquid gas system and b. a liquid-liquid system where the different contact lines represent the various adhesion forces.

area with the other phases. So, the force balance can be expressed as balance of interfacial tensions in the three-phase contact line, as can be seen in Figure 9.3, resulting in the following equation (known as Young's equation):

$$\sum f_x = \sigma_{so} - \sigma_{sw} + \sigma_{wo}\cos\theta = 0 \tag{9.1}$$

or commonly written as:

$$\sigma_{so} - \sigma_{sw} = \sigma_{wo}\cos\theta \tag{9.2}$$

Where $\sum f_x$ are the forces in horizontal plane, σ_{so} is the surface tension between the solid and oil, σ_{sw} is the surface tension between the solid and water, $\sigma_{wo}\cos\theta$ is the horizontal component of the interfacial tension between (in this example) the oil and water, σ_{wo} is the surface tension between oil and water, and θ is the contact angle measured through the densest phase (water) at the 3-phase contact line, where water, oil, and solid meet.

This **three-phase contact line** is the line along which the net forces of adhesion and cohesion act. The contact angle cannot be seen independent from the interfacial tension. Figure 9.3 illustrates this, assuming σ_{sw} and σ_{so} remain constant, that with a higher interfacial tension σ_{wo2} (left) the contact angle increases towards 90°.

The term $\sigma_{so} - \sigma_{sw}$ is also called adhesion tension A_T, which can then also be expressed as $\sigma_{wo}\cos\theta$. A positive adhesion tension A_T indicates that the phase that forms the droplet (here water) preferentially wets the solid surface (water wet). An A_T of zero indicates that both phases have an equal affinity for the surface (neutral system). A negative A_T indicates that the surrounding phase (oil in this case) wets the solid surface (oil wet). Therefore, the magnitude of the adhesion tension determines the ability of the wetting phase to adhere to the solid and to spread over the surface of the solid. Note that the adhesion forces are dependent on the surface properties, so coatings on the solid surface can significantly alter the wettability behavior of a system. This occurs during the aging process, for example where oil components absorbed into the rock can alter the wettability from water to oil wet.

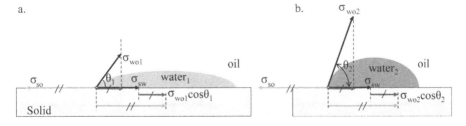

FIGURE 9.3 The effect of interfacial tension on the force balance and resulting contact angle in an water-oil-solid system, with a. showing a smaller interfacial tension than in b., ($\sigma_{wo1} < \sigma_{wo2}$) and $\sigma_{sw1} = \sigma_{sw2}$, $\sigma_{so1} = \sigma_{so2}$.

9.3   WETTABILITY CLASSIFICATION

Within reservoir engineering context, using as a basis oil and brine as the two fluid phases and rock mineral as the solid phase, wettability is classified into four different groups based on the static contact angle measured:

Water-wet: In this type of reservoir, the rock surface exhibits a preference for water coating. Consequently, the rock manifests a strong attraction to water, facilitating water's broad spreading across the surface (as illustrated in Figure 9.4a). With oil as the lighter phase, this implies that the contact angle will be smaller than 90°, indicating the water's tendency to spread across the surface.

1. **Intermediate wet or neutral wet**, where the rock surface shows an almost equal preference to be coated by either of the fluids, oil or water (as depicted in Figure 9.4b). This equilibrium results in a contact angle of approximately 90°, signifying the surface's equal affinity for both oil and water.
2. **Oil wet**: In this situation, the rock exhibits a preference for coming into contact with oil, in direct contrast to the water-wet scenario (as illustrated in Figure 9.4c). Consequently, the contact angle surpasses 90° as the oil-water interface inclination is towards oil rather than water.
3. **Mixed-wet (fractional wet)**: In this category, certain portions of the rock favor contact with oil, while other segments favor contact with water (as depicted in Figure 9.4). Mixed wettability can also be denoted as fractional wettability, and the contact angle will display variations contingent upon the specific region of the rock.

When contact angle is less than 90°, the surface is called **hydrophilic**, and when the contact angle is above 90°, the surface is called **hydrophobic**. In the case of fluids other than oil and water, the fluid that has the highest affinity to wet the solid surface is the **wetting phase** and the fluid with least affinity is **non-wetting phase**.

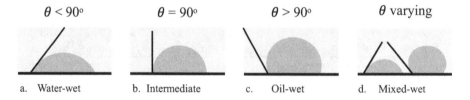

| $\theta < 90°$ | $\theta = 90°$ | $\theta > 90°$ | θ varying |

a.   Water-wet   b. Intermediate   c.   Oil-wet   d.   Mixed-wet

FIGURE 9.4   The four different classes of wettability based on the static contact angle, measured through the denser droplet.

Based on the contact angle, a **wetting index WI** is defined as $cos\theta$, so contact angles can be defined as in Table 9.1:

TABLE 9.1
The Wetting Index for Different Contact Angles

Wetting Index	Contact Angle (θ)	Wetting Condition
1.0	0°	Completely water wet
0.0	90°	Intermediate (neutral) system
−1.0	180°	Completely oil wet

The limits of the scales are not definite, since a system with contact angle in the range of about 70° to 110° is considered neutral. Other scaling systems are often used, and information is to be given on the definition used in each case.

The wettability of a reservoir rock system will depend on the following factors:

- Reservoir rock material and pore geometry.
- Geological mechanisms (accumulation and migration).
- Composition and amount of oil and brine.
- Physical conditions: Pressure and temperature.
- Mechanisms occurring during production, for instance., change in saturations, pressure, and composition.

Note that it is difficult to make a general model of wettability including all these factors. Although a lot of work has been done on wettability, it is not fully understood how the wettability of a porous rock surface is composed.

EXERCISE 9.1

Briefly describe what the interfacial tension and contact angle are. Make necessary calculations and draw a drop of water on a fixed surface surrounded by oil with the following interfacial tensions:

$$\sigma_{ow} = 30 \times 10^{-3} \frac{N}{m}$$

$$\sigma_{os} = 45 \times 10^{-3} \frac{N}{m}$$

$$\sigma_{ws} = 70 \times 10^{-3} \frac{N}{m}$$

What kind of wettability does this system have?

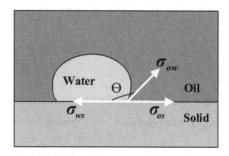

FIGURE 9.5   Oil wet system shown by water droplet placed on a flat, solid surface surrounded by oil.

SOLUTION

At the interface between two phases there is a contracting force that tries to minimize the area of the interface. The interfacial tension is a quantitative measure of this force per unit length $\left(\dfrac{N}{m}\right)$. When a liquid drop is placed on a flat, solid surface, a contact angle will be formed at equilibrium. This is the angle between the tangent of the liquid surface at the three-phase point and the solid surface, measured through the denser phase.

$$\sigma_{os} = \sigma_{ws} + \sigma_{ow} cos\theta$$

$$cos\theta = \frac{\sigma_{os} - \sigma_{ws}}{\sigma_{ow}}$$

$$\frac{45 - 70}{30} = cos\theta \Rightarrow \theta = 146°$$

The system is oil wet, see Figure 9.5.

9.4   LABORATORY MEASUREMENT OF WETTABILITY

Whether a rock is oil or water wet generally depends on the degree of adsorption of oil components into the surface of the mineral. Depending on the reservoir conditions and oil/brine composition, this might happen. On the other hand, the distribution of the oil saturation influences the distribution of the wettability throughout the pores. This is different considering just a prepared flat surface, which makes the representation of contact angle measurements on rock surfaces questionable.

The contact angle measurement on a flat surface does not cover the dynamic differences in pores filled with oil based on capillary forces. Therefore, other methods, invasive methods, were introduced for quantification of the wettability in porous rocks.

For wettability measurements, one needs to use either cores that hold the original reservoir wettability without impact of drilling fluids or the cores with precipitation of oil and salt components. In general, after the cleaning process as discussed in

Section 9.5, an attempt is made to restore the original wettability by an aging process. The wettability tests are performed on these samples.

Wettability in rocks can be assessed through either direct or indirect methods. Direct measurement techniques encompass quantification by e.g., the determination of contact angle, Amott index (Amott 1959), and USBM method. Indirect methods to obtain qualitative information on the wettability involve the assessment of imbibition capillary pressure and waterflood relative permeability. The subsequent section will delve into the details of direct measurements on core plug samples, while the techniques for indirect measurements will be explained in their respective chapters.

No satisfactory method exists for direct measurements of wettability in the reservoir, and therefore it is necessary to estimate the wettability from laboratory measurements.

The most common methods for measuring wettability on core samples, which will be discussed in more detail, are:

- The Amott method (imbibition and displacement or imbibition only).
- The USBM method.

The test developed by Amott appears to be the most accepted and widely used test in the oil industry. Other less common direct methods are:

- Measurements of nuclear magnetic relaxation rate.
- Measurement of wetting-dependent dye absorption.

9.4.1 CONTACT ANGLE METHOD

When a liquid drop is placed on a flat surface surrounded by another fluid, the degree of wetting of this solid interface by liquids is usually measured by the contact angle θ that a liquid-liquid interface makes with a solid, rather than the adhesion tension and the effect of the interfacial tension. The contact angle measured in static conditions in equilibrium is defined as the static contact angle.

When a liquid swiftly traverses a surface, the contact angle can deviate from its static equilibrium value. The **advancing contact angle** is determined while the front of the droplet is advancing, whereas the **receding contact angle** is measured when the droplet front is receding. On an ideally uniform surface, these two values closely align. However, in practice, the observed contact angle tends to vary based on the direction in which the contact line moves. The advancing contact angle tends to increase as the speed of motion increases, while the receding contact angle tends to decrease. This variability is known as **contact angle hysteresis**, quantified as the disparity between the advancing and receding contact angles. Contact angle hysteresis predominantly arises from factors such as the chemical and topographical irregularities of the surface (surface roughness), the absorption of impurities from the solution onto the surface, or changes in the surface caused by solvent-induced swelling, rearrangement, or modification.

Direct measurement of contact angle hysteresis is not feasible; instead, it is assessed by measuring the advancing and receding contact angles. To illustrate, imagine a water droplet on an inclined surface (see Figure 9.6a). In the absence of

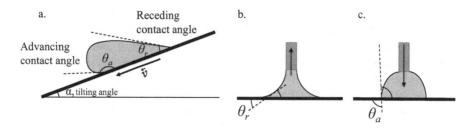

FIGURE 9.6 Advancing and receding contact angle using: (a) moving interfaces on a tilted plate or (b) on a flat plate under removal of liquid or (c) on a flat plate adding liquid to the droplet.

contact angle hysteresis, the droplet would immediately start sliding upon impact due to gravity's downward pull. However, contact angle hysteresis allows some droplets to remain adhered even to vertical surfaces, such as window glass. As the surface tilt increases or raindrop size on a window glass expands, a point is reached where gravity overcomes the hysteresis, prompting the droplet to move. At this point, the measurement of the advancing and receding angles is possible. The advancing contact angle indicates the surface's maximum attainable contact angle, while the receding contact angle signifies the minimum angle at a defined speed.

In the following paragraphs, different variations of the contact angle measurement will be discussed before the challenges will be addressed regarding why other methods for wettability determination also have been developed specifically for porous media.

A variation to the static contact angle measurement is the dynamic contact angle measurement. In this case images are made of moving droplets as seen in Figure 9.6a. Alternatively a droplet is created on a flat surface, where under constant flow rate fluid is added to the droplet (b) respectively removed (c) using a syringe which is in capillary contact with the droplet. With addition of fluid, the contact angle formed is the advancing contact angle, where, when fluid is removed, the contact angle measured is the receding contact angle.

The contact angle, denoted as θ, quantitatively assesses the level of wetting between a liquid and a solid surface. Geometrically, it is the angle established by a liquid at the point where three phases intersect: The liquid, gas (or in certain cases, another liquid), and solid. At this three-phase juncture, three distinct forces come into play. These forces are present at the contact point where the solid, liquid, and gas (or liquid) meet. Contact angles can be categorized into static, dynamic, and contact angles adjusted for surface roughness.

STATIC CONTACT ANGLE METHOD

Contact angle measurement is a classical method widely used in chemical engineering to derive the wettability in three-phase system (gas/liquid/solid or liquid/liquid/solid). The method was introduced into petroleum engineering more than 50 years ago. This method is used to determine reservoir formation wettability. The imaging method is easily carried out in the laboratory and gives you a clear understanding of the wetting mechanism in oil-water-rock system.

Generally, sandstone formation and carbonate formation are represented by small, polished quartz and carbonate blocks, respectively. A small drop (2–3 mm^3) of water is laid on the smooth surface of rock, which has previously been submerged in an oil-filled transparent cell. Then, the enlarged image of the water drop is obtained by photographing. The dimensions of the drop image are used to calculate the contact angle in the system.

The measurement of contact angles is based on Young's equation. When placing a drop of liquid on a solid surface, a finite angle of contact will in most cases be observed. However, complete spreading may also occur and then Young's equation ceases to hold.

The measurement of contact angle is simple in concept, but in practice it is an extraordinarily complex topic, and the interpretation of results is not straightforward. The measurement puts up severe demands to the cleaning procedures and the preparation of fluids and solid surface. The contact angle between oil, water, and a solid surface will depend on the following:

- Crude oil composition.
- Surface electric properties (pH and salt content of water).
- The solid surface itself.
- Roughness and heterogeneity of the solid surface.
- Dynamic effects.
- Pressure and temperature.

In practice, in a porous material it is even more complex because of different pore shapes and complex mineralogy.

Contamination will affect the contact angle measurement, so the cleaning procedures are critical. Most contact angle measurements are performed at room condition, and this is often due to the time-consuming cleaning procedures necessary for measurement at high pressure and temperature and complex and costly apparatus needed.

Contact angle measurements can also be performed at reservoir conditions, which is one of the main advantages in a crude oil-brine-rock system, as it has been shown that temperature may change the wettability significantly.

Static contact angles are determined when a droplet is stationary situated on a surface, and the intersection of the three phases remains fixed, often referred to as **sessile drop method or goniometer**. Among the various wettability measurements, static contact angles are the most commonly employed. The measurement of static contact angles relies on Young's equation, which assumes that the interfacial forces are in a thermodynamically stable state. So, they are particularly suitable for relatively even and uniform surfaces. Additionally, static contact angles serve as the basis for establishing the surface free energy (i.e., the solid's surface tension) of the substrate. This method provides a swift, straightforward, and quantifiable assessment of wettability.

Practically conducting sessile drop measurements requires the presence of a camera to make an image of the droplet outline, a light source for parallel backlighting such that the contrasts and droplet outline are enhanced, and a means to apply the droplet onto the surface. Commonly these components are assembled into a single apparatus known as an optical tensiometer, ranging from manual setups to fully automated systems. Accompanying software using fitting algorithms is typically employed to promptly extract contact angle data following the deposition of the droplet.

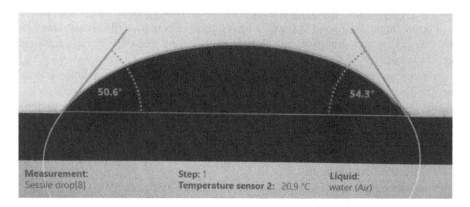

FIGURE 9.7   Sessile drop method of a water droplet in air on a quartz crystal.

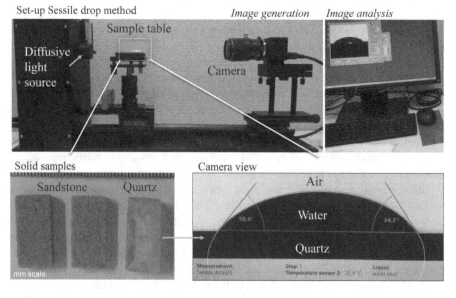

FIGURE 9.8   Sessile drop method applied using quartz and a water droplet in air. The bottom right shows the image generated and contact analysis on solid samples used as seen at bottom left.

For the sessile drop method, the density of the droplet needs to be higher than the surrounding fluid and the surrounding fluid needs to be transparent. The inversed system is called **the captive bubble method,** where the lighter phase is used to generate the bubble/droplet placed underneath the surface. Buoyancy forces the droplets against the solid surface. The choice of using the sessile drop or captive bubble method depends on the transparency of the fluids, the available quantities of fluids, and HSE implications of use of the fluids. In an oil water system, having water as surrounding phase is safer, and as it is a transparent fluid the droplet can be visualized. Oils are not always transparent, especially crude oils, and evaporation can cause health problems and an increased risk of fire.

PROCEDURE SESSILE DROP METHOD

Based on HSE risks due to liquid evaporation or spills, consider placement of the set-up in the fume hood.

1. Prepare the solid samples using the appropriate cleaning and aging procedure.
2. Fill the cell with the oil phase and then place the prepared sample in the cell.
3. Carefully place one or more water drops onto the surface with a syringe.
4. Adjust the imaging system until a drop is in focus (a clear outline of the image appears on the screen) and the droplet position shall be in line with the camera and light source.
5. Take an image of the droplet on the surface.
6. Determine on the image the contact angle at the left and right side of the bubble.

In case of manual analysis:

    a. Determine the three-phase contact point on both sides of the droplet, by
       i. Finding the interface of the solid and
      ii. determine where the liquid-liquid interface contacts the flat surface.
    b. Draw a tangent from the three-phase contact point along the liquid-liquid interface.
    c. Determine the angle between the interface of the solid and the tangent through the densest phase.
7. Repeat this procedure for multiple droplets to obtain a quantification of the statistical variation in the measurements.

Finish

8. Dispose of the fluid in the container and syringe according to regulations and clean the container accordingly. Store the rock/mineral sample for cleaning.
9. Dispose of the syringe appropriately.

9.4.2   EXPERIMENT 9.1: STATIC CONTACT ANGLE
MEASUREMENT

Compare minerals surfaces with rock surfaces composed of similar minerals.

Objective: Determine the contact angle at room temperature on different surfaces for Exxcol 60 as common model oil and a) 3wt% NaCl solution and b) 1wt% sodium dodecyl (SDS, surfactant) solution (or any other available surfactant) in 3wt% NaCl solution. The surfaces used should be quartz crystal, polished Berea sandstone, carbonate crystal, and polished reservoir carbonate core. Consider surface tension, interfacial tension, and the sessile versus captive droplet method. Create your own experimental plan.

Phase 1	Phase 2	Solid
Exxsol D 60	3wt% NaCl solution	Quartz crystal
air	3wt% NaCl solution + 1wt% SDS	Quartz sandstone (e.g. Berea)
	air	Carbonate crystal
		Carbonate rock

QUESTIONS

Determine the wettability for the different combinations of liquids, air-liquid, and rocks.

Determine whether there is a difference between the results of quartz and Berea sandstone (> 95% quartz) and the carbonate crystal and rock sample, considering e.g. difference in presence of porosity and surface roughness.

Determine how the addition of SDS affects the results.

CALCULATIONS AND REPORT

TABLE 9.3
Table for Data Contact Angle (CA) Measurement Using Imaging Method Calculation; Example to Extend for Number of Systems and Droplets Measured

System	Sample Dimension, mm		Liquid Phase		CA, Degree		Comments	Average CA, Degree	System Wettability
	H	D	Droplet/ Bubble	Surrounding Phase	Left CA	Right CA			
Sandstone (quartz)					1.				
					2.				
					3.				

DYNAMIC CONTACT ANGLE METHOD

A variation to the static contact angle measurement is the dynamic contact angle measurement. In this case images are made of moving droplets as seen in Figure 9.6. Alternatively a droplet is created on a flat surface, where under constant flow rate fluid is added to the droplet (b) respectively removed (c) using a syringe which is in capillary contact with the droplet. With addition of fluid the contact angle formed is the advancing contact angle, where when fluid is removed the contact angle measured is the receding contact angle.

IN-SITU CONTACT ANGLE MEASUREMENT BY PORE SCALE IMAGING

With the development of 3D imaging, the possibility arises to also measure in-situ at pore level contact angles when the resolution is high enough (Andrew et al. 2014). From the 3D images, from e.g. micro CT scanning, contact angles are derived from the three-phase contact line using specifically developed algorithms for segmentation and surface smoothing (Khanamiri et al. 2018). Micro-CT scanning provides an image resolution high enough to identify pores. However, the image resolution is not ideal for contact angle determination, with minimally 1–3 μm voxel size, especially considering pore sizes are one-tenth of μm. Based on the image resolution, the algorithms include a surface smoothing step before the normal of the two surfaces is determined, which meet at the three-phase contact line. The contact

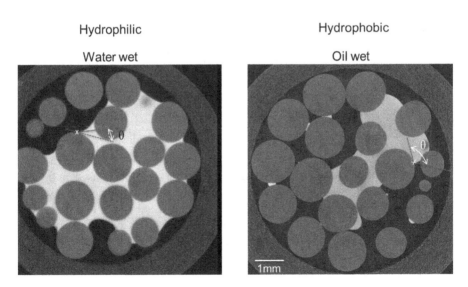

FIGURE 9.9 A horizontal cross-section of a vertical glass bead pack, partially saturated with cesium chloride solution (white) and air (black), using a micro-CT scan. Courtesy of T. Vukovic (Vukovic et al. 2023).

angle can then be determined using the normal of the two surfaces, as sketched in. The in-situ obtained contact angles display in general a significant spread, based on both physical limitations as well as image analysis errors. For more information on the image analysis, see Chapter 12. The contact angles are likely to be pinned, ranging from an advancing to receding angle, as the mineral surfaces are not smoothed. Additionally, mineralogy can vary when using original reservoir rocks, and depositions of clays or minerals can affect the surface interactions on presence of different minerals, so contact angles can show a large spread, even without considering potential differences in wetting regions due to variations in initial oil distribution after oil migration.

CHALLENGES IN CONTACT ANGLE MEASUREMENTS

The measurement of contact angle is simple in concept, but in practice it is an extraordinarily complex topic, and the interpretation of results is not straightforward (Deng et al. 2018; Andrew et al. 2014). The measurement puts up severe demands for the cleaning procedures and the preparation of fluids and solid surface.

The method is based on obtaining an equilibrium between the three interfacial forces. Pinning of the three-phase contact line due to surface roughness makes measurement erroneous. If the surface is not naturally flat, the solid surface must therefore be carefully polished.

Measurements on mineral surfaces are often performed on the natural plane of cleavage.

Using a porous material is even more complex because of varying mineralogy and due to the presence of pores as will be discussed further later. Generally, sandstone formation and carbonate formation are represented by small, polished quartz and carbonate crystal blocks, respectively, as seen in Figure 9.8 (bottom left).

Contamination will affect the contact angle, so the cleaning procedures are critical.

Finally, contrast is important as well as image resolution to determine the three-phase contact point on the image and tangent accurately.

9.4.3 Amott Method

Amott (Amott 1959), presented a method for measuring wettability based on spontaneous and forced displacement. First it is necessary to define two important reservoir engineering terms: Imbibition and drainage. Generally, imbibition is a process in which wetting phase saturation increases while drainage refers to a process in which non-wetting phase saturation increases. As an example, for a water wet reservoir, water injection is an imbibition process while gas injection is a drainage process.

EXPERIMENTAL PROCEDURE

The Amott test basically consists of natural and forced displacement of the fluid in the sample. In general, the Amott test involves four displacement steps. The objective of the experiment is to measure the amount of displaced fluid in a core sample, by spontaneous imbibition and forced displacement using two immiscible phases. Here oil and water will be used as example.

The experimental procedure for Amott test is as follows:

1. Select a clean and dry sample and measure physical properties such as length, diameter, and dry weight.
2. Saturate the core sample with 100% water. Measure the saturated core weight and calculate pore volume of the sample.
3. Displace the mobile water with oil to reach "initial water saturation" S_{wi} and measure the saturated weight, then compare the S_{wi} from the production data with the weight measurement.
4. Immerse the core sample in water in the Amott cell (Figure 9.10) to observe the spontaneous displacement of water by oil. The water is allowed to imbibe into the core sample, displacing oil out of the sample until equilibrium is reached (Figure 9.10a). The volume of oil displaced is measured, which represents the volume of water imbibed (V_{ws}).
5. The core sample is then removed and the remaining oil in the sample is forced down to residual saturation by displacement with water. This may be done either in a centrifuge or by being displaced with a pump in a sealed core holder. The volume of oil displaced may be measured directly or determined by weight measurements (V_{wf}).

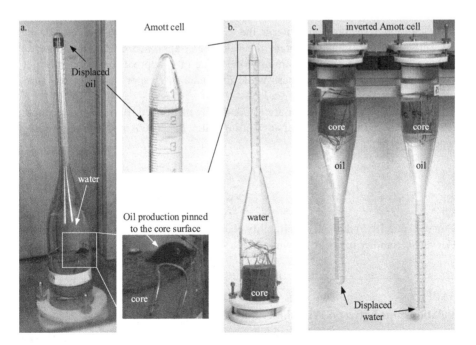

FIGURE 9.10 Imbibition cell with oil saturated core plug surrounded by water, using respectively crude oil a. and model oil b. and water-saturated core plug surrounded by oil c. (courtesy of Stratum).

6. The core, now saturated with water at residual oil saturation, is placed in an imbibition cell and surrounded by oil. The oil is allowed to imbibe into the core, displacing water out of the sample (Figure 9.10b). The volume of water displaced is measured, which is equal to volume of oil imbibed (V_{os}).
7. The core is removed from the cell after equilibrium is reached, and the remaining water in the core is forced out by displacement in a centrifuge or flooding using a core holder. The volume of water displaced is measured (V_{of}).

To determine the equilibrium time for spontaneous displacement in the Amott test, the production versus time is to be plotted. Once production ceases and reaches a stable equilibrium value, the spontaneous imbibition is finalized. For reservoir conditions e.g., the procedure described by Cuiec (Cuiec 1995) can be applied.

DATA ANALYSIS

By recording all volumes produced, it is possible to calculate water and oil wettability-index by the following equations:

$$I_o = V_{ws}/(V_{ws} + V_{wf}) \tag{9.3}$$

$$I_w = V_{os}/(V_{os} + V_{of}) \tag{9.4}$$

In addition to the previous criteria, the value of I_o approaching 1 indicates strong oil wetness of the sample, whereas the value of I_w approaching 1 strongly shows water-wetness of the sample. Similarly, the value of I_o approaching zero indicates weak preference for water-wetness while the value of I_w approaching zero shows weak water-ness of the sample.

By following the same procedure as previously mentioned, Boneau and Clampitt (1977) introduced the Amott-Harvey wetting index I_{AH}. The Amott-Harvey wetting index can be determined as:

$$I_{AH} = I_w - I_o = \frac{V_{ws}}{V_{ws} + V_{wf}} - \frac{V_{os}}{V_{os} + V_{of}} \tag{9.5}$$

Cuiec (Cuiec 1995) defined a wettability classification based on the range of Amott-Harvey wettability index as shown in Table 9.5.

TABLE 9.4

Relationships between Wettability and Amott Wettability Index

Wettability Index	Preferentially Water Wet	Neutral Wet	Preferentially Oil Wet
I_o	0.0	0.0	Positive
I_w	Positive	0.0	0.0

TABLE 9.5
Cuiec's Wettability Classification for Amott-Harvey Test

I_{AH} Range	Wettability
+0.3 to +1.0	Water-wet
+0.1 to +0.3	Slightly water-wet
−0.1 to 0.1	Neutral
−0.3 to −0.1	Slightly oil wet
−1.0 to −0.3	Oil wet

During the spontaneous imbibition, the wetting phase invades to reduce the capillary pressure to zero, where in the forced displacement the capillary pressure increases in magnitude again.

A drawback is that the tests are difficult to perform at reservoir pressure and temperature and droplets of produced fluids pinned to the rock surface can affect the accuracy of the production estimations.

For screening purposes, for example chemical influence, sometimes only the spontaneous imbibition test with brine (part 1–3) is performed to compare the production rate and recovery factor, which gives an idea about wettability of the sample.

9.4.4 EXPERIMENT 9.2: WETTABILITY DETERMINATION USING AMOTT METHOD

Objective: Determine Amott-Harvey wetting index for a cleaned sandstone core at S_{wi} and room conditions. Use 3wt% NaCl and Exxsol D60 as brine and oil phase.

Beside production data, also record temperature to determine that expansion of fluids did not cause additional production. Estimate the error for each reading, also taking into account the number of droplets pinned to the core surface, which are produced but not accounted for in the production volume.

PROCEDURE

1. Saturate the core with brine.
2. Use the centrifuge (see Chapter 10, capillary pressure) or core flooding to flood the core with Exxsol D60 to obtain S_{wi}. Record the amount of brine produced.
3. No aging is required.
4. Place the core in an Amott cell submerged in brine and record every day the oil production from the core till no additional production has been recorded for a week. Record time, the amount of produced oil, and temperature.
5. Place the core in a flooding rig to flood the core with brine to (1-S_{or}). At this stage there is no oil production anymore and ΔP is constant. Record the amount of produced oil and temperature.
6. Repeat point 4 submerging the core in oil and point 5 using oil as flooding fluid.

CALCULATIONS AND REPORT

The following is an example to show how the Amott-Harvey wettability index can be determined from experimental data. Note that this data is reported without error range in the produced liquid volumes. This shall be estimated.

EXERCISE 9.2

The following table provides the displacement data for Amott wettability test on three cores from a North Sea reservoir:

TABLE 9.6
Displacement Data for Amott Wettability Test

Core No	Displacement by Oil (ml)		Displacement by Oil (ml)	
	Spontaneous	Forced	Spontaneous	Forced
1	0.05	1.25	0.81	0.85
2	0	1.69	0	0.97
3	0.47	0.53	0.01	0.59

Calculate the Amott-Harvey wettability index and determine the wetting state of each core.

SOLUTION

$$I_o = \frac{V_{ws}}{V_{wt}} \qquad I_w = \frac{V_{os}}{V_{ot}} \qquad I_{AH} = I_w - I_o$$

Table 9.7 shows the calculated Amott-wettability indices.

TABLE 9.7
Amott-Harvey Wettability Index Calculation

Core No.	I_o	I_w	I_{AH}	Wettability State
1	0.038	0.48	0.45	Water wet
2	0.00	0.00	0.00	Neutral
3	0.47	0.016	−0.45	Oil wet

9.4.5 USBM METHOD

Donaldson et al. (Donaldson et al. 1969) developed a method, known as the U.S. Bureau of Mines-method (USBM) to measure average wettability of a core sample. The method is based on a correlation between the degree of wetting and the work involved in displacing the fluids. This is represented by the areas under the

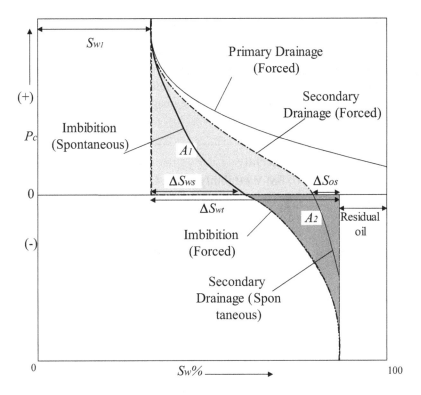

FIGURE 9.11   USBM data: Capillary pressure (Pc) curve to express wettability (water-wet) as "work needed to induce a saturation change" represented by the surface area ratio A1/A2 (modified from Morrow (Morrow 1990).

capillary pressure curves, further explained in Chapter 10 and in Tiab (Tiab and Donaldson 1996). The method employs the two areas under the oil-water capillary pressure curves obtained from the centrifuge method. Figure 9.11 shows full capillary pressure curves for drainage and imbibition and the areas A_1 and A_2 measured in the USBM method. In fact, the areas under the oil injection curve (A_1) present the energy needed for displacement of water by oil and the area under the water injection (A_2) curve represents the energy needed to displace the oil by water. One can notice that displacement of the wetting phase by the non-wetting phase (drainage process) requires more energy than displacement of the non-wetting phase by the wetting phase (imbibition process). The USBM wettability index is then calculated from the ratio of the area under the two pressure curves according to the following equation:

$$I_{USBM} = \log\left(\frac{A_1}{A_2}\right) \quad (9.6)$$

where A_1 is the area under the secondary drainage capillary pressure curve and A_2 is the area under the primary imbibition capillary pressure curve. Figure 9.11 clearly shows different steps involved in the USBM method.

Table 9.8 shows criteria for USBM wettability index.

Figure 9.12 shows examples of typical curves for water wet, oil wet, and neutral wet samples (Donaldson et al. 1969). A positive USBM index shows water wetting, the negative index shows an oil-wetting state, and a value of close to zero shows neutral wettability state. The larger the absolute value, the greater the wetting preference. In water-wet sample, the area under drainage capillary pressure curve (A_1)

TABLE 9.8

USBM Wettability Index Criteria

I_{USBM}	Wettability
> 0.0	Water wet
< 0.0	Oil wet
Near 0.0	Neutral

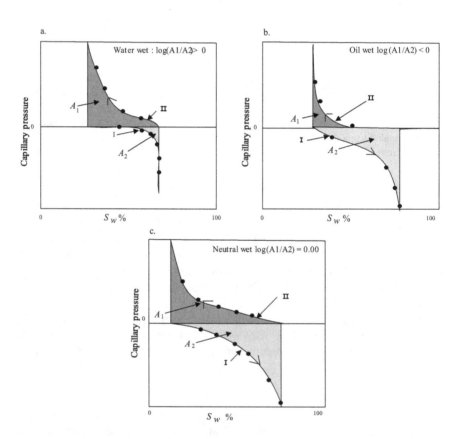

FIGURE 9.12 USBM examples for different wettability states. Curve I shows the oil injection and II the water injection (modified from Donaldson et al. (Donaldson et al. 1969)).

is much larger than the area under force imbibition capillary pressure curve (A_2). This is quite reasonable since in a water-wet sample, water tends to stick to the pore wall, hence the non-wetting phase needs high energy and work to displace the wetting phase. Conversely, in the situation of an oil wet sample, the area beneath the secondary drainage curve (A_1) is significantly smaller than the area beneath the force imbibition curve (A_2).

EXPERIMENTAL PROCEDURE

To determine wettability state of core plugs using the USBM method, the following procedure is suggested:

1. A core sample with a diameter of 1" or 1.5" and length < 5 cm should be selected, based on the dimensions of the core cups in the centrifuge.
2. Physical properties of the sample including porosity, permeability, and pore volume should be determined prior to the test.
3. Sample should be saturated with wetting phase, for instance, water.
4. Follow the test procedure for capillary pressure for drainage described in Section 10.9.3 to obtain primary drainage curve.
5. Once the sample reaches S_{wi}, place it into the Amott cell to conduct spontaneous imbibition.
6. Follow the test procedure for imbibition capillary pressure described in Section 10.9.3 to obtain force imbibition curve.
7. Once the sample reaches S_{or}, place it into the Amott cell and fill it with oil to conduct spontaneous drainage.
8. Follow the test procedure for drainage capillary pressure described in Section 10.5.3 to obtain forced secondary drainage curve.

EXERCISE 9.3

The following data are recorded (Table 9.9):

TABLE 9.9
Recorded Data

Pore volume of the core plug [cm³]	8.0
Initial water saturation	0.20
Spontaneous imbibition of water [cm³]	2.5
Forced displacement of water [cm³]	2.0
Spontaneous uptake of oil [cm³]	0.8
Forced displacement of oil [cm³]	3.7

Calculate Amott-index. Calculate all the end point saturations and outline all in a capillary pressure curve.

SOLUTION

$$S_{wi} = 0.20$$

$$S_{w2} = 0.2 + \frac{2.5}{8.0} = 0.51$$

$$S_{w3} = 0.51 + \frac{2.0}{8.0} = 0.76$$

$$S_{w4} = 0.76 - \frac{0.8}{8.0} = 0.66$$

$$S_{w5} = 0.66 - \frac{3.7}{8.0} = 0.20$$

According to Equation 9.5:

$$WI_{Amott} = \frac{2.5}{2.5+2} - \frac{0.8}{3.7+0.8} = 0.37$$

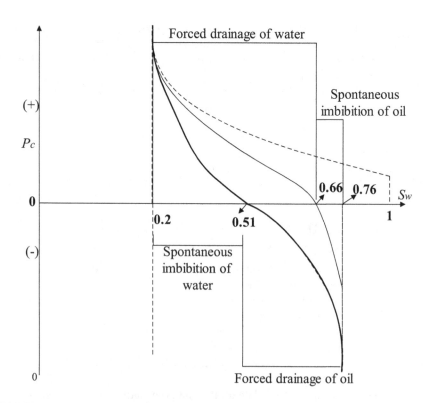

FIGURE 9.13 Capillary pressure versus water saturation.

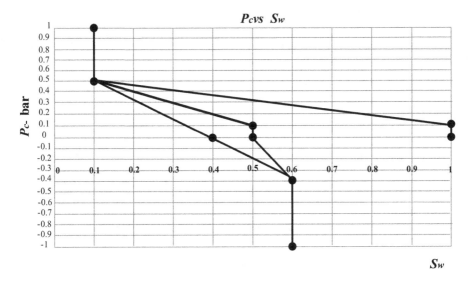

FIGURE 9.14   Capillary pressure versus saturation. Data for Exercise 9.4.

Most experimental evidence indicates that the irreducible saturation obtained by primary drainage is the same as that obtained by secondary drainage. This is a requirement for the analysis of the data. When the residual saturation is the same, the imbibition after secondary drainage will follow exactly the imbibition curve obtained after primary drainage. So, the USMB experiment can be restarted after the initial index determination (using primary imbibition and secondary drainage), for example, doing secondary imbibition and tertiary drainage resulting in the same index. This effect is called capillary pressure hysteresis and is a result of the different mechanisms governing filling/emptying of pores with a non-wetting or a wetting fluid, respectively.

EXERCISE 9.4

In Figure 9.14 a complete capillary loop is given for an oil/water/core sample system. Make a sketch of the following plot and indicate with arrows the direction of the saturation changes and give the curves appropriate names. Calculate the Amott and USBM indexes based on the values from the following capillary pressure curves.

SOLUTION

Here are all the names mentioned in Figure 9.11 are placed on this simlified capillary versus saturation loop.

$$WI_{Amott} = \frac{\Delta S_{ws}}{\Delta S_{wt}} - \frac{\Delta S_{os}}{\Delta S_{wt}} = \frac{0.3}{0.5} - \frac{0.1}{0.5} = 0.4 \Rightarrow \text{Slightly water wet}$$

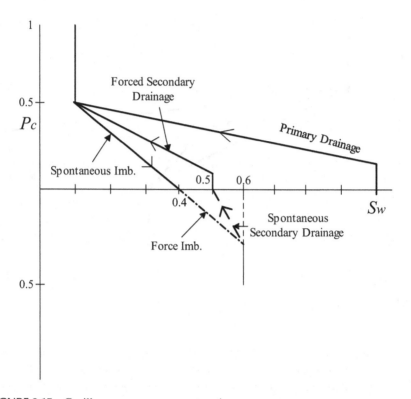

FIGURE 9.15   Capillary pressure versus saturation.

$$WI_{USBM} = log \frac{A_1}{A_2}$$

$$A_1 = 0.4 \cdot 0.1 + \frac{0.4 + 0.4}{2} = 0.12$$

$$A_2 = \frac{0.1 \cdot 0.4}{2} = 0.02$$

$$WI_{USBM} = log \frac{0.12}{0.02} = 0.78 \Rightarrow \text{Slightly water wet}$$

9.5   WETTABILITY RESTORATION

In Chapter 3 the core preservation and cleaning were discussed. Some definitions relating to the state of a core sample are used for SCAL (Section 1.5) in respect to the preservation, cleaning, and wettability.

Native state cores: The original wettability of the core sample has attempted to be maintained during coring operation by using fluids not supposed to influence wetting properties. The cutting of such cores also requires the use of proper drilling fluid.

Fresh cores: Those that have undergone normal exposure to drilling fluids but have been preserved upon recovery at the surface to prevent loss of fluid or exposure to oxidation. Native-state and fresh cores are not cleaned or dried prior to testing. The use of these cores will not require a wettability restoration.

Restored state cores: Cores to which a procedure is applied to restore the original wettability state of the core. This procedure should be carried out after cleaning and measuring permeability and porosity as part of the RCAL program. The restored state condition is often erroneously referred to as samples that have only been cleaned and saturated with any fluids but not aged at reservoir conditions before the experiment starts. Restoration of wettability is called aging, and it can be done in a static manner or dynamically. The principle is based on exposing the rock or individual minerals under reservoir conditions to oil, stimulating oil adsorption to the specified minerals. With the increased interest of changing wettability to more water wet conditions by injection of low-salinity water (Pinerez et al. 2020; Shariatpanahi et al. 2012), multiple studies have been performed to understand the principles of changing wettability from oil- to water wet. This additionally questions how to establish the original wettability in the first place. As mentioned in Section 9.1 in general, the wettability and wettability distribution of a reservoir rock system will depend on the following factors:

- Reservoir rock minerals.
- Pore geometry and roughness.
- Composition and saturation of crude oil and brine.
- Geological mechanisms (accumulation and migration).
- Physical conditions: pH, pressure, and temperature.
- Mechanisms occurring in the reservoir during production, for example, change in saturations, pressure, and composition.

The factors affecting wettability and wettability distribution are likely important to consider in the wettability restoration. Beside the oil composition and rock minerals with preferred adsorption for polar oil components, specifically clays, the brine concentration and composition, temperature, and pH are found to influence whether a rock wettability is changed to an oil wet state.

For the purpose of wettability restoration, parameters to consider are:

- Choice of sample.
  - Rock sample.
    - History of sample (drilling, transport, storage).
    - Shape (contact angle measurement or core experiments).
  - Minerals.
- Core cleaning.
  - Liquids used.
  - Cleaning method (hard versus soft).
- Core wettability restoration.

- Liquid samples (oil composition, brine composition).
- Temperature.
- Duration.
- Initial water saturation.

MILD SAMPLE CLEANING FOR WETTABILITY RESTORATION

The traditional cleaning, as described in Chapter 3, has the aim to remove all adsorbed oil components from the surface to create a perfect water-wet state. This is considered "hard cleaning", where the samples are washed very vigorously using toluene and methanol. There are indications that the methods do not always achieve a full water wet state; it is suspected circulation of oil soluble solvents through the sample can redistribute oil components and make the core oil wet. Alternatives have been considered by using mixtures such as heptane and kerosene, which do not desorb the very long-chain organic molecules adsorbed to the surface of the rock (Pinerez et al. 2020; Shariatpanahi et al. 2012). This is named "soft or mild cleaning", with the aim to preserve some of the sample's "acquired" wettability in the reservoir as a basis for the aging process.

AGING PROCEDURE

The technique of aging is under development. Traditionally aging is performed statically, where the more recent technique applied is based on aging the core under dynamic conditions (Fernø et al. 2010). Both techniques will be described later. Static aging has the advantage of performing the aging process on several cores in parallel, where dynamic aging is quicker, but in general more crude oil is needed for the aging process.

STATIC AGING

Hereby the core is statically exposed to oil under high temperatures, by saturating the core at S_{wi} and submerging the core plug in crude oil for a long duration.

PROCEDURE

1. Saturate the core with oil at initial water saturation (core flooding or centrifuge).
2. Submerge the core plug in a vessel containing crude oil.
3. Place the vessel with a small ventilation hole in the oven and increase the temperature.
4. Leave the vessel under high temperature (60–80°C advised) for several days or weeks.

   Regular resistivity measurements can be performed to monitor the wettability change (Section 6.5.6) and can be used to determine when the process of aging is finalized. With aging the resistivity will increase as the conductive water film along the grain surface is reduced.

5. Once the aging process is finished, turn off the oven and let the vessel cool down before the core plug is taken out for further use.

Note that this method is also easily suitable for other geometries e.g. plates to be used for contact angle measurements.

DYNAMIC AGING

During dynamic aging, the core is exposed to oil through a flooding process at a low flow rate under temperature. The difference with static aging is the continuous supply of new oil components enabling the aging to proceed quicker but more oil volume needed to complete the process.

PROCEDURE

1. Saturate the core with oil at initial water saturation (core flooding or centrifuge).
2. If not already placed in the core holder; mount the core in the core holder with the flow lines filled with oil, avoiding air in the inlet flow lines.
3. Set sleeve pressure. The use of sleeve pressure consequently results in an outer face that is not aged. Contact angle measurement on the core surface cannot be performed then.
4. Start the flow of oil (recommended 1.5 ml/hr)
5. In case needed, increase the back pressure to the preferred pressure.
6. For determination of the necessary duration, the change in pressure drop over the core can be monitored during the flood. Becoming more oil wet, the pressure should increase slightly due to increased flow resistance along the grain walls. In-situ resistivity measurement can be used also here to monitor the increase in resistivity.

9.6   WETTABILITY ALTERATION

With the increased understanding of the effect of wettability on flow properties and ongoing research on improved conditions for reservoir production or storage, manipulation of wettability conditions is a focus of study. For example, low salinity flooding is a largely studied topic, whereby the saline concentration as well as the composition of injection water is being manipulated to change the wettability from oil to water wet, for accelerated production (Bartels et al. 2019). Similarly, to change wetting conditions, for example in fractured oil wet carbonates, surfactants are being studied, which can absorb to the rock surface or interact with the absorbed oil components (AlZaabi et al. 2023) and change the wetting state of the rock. Also, for facilitation of wettability studies in the laboratory, surface coatings are applied to obtain reproducible wetting conditions, like silicon coatings (Vukovic et al. 2023).

It lies beyond the scope of this book to fully address all known details and we refer to literature for further information.

REFERENCES

AlZaabi, A., Arif, M., Ali, M., Adila, A., Abbas, Y., Kumar, R.S., Keshavarz, A., Iglauer, S. 2023. "Impact of carbonate mineral heterogeneity on wettability alteration potential of surfactants." *Fuel 342.*

Amott, E. 1959. "Observations relating to the wettability of porous rock." *Transactions of the AIME 216* 156–162.

Andrew, M., Bijeljic, B., Blunt, M.J. 2014. "Pore-scale contact angle measurements at reservoir conditions using X-ray microtomography." *Advances in Water Resources 68* 24–31.

Bartels, W.-B., Mahani, H., Berg, S., Hassanizadeh, S.M. 2019. "Literature review of low salinity waterflooding from a length and time scale." *Fuel 236* 338–353.

Boneau, D.F. and Clampitt, R.L. 1977. "A surfactant system for the oil wet sandstone of the North Burbank Unit." *Journal of Petroleum Technology 29* 501–506.

Cuiec, L. 1995. "Wettability laboratory evaluation under reservoir conditions: A new apparatus." *Society of Core Analysts 9529.*

Deng, Y., Xu, L., Lu, H., Wang, H., Shi, Y. 2018. "Direct measurement of the contact angle of water droplet on quartz in a reservoir rock with atomic force microscopy." *Chemical Engineering Science 177.*

Donaldson, E.C., Thomas, R.D., Lorenz, P.B. 1969. "Wettability determination and its effect on recovery efficiency." *SPEJ 9* 13–20.

Fernø, M.A., Torsvik, M., Haugland, S., Graue, A. 2010. "Dynamic laboratory wettability alteration." *Energy and Fuels 24* 3950–3958.

Khanamiri, H.H., Berg, C.F., Slotte, P.A., Schlüter, S., Torsæter, O. 2018. "Description of free energy for immiscible two-fluid flow in porous media by integral geometry and thermodynamics." *Water Resources Research 54* (11) 9045–9059.

Morrow, N.R. 1990. "Wettability and its effect on oil recovery." *Journal of Petroleum Technology* 1476–1484.

Pinerez, I., Puntervold T., Strand S., Hopkins P., Aslanidis P., Yang H.S., Sundby Kinn M. 2020. "Core wettability reproduction: A new solvent cleaning and core restoration strategy for chalk cores." *Journal of Petroleum Science and Engineering 195.*

Shariatpanahi, S.F., Strand, S., Austad, T., Aksulu, H. 2012. "Wettability restoration of limestone cores using core material from the aqueous zone." *Petroleum Science and Technology 30* (11) 1082–1090.

Tiab, D., Donaldson, E.C. 1996. *Petrophysics: Theory and Practice of Measuring Reservoir Rock and Fluid Transport Properties.* Houston: Gulf Publishing Company.

Vukovic, T., Røstad, J., Farooq, U., Torsæter, O., van der Net, A. 2023. "Systematic study of wettability alteration of glass surfaces by dichlorooctamethyltetrasiloxane silanization—a guide for contact angle modification." *ACS Omega 8* (40) 36662–36676. http://doi.org/10.1021/acsomega.3c02448.

10 Capillary Pressure

10.1 INTRODUCTION

Capillary pressure is responsible for holding the fluid inside a porous medium. The phenomenon is like rise of a liquid in a capillary tube or a cloth (see Figure 10.1), which is due to the surface tension between the liquid and solid surface and contact angle. In our daily life, an oil droplet will be at the top of a beaker or bowl containing water showing the effect of gravity (density difference between oil and water). However, in the same situation when the beaker or bowl is filled with a porous medium, the upward movement of the oil droplet will be blocked while traveling from the bottom to the top of the surface, which shows existence of capillary pressure.

Chapter 8 addressed the interfacial tension that exists *between two fluids* that form a curved interface, which can be described by the Young–Laplace equation dependent on the pressure difference over the interface. In Chapter 9 the wettability was described, defining the surface affinity of the surface, where the contact angle is dependent on the force balance *between the two fluids and the rock*. A method was presented there where wettability of porous media was measured using spontaneous imbibition driven by capillary pressure minimalization. This chapter focuses on capillary forces and pressure occurring in porous media when

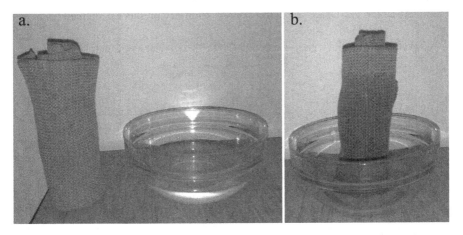

FIGURE 10.1 Spontaneous uptake of liquid in a rolled-up cloth caused by capillary forces. The dry cloth in a. shows in b., when placed in fluid, a fluid rise much higher than the fluid level.

DOI: 10.1201/9781003382584-10

multiple phases are present, combining the influence of wettability and interfacial tension.

The concept of capillary pressure has many applications in fluid flow in porous media and fluid storage in subsurface reservoirs. For example, it determines the initial fluid distribution in the reservoir, and depending on the dominating forces in the field during flooding, it can affect the flooding pattern preferential flow in fractures in carbonates and recovery factors. It can affect the storage capacity when small pores cannot be invaded. The opposite is preferred as a sealing property of cap rocks, which is largely dependent on the entry capillary pressure. The larger it is, the less likely it is for hydrogen or CO_2 to penetrate and leak off.

10.2   DEFINITION

When two immiscible fluids are in contact in the porous structure, a discontinuity in pressure exists across the interface separating them. Each fluid phase, either wetting or non-wetting, has a different pressure than the other. The difference in pressure between these two phases is called capillary pressure, which is pressure in the non-wetting phase minus the pressure in the wetting phase:

Capillary pressure = pressure in the non-wetting phase – pressure in the wetting phase.

or

$$P_c = P_{non-wetting} - P_{wetting} \tag{10.1}$$

Capillary pressure can be defined depending on the phases present in the reservoir. For an oil-water, gas-water, or gas-oil system, capillary pressure is defined as:

- Water-oil capillary pressure denoted by P_{cwo} (in water-wet media):

$$P_{cwo} = P_o - P_w \tag{10.2}$$

- Water-oil capillary pressure denoted by P_{cwo} (in oil wet media):

$$P_{cwo} = P_w - P_o \tag{10.3}$$

- Gas-water capillary pressure denoted by P_{cwg}:

$$P_{cwo} = P_w - P_o \tag{10.4}$$

- Gas-oil capillary pressure denoted by P_{cgo}:

$$P_{cgo} = P_g - P_o \tag{10.5}$$

The capillary pressure can also be defined as result of the curvature of fluid interfaces in a capillary, according to the well-known Laplace equation discussed in Chapter 8:

$$P_c = \sigma \left(\frac{1}{r_1} + \frac{1}{r_2} \right) \cos \theta \tag{10.6}$$

10.3 CAPILLARY RISE IN A TUBE

Figure 10.2 shows the elevation of water in a single capillary tube with a narrow internal diameter that is placed in a large open beaker filled with water.

The water rises in capillary tube due to the attractive forces existing between the tube and the water. The attractive forces, known as adhesion tension, draw the liquid up the tube's wall. The height liquid rises up in the capillary tube depends on the balance between the force acting to pull the liquid upward and the weight of the column of the liquid acting downwards. These opposing forces can be expressed as:

$$\text{Force up} = A_T \cdot 2\pi r \tag{10.7}$$

$$\text{Force down} = \pi r^2 \Delta \rho g h \tag{10.8}$$

Where A_T is the adhesion tension, r the radius of the capillary tube, h the height of the capillary rise, $\Delta \rho$ the density difference between the two phase (in the case of gas and liquid this can be approached as just the density of the liquid) and g the gravity force.

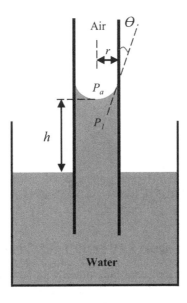

FIGURE 10.2   Liquid rise in a single capillary tube.

By equating the previous equations:

$$A_T \cdot 2\pi r = \pi r^2 \Delta \rho g h \qquad (10.9)$$

or

$$h = \frac{2\pi r A_T}{\pi r^2 \Delta \rho g} = \frac{2 A_T}{r \Delta \rho g} \qquad (10.10)$$

by replacing adhesion tension with $\sigma \cos\theta$:

$$h = \frac{2\sigma \cos\theta}{r \Delta \rho g} \qquad (10.11)$$

Note that the height of the fluid is inversely proportional dependent on the density difference and pore radius and linearly changing with interfacial tension. So, in case of two liquids, the density difference reduces compared to a gas and liquid and the fluid level height will be lower, where a higher interfacial tension will give a larger fluid rise in the tube.

On the other hand, the pressure at the top of the liquid column is equal to the pressure at the bottom minus the pressure due to the head of liquid h:

$$P_1 = P_a - \rho g h \qquad (10.12)$$

or

$$P_a - P_1 = \rho g h = P_c \qquad (10.13)$$

Combination of equations 10.11 and 10.13 leads to the capillary pressure equation in terms of surface forces, wettability, and capillary size:

$$P_c = \frac{2\sigma \cos\theta}{r} \qquad (10.14)$$

Where the P_c is the capillary pressure, σ the surface tension, θ the contact angle, and r the capillary radius. Thich concept is applied in the capillary rise method, to determine the interfacial tension, with knowing the contact angle and radius or having a reference fluid.

10.4 CONCEPT OF CAPILLARY FORCES IN POROUS MEDIA

Figure 10.3 shows a series of capillary tubes with varying diameter immersed in water. The figure clearly shows different air/water interface levels, induced due to the size of each capillary tube. The capillary model shown in the figure is

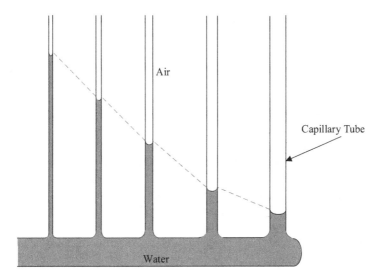

FIGURE 10.3   Liquid rise in capillary tubes.

somewhat analogous to the underground porous media when reservoir rock consists of different pore sizes, filled with water and oil or gas. In a low permeability reservoir, the pore throats are small, resulting in a network of very fine capillaries and hence high capillary pressure. On the contrary, in a high permeability reservoir the pore throats are larger, resulting in a network of larger capillaries and hence lower capillary pressure.

Similarly in porous media, in smaller pores the capillary forces act stronger on the wetting fluid to wet the rock surface, counteracting the gravitational forces. This will affect the saturation profile at the interface between the two immiscible phases, creating a transition zone, further discussed in Section 10.8.

10.5   CAPILLARY PRESSURE CURVES OF DRAINAGE AND IMBIBITION

Drainage is a fluid flow process in which the saturation of the non-wetting phase increases while saturation of the wetting phase decreases. Some examples of drainage processes are:

- Hydrocarbon movement millions of years ago to fill the pore space and displace the original water in water wet reservoir.
- Waterflooding in an oil wet reservoir.
- Gas injection.
  - CO_2 injection in aquifer (immiscible).
  - Hydrogen injection for storage.
- Gas cap expansion as reservoir pressure decreases.

Imbibition is the opposite fluid flow process of drainage, where the saturation of the wetting phase increases while saturation of the non-wetting phase decreases. Some examples of imbibition processes are:

- Waterflooding an oil reservoir in a water wet rock.
- Hydrogen recovery from storage reservoir.
- Accumulation of oil in an oil wet reservoir.

Thus, as the invading fluid needs to have a higher pressure, based on Equations 10.2–10.5 the capillary pressure may have either positive or negative values.

Considering a reservoir for storage of natural gas or oil, initially the porous medium is completely saturated with water and considered water wet. Due to migration driven by gravity force, the immiscible lighter phase would like to be on top – in general the non-aqueous phase. As the system is water wet, if the water is to be forced out, a pressure equal to the capillary pressure of the pores must be applied to start to expel the water.

In other words, a capillary pressure greater than zero is required for the non-wetting phase to invade into the rock. The minimum pressure is so called "entry pressure" or "displacement pressure". As the capillary pressure is inversely proportional to the radius (see Equation 10.14) the entry pressure is related to the largest pore diameter as:

$$P_e = \frac{2\sigma cos\theta}{r_{max}}$$ (10.15)

Depending on the capillary pressure up till Pc^*, the pores with a radius larger than r^* can be drained by the non-wetting phase, as all the larger radii can be accessed as well, as they will need a lower capillary pressure to enter.

$$P_c^* = \frac{2\sigma cos\theta}{r^*}$$ (10.16)

For imbibition this will be the opposite; smaller pores will be invaded first as capillary pressure will reduce, which will occur first in the smallest pores with the highest capillary forces. It can be imagined as a tense stretched elastic film, which will release tension first where the tension is highest, which is in the smallest pores. Therefore, the drainage process begins only when an entry pressure greater than zero is applied, which is related to the largest pores, while the imbibition process occurs spontaneously and does not require an entry pressure.

Reservoir rocks have different pore radii, and a particular capillary pressure will be associated with a specific set of pores having the same pore radius. Lower capillary pressure will displace fluid in the bigger pores while higher capillary pressure will displace fluid in the even smaller pores. The amount of fluid displaced can be correlated to the core phase saturations. The graphical relationship that relates the apparent capillary pressure to the average water saturation in a given pore system is called a capillary pressure curve.

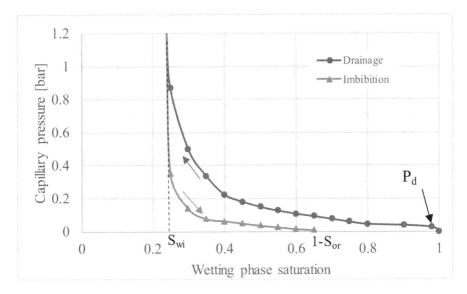

FIGURE 10.4 Typical shape of capillary pressure curve.

Figure 10.4 shows typical shapes of capillary pressure curves including both drainage and imbibition processes in an oil/water system where water is the wetting phase. The relationship between capillary pressure and saturation is not unique but depends on the saturation history of the system. Typically, capillary pressure is plotted on a linear X-Y axis. In the laboratory, normally capillary pressure is measured as a function of the wetting phase saturation.

All laboratory experiments are designed to mimic or simulate the saturation history of the reservoir. In other words, the drainage capillary process, which is displacing the wetting phase with the non-wetting phase, primarily simulates the process that occurred in the reservoir at a geological time millions of years ago. This process may continue to displace the wetting phase with the non-wetting phase until reaching irreducible water saturation (S_{wi}). The core sample then mostly contains a non-wetting phase in the large pores and a wetting phase in the small pores and as coating on the pore walls.

The imbibition process may then start by spontaneous imbibition of water into the porous medium, which is like the process in which oil is displaced by water in a water-wet rock. The process is also like what happens for a single core sample placed in an Amott cell to measure spontaneous imbibition of water into the rock sample. This process continues until reaching a certain saturation at zero capillary pressure. The imbibition process may continue by forcing the wetting phase to displace the non-wetting phase until reaching residual oil saturation (S_{or}). This is referred to as forced or negative imbibition. The core sample may then experience spontaneous drainage and secondary drainage through displacing the wetting phase with the non-wetting phase either by spontaneous or forced drainage, which are called spontaneous drainage and secondary drainage.

The most essential elements of the capillary pressure curve are:

- **Irreducible water saturation** S_{wi}. The volume of the wetting phase retained at the highest capillary pressure during the drainage process, where the wetting phase saturation is independent of further increases in the externally measured capillary pressure.
- **Residual oil saturation** S_{or}: The volume of the non-wetting phase that remains entrapped during the imbibition process when externally measured capillary pressure is decreased from a high positive value to a large negative value.
- P_d or P_{th}: Displacement pressure or entry pressure, which is the minimum required pressure to enter the non-wetting phase into the pores.
- **Primary drainage curve**: The capillary pressure–saturation relationship characteristic of the displacement of the wetting phase from 100% saturation to the irreducible saturation.
- **Imbibition curve**: Contains spontaneous imbibition, which is characteristic of the displacement of the non-wetting phase by the wetting phase from the irreducible wetting saturation to the residual non-wetting saturation.

EXERCISE 10.1

Calculate the entry pressure for CO_2 and formation water for the following geometries:

1. Cylindrical pore with diameter: $d = 35 \ \mu m$
2. Linear fracture, with fracture opening: $d = 15 \ \mu m$
   The interfacial tension is: 34.2 mN/m.

In which geometry will CO_2 first penetrate?

SOLUTION

To calculate the entry pressure, we use the Equation 10.9:

Cylindrical pore. The CO_2/water meniscus is a hemisphere:

$$R_1 = R_2 = \frac{35 \cdot 10^{-6}}{2} = 17.5 \cdot 10^{-6} \, m$$

$$p_e = 34.2 \cdot 10^{-3} \, \frac{N}{m} \cdot \frac{2}{17.5 \cdot 10^{-6} \, m} = \underline{3,909 \, Pa}$$

Linear crack. The CO_2/water meniscus is a half-cylinder:

$$R_1 = 7.5 \cdot 10^{-6} \, m \quad R_2 = \infty$$

$$p_e = 34.2 \cdot 10^{-3} \, \frac{N}{m} \cdot \left(\frac{1}{7.5 \cdot 10^{-6} \, m} + \frac{1}{\infty} \right) = \underline{4,560 \, Pa}$$

CO_2 will most easily penetrate the cylindrical pore.

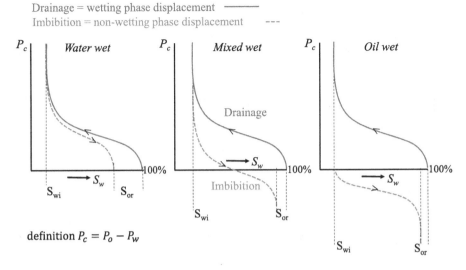

FIGURE 10.5 Effect of wettability on capillary pressure curves: for a water wet, mixed wet, and oil wet system. The level of Pc changes and the end point, S_{or}, as well.

10.6 EFFECT OF WETTABILITY ON CAPILLARY PRESSURE

With a different wettability, the definition of the wetting and non-wetting phases changes – and so the capillary pressure calculation – as Equations 10.2 and 10.3 show. The aging process in the field or in the lab after primary drainage can result in wettability change making the imbibition, secondary drainage, and all cycles afterwards. To compare capillary pressure curves depending on wettability, the capillary pressure definition is fixed as $P_c = P_o - P_w$, independent of the wetting state. Figure 10.5 clearly shows the effect of wettability on capillary pressure curves for primary imbibition and secondary drainage. This illustrates that the Amott curves (Amott 1959) represent the capillary pressure curves; see Chapter 9. The spontaneous invasion of fluids is driven to reduce the capillary pressure to zero; this happens as mentioned when introducing the wetting phase, where no energy is needed. This, however, only displaces the non-wetting fluid to a capillary pressure of zero. In the case of mixed/neutral wet, where pores are particularly oil or water wet, only pores filled with non-wetting fluid are invaded by the wetting fluid. The corresponding equilibrium saturation is determined by the volumes of the wetting and non-wetting pores and can be derived from the capillary pressure curve, at $P_c = 0$.

10.7 EFFECT OF PORE SIZE DISTRIBUTION ON CAPILLARY PRESSURE

In case the porous media would only consist of one pore size radius, there would be one capillary entry pressure at which all pores would be invaded and saturation would drop at one Pc from 1 to 0. The capillary pressure curve will then look

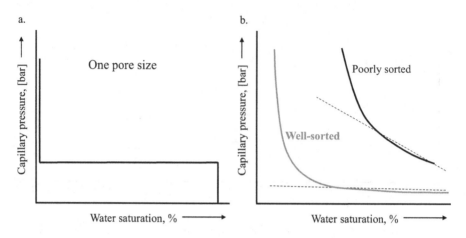

FIGURE 10.6 a. Theoretical capillary pressure curve from a porous media with one pore size radius, compared to a porous media well- and poorly-sorted in b.

like Figure 10.6a. From Equation 10.14 it is known that the Pc scales with $1 / r$. When the pore radius distribution is wider, generally occuring when the grains are poorly sorted, then the water saturation change happens over a larger range of entry capillary pressures of the different pores. No clear plateau can be recognized. Translating this to the distribution of pore sizes, the pore size distribution is more homogenous; if the grains are more homogeneous and well sorted. The capillary pressure curve shows then a clear plateau. If the pore size is more heterogeneous, there are more entry pressures and no plateau or significant jump in water saturation at a small range of capillary pressures is visible. This is illustrated in Figure 10.6b, showing an example of a well- and poorly sorted rock. Similarly, a low permeability (refer to Kozeny–Carman theory in Chapter 7) will give higher capillary pressures to invade pores to reach similar water saturations. In the low permeable rock more not-so-invadable small pore sizes are present, such that the initial water saturation S_{wi} is higher than for high permeable rock, as can be seen in Figure 10.7.

So, based on the comparison of capillary pressure curves, indications on permeability and grain sorting can be derived. This needs verification with the geological information from the core and thin section analysis.

Mercury injection is the best method to determine the pore size distribution in a rock sample; see further Section 10.9.3.

10.8 EFFECT OF CAPILLARY PRESSURE ON INITIAL FLUID DISTRIBUTION

As described earlier, natural reservoir rock is filled by the phases migrating upwards from the source rock. In the oil or gas migration process in the reservoir, where gravitation forces are driving the flow, a drainage process takes place, with the oil or gas invading the water wet reservoir. This process stops when the migration ceases.

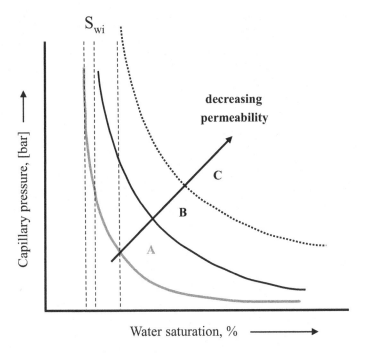

FIGURE 10.7   The effect of permeability on the capillary pressure curve.

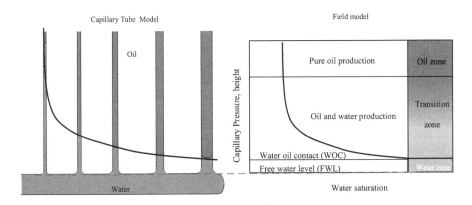

FIGURE 10.8   Fluid distribution based on the capillary pressure curve.

The capillary pressure will determine the static fluid distribution in the reservoir (Figure 10.8).

Capillary pressure is zero at 100% water saturation. The height of zero capillary pressure is defined as the free water level. The highest level with 0% non-aqueous phase (shallowed depth) is based on the entry pressure and not necessarily the point of zero capillary pressure. The latter is called the contact, for example, oil-water contact, oil-gas contact, or gas-water contact. Equation 10.11 represents the height above the free water level. If the free water level is known as well as the entry pressure of the rock—or the largest pore size r_{max}—the height of the contact with water can be determined by:

$$\text{Depth phase contact (OWC)} = \text{free water level height} - \frac{2\sigma\cos\theta}{r_{max}\left(\rho_w - \rho_o\right)g} \qquad (10.17)$$

Where FWL = 0 m and the depth axis is positive upwards.

Consider the hydrostatic pressure of a liquid x with density r dependent on the elevation z. The pressure gradient along the column can be described as follows:

$$\frac{dP_x}{dz} = \rho_x g \qquad (10.18)$$

See Figure 10.9 for an example with water and gas. The gradient for water is much larger than for gas. By multiple pressure measurements in the well (black dots) the phase pressure gradients can be determined and by extrapolation the free water level can be determined, where both pressures are equal and $Pc = 0$.

Using the gradients for the phases present in the reservoir, e.g., for an oil-water system, the capillary pressure, if fluid columns are continuous in the reservoir, becomes:

$$\frac{dP_c}{dz} = \Delta\rho g \qquad (10.19)$$

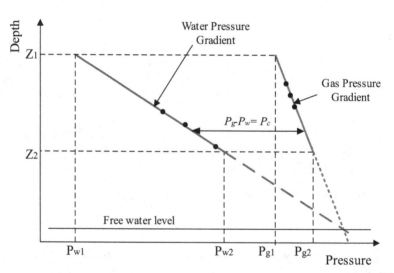

FIGURE 10.9 Pressure gradients for water and gas and the corresponding capillary pressure gradient in this field.

The imposed capillary pressure gradient due to the hydrostatic column determines which pore radii can be invaded by the non-wetting phase, depending on the elevation z. The volume of the invaded pore radii can be linked to the saturation of the porous medium.

The capillary pressure can be related to the height z, such that the elevation above the free water contact can be related to the reservoir's saturation. This is the so-called *saturation-height function*, which can be created describing the transition zone based on the capillary pressure curve of the reservoir. Saturation distribution in the transition zone in various reservoirs is discussed in reservoir engineering textbooks (Amyx 1960; Dake 1978; Zolotukhin and Ursin 2000).

$$z\left(S_w\right) = \frac{P_c\left(S_w\right)}{\Delta \rho g} \tag{10.20}$$

So, from the free water level upwards, the induced capillary pressure increases, resulting in a reduction of the wetting phase saturation and increase of non-wetting saturation, till the irreducible saturation has been reached.

EXERCISE 10.2

1. Fluid pressure in a reservoir is measured as a function of depth as given in the following table:

TABLE 10.1

Fluid Pressure as a Function of Depth

Depth (m)	Pressure (10^6 Pa)
2,475	20.07
2,500	20.13
2,525	20.18
2,550	20.24
2,575	20.40
2,600	20.67
2,625	20.95
2,650	21.20

   a. Calculate the density of the fluids in the reservoir and specify the fluid types. Determine the depth of the fluid contact.
   b. Define capillary pressure and calculate capillary pressure at depth of 2,550 m.

2. An oil reservoir has an area of coarse-grained sandstone (zone 1) and an area of fine-grained sandstone (zone 2). Due to capillary conditions each reservoir has a zone over the water-oil contact (free water level) where water saturation is approximated 100% and only water is produced. In

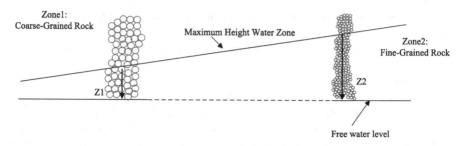

FIGURE 10.10   Exercise 10.2 oil reservoir cross section.

Figure 10.10 this zone is limited by the water-oil contact and an inclined surface.

a. Calculate the height difference $\Delta Z = Z_2 - Z_1$. Which assumption must be made for the water zone level between zones 1 and 2 to be as indicated in Figure 10.10?

b. Assume that the sandstone in zones 1 and 2 is completely water wet with cylindrical pores. Calculate the radius of the largest pores in zones 1 and 2.

Data:
Capillary entry pressure in zone 1: $P_{e1} = 6 \cdot 10^3 Pa$
Capillary entry pressure in zone 2: $P_{e2} = 60 \cdot 10^3 Pa$
Water density: $\rho_w = 1040\,kg/m^3$
Oil density: $\rho_o = 700\,kg/m^3$
Water-oil interfacial tension: $\sigma = 30 \cdot 10^{-3}\,N/m$

SOLUTION

1. a. Upper fluid density:

$$\rho = \frac{\Delta P}{g\Delta Z} = \frac{(20.24 - 20.07)10^6}{9.81 \cdot 75} = 231\ kg/m^3$$

This low density indicates that the fluid is gas.
Water density:

$$\rho_w = \frac{\Delta P}{g\Delta Z} = \frac{(21.2 - 20.4)10^6}{9.81 \cdot 75} = 1087\ kg/m^3$$

According to the plot, Figure 10.11, the gas-water contact is read at 2,565 m.

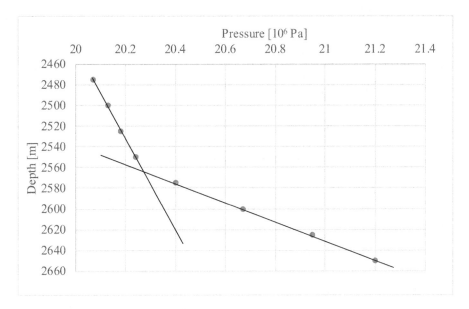

FIGURE 10.11   Pressure versus depth.

b. $P_c = P_g - P_w = 20.24 \cdot 10^6 - (20.24 \cdot 10^6 \, 0.27 \cdot 10^6) = 0.11 \cdot 10^6$ Pa

3.   a. $P_{e1} = (\rho_w - \rho_o) g z_1$

$$P_{e2} = (\rho_w - \rho_o) g z_2$$

$$z_2 - z_1 = \frac{P_{e2} - P_{e1}}{(\rho_w - \rho_o) g} = \frac{(60 - 6) \cdot 10^3}{(1040 - 700) \cdot 9.81} = 16 \text{ m}$$

Assumption: the entry pressure, pe, increases linearly from zone 1 to 2. That is, uniform change of the rock type from coarse to fine grain.

b. $P_{e1} = \dfrac{2\sigma \cos\theta}{r_1} \Rightarrow r_1 = \dfrac{2\sigma \cos\theta}{P_{e1}}$

$$r_1 = \frac{2.3 \cdot 10^{-3}}{6 \cdot 10^3} = 10 \cdot 10^{-6} \text{ m} = 10 \text{ μm}$$

$$r_2 = \frac{2.3 \cdot 10^{-3}}{60 \cdot 10^3} = 1.0 \cdot 10^{-6} \text{ m} = 1.0 \text{ μm}$$

EXERCISE 10.3

A well has free water level at 2,093.1 m and gas-oil contact (when capillary pressure between oil and gas is zero) at 2,079 m. Create a plot showing the pressure in the gas phase, oil phase, and water phase with respect to depth. (P_g, P_o, P_w, vs. D).
The following data is provided:

- Water is the wetting phase.
- The pressure in the water phase at 2,088 m is $21.38 \cdot 10^6$ Pa
- Fluid densities are assumed constant in the reservoir:

$\rho_w = 1{,}041$ kg/m³
$\rho_o = 657$ kg/m³
$\rho_g = 197$ kg/m³

Determine the capillary pressure between gas and water at a depth of 2,073.6 m.

SOLUTION

Pressure difference across the interface between two immiscible fluids:

$$P_c = P_{nw} - P_w$$

Where P_{nw} is the non-wetting phase pressure and P_w is the wetting phase pressure.

$D_{o/w} = 2093.1$ m   Free water level
$D_{g/o} = 2079.0$ m   Gas-oil contact
$P_{w,\,2088} = 21.38 \cdot 10^6$ Pa

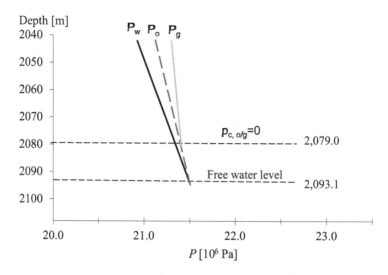

FIGURE 10.12   Depth versus pressure, Exercise 10.3.

Pressure calculations:

$$P_{w,2093.1} = \left(21.38 \cdot 10^6 + 1041 \cdot 9.81(2093.1 - 2088)\right) Pa = 21.43 \cdot 10^6 \, Pa$$

$$P_{o,2079.0} = \left(21.43 \cdot 10^6 + 657 \cdot 9.81(2079.0 - 2093.1)\right) Pa = 21.34 \cdot 10^6 \, Pa$$

$$P_{g,2073.6} = \left(21.34 \cdot 10^6 + 197 \cdot 9.81(2073.6 - 2079.0)\right) Pa = 21.33 \cdot 10^6 \, Pa$$

$$P_{c,2073.6\,gw} = P_{g,2073.6} - P_{w,2073.6}$$

$$= 21.33 \cdot 10^6 \, Pa - \left[21.38 \cdot 10^6 + 1041 \cdot 9.81(2073.6 - 2088)\right] Pa = \underline{9.8 \cdot 10^4 \, Pa}$$

EXERCISE 10.4

The following oil-water capillary pressure data for two core samples are measured in the laboratory. Data is representative of the reservoir conditions.

A reservoir consists of layers of alternatingly high and low permeability. Free water level, the permeabilities, and the thicknesses of the layers are given in Figure 10.13. The density of the water and the oil is 1,010 kg/m³ and 810 kg/m³ respectively and g is 9.81 m/s². Calculate capillary pressure and saturation in the middle of each rock layer.

TABLE 10.2
Measured Oil-Water Capillary Pressure Data
for Two Core Samples

Sample No.1 (high permeability) $K_1 = 1 \, \mu m^2$		Sample No.2 (low permeability) $K_2 = 0.1 \, \mu m^2$	
P_c (kPa)	S_w	P_c (kPa)	S_w
3.4	1	10.2	1
3.9	0.90	12.2	0.90
5.1	0.80	13.6	0.60
6.1	0.40	14.7	0.32
7.5	0.20	15.3	0.28
10.2	0.13	18.7	0.20
13.6	0.12	23.8	0.18
17	0.12	34.0	0.18

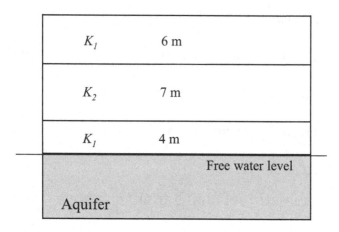

FIGURE 10.13   Schematics of the layered reservoir.

SOLUTION

$$P_c = \Delta \rho g h$$

In the middle of the first layer: $h = 2$ m $\Rightarrow$ $P_c = 200 \cdot 9.81 \cdot 2 = 3,924\,Pa$
From P_c measurement table (K_1): $S_w = 0.9$
In the middle of the second layer: $h = 7.5$m $\Rightarrow$ $P_c = 200 \cdot 9.81 \cdot 7.5 = 14,715\,Pa$
From P_c measurement table (K_2): $S_w = 0.32$
In the middle of the third layer: $h = 14$ m $\Rightarrow$ $P_c = 200 \cdot 9.81 \cdot 14 = 27,468\,Pa$
From P_c measurement table (K_1): $S_w = 0.18$

The relation between fluid saturation and capillary pressure in the reservoir is a function of pore sizes, wettability, interfacial tension, and fluid saturation history (drainage and imbibition). Based on laboratory measurements of capillary pressure, it is possible to convert those into reservoir capillary pressure. From these values fluid saturations in the reservoir can be evaluated. This will be further addressed in Section 10.10.

10.9   EXPERIMENTAL METHODS

Laboratory capillary pressure measurements are limited to three different methods: porous plate or restored state method, centrifuge method, and the mercury injection method. These methods are discussed in the following.

10.9.1   Porous Plate Method

The porous plate method is one of the most accurate and simplest methods that can be used for measurement of both drainage and imbibition process using different pairs of fluids. The method is called semi-permeable membrane and restored state in

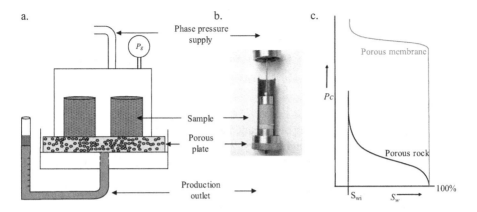

FIGURE 10.14 a. Schematics of the porous plate equipment for multiple core samples. b. A porous plate set-up in a core holder to allow confinement (courtesy of Stratum). c. The porous plate needs a higher entry pressure for the injection phase than porous rock.

older books. The method is slow and one full curve may take up to 40 days or more to obtain depending on how permeable the sample is. However, equipment needed for this method is simple and inexpensive and the work needed is limited to some volume reading during the process. Several samples may be placed in one chamber. Then the samples must be removed to weigh them separately between each pressure increase. Preferably, one sample should be run in an assembly of one-sample cells. Then it is not necessary to decrease pressure between each reading. The schematic of the experimental setup is shown in Figure 10.14.

It simply consists of source (either gas or liquid), connection lines, sample chamber, contact material, porous plate, and beret for production measurement. The main component of the apparatus is the porous plate ceramic, which limits the applied pressure in the test to its entry pressure and imposes constraints on the maximum applicable pressure. In fact, the porous plate must have a capillary entry pressure much higher than the core sample, which is the maximum applicable pressure and should not be reached during experiment (see Figure 10.14). The porous plate is strongly water wet and has smaller pore throats than most of the core samples. Once it is saturated with brine (wetting phase), air or oil (no-wetting phase) cannot enter the brine-saturated pores in the plate until the entry pressure is reached. However, brine as a wetting phase can pass through the plate, to the atmosphere. The functionality of the porous plate is comparable to the preferred capillary pressure curve of the cap rock, with a high entry pressure such that the cap rock is sealed for the non-wetting phase, which is to remain captured in the reservoir.

The porous plate method is regarded as the standard method against which all other methods are compared. Routinely only the drainage curve is measured, but with appropriate modifications the imbibition curve may be determined in the same manner. The weakness, as with all the other methods, is the transformation of data to reservoir conditions, which is discussed in Section 10.10.

PROCEDURE FOR POROUS PLATE

The test procedure given here is specialized for air-water system in a water-wet sample, starting with a 100% water-saturated sample, and knowledge of the dry and wet weight needs to be available.

PROCEDURE

1. Measure the bulk volume of a cleaned and dried core sample by measuring length and diameter. Weigh the dry core, W_{dry}.
2. Saturate the core sample and porous plate with brine/water with a known density of ρ_w. Weigh the saturated core, W_{sat}. Calculate pore volume of the sample.
3. Place saturated sample on porous plate. To ensure full capillary contact between the porous plate and core sample, a tissue paper wet by water is needed to place between the sample and the porous plate.
4. Start to apply pressure as low as possible. The applied pressure should be at a minimum to be able to obtain the entry pressure.
5. Start to monitor production as a function of time from a graduated pipette or alternatively by removing the plug and weighing in each step. The criterion for starting the next step is reaching equilibrium in which no production occurred (Figure 10.15). The system should undergo the next pressure after reaching the final equilibrium production in each step.
6. Production might be slow and take several hours or days to reach the plateau. Typically, 8 to 10 points are needed to reach the full capillary pressure curve and about 1–3 days' time for each pressure point (depending on sample permeability).
7. The maximum pressure applied on the core sample should not be higher than "porous plate entry pressure", which normally can be found from data provided by the manufacture.
8. At the final stage, while the sample is in S_{wi} condition, remove it from the apparatus and weigh.

DATA INTERPRETATION, QUALITY CHECK, AND SOURCES OF ERRORS

Since water is the wetting phase in both the porous plate and core sample, the capillary pressure at each step is equal to applied pressure. At each applied pressure, the desaturation process continues until no more water production is observed or weight loss is constant. The quality check of the experiment can be performed in two ways:

1. At each step or after the last step, by comparing the produced amount of water (V_1), with volume calculated by weight difference divided by the density (V_2).
2. Determine the amount of water (S_{wi}) by using the Dean-Stark extraction method and compare it with V_1 and V_2. The final S_{wi} (V_3) from Dean-Stark should match the final water production recorded from graduated burette (V_1) and volume calculated from the weight difference (V_2).

Loss of capillary contact between core sample and porous plate can be a source of error.

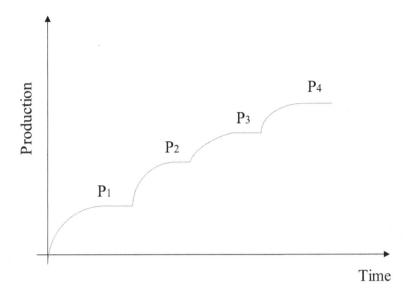

FIGURE 10.15 Production versus time dependent on the pressure in a porous plate experiment.

TABLE 10.3
Capillary Pressure Measurement Using Porous Plate Method Calculation Table

No. $_{(i)}$	$P_{c(i)}$, H_{water} (cm)	$W_{wet(i)}$ (g)	$W_{water(i)}$ (g)	$S_{w(i)}$ (fraction)	$r_{(i)}$ (µm)	$\Delta W_{(i)} / W_{water}$
0						
1						
...						
10						

10.9.2   EXPERIMENT 10.1: POROUS PLATE

Objective: Determination of the capillary pressure curve for a core sample using model oil and brine, at room temperature.

Follow the procedure as presented previously.

Procedure for calculations and reporting:

1. Calculate and fill in the data form in Table 10.3.
2. Plot capillary pressure curve ($S_w - P_c$) and pore size histogram (distribution).

Data form:

Core No. D:____cm, L:____cm, W_{dry}:____ g, W_{sat}: ____ g, ϕ =____

Where:

$P_{c(i)}$ = capillary pressure of ith measurement, cm of water, reading of U-manometer

W_{wet} = core weight of ith measurement, g

$S_{w(i)} = (W_{wet(i)} - W_{dry})/W_{water}$, i^{th} water saturation of $P_{c(i)}$

$\quad\quad W_{water} = W_{sat} - W_{dry}$, or $\quad W_{water} = \phi \cdot V_{bulk} \cdot \rho_{water}$

$r_{(i)} = 2\sigma_{g-w}/Pc(i)$, radius corresponding to $P_{c(i)}$

$\quad\quad\quad \sigma_{g-w}$ = 72.8 dynes/cm (10^{-3}N/m), interfacial tension of air-water at 20°C

$\quad\quad\quad P_{c(i)} = 981 \cdot H_{water}$, dynes/cm^2

$\Delta W_{(i)}/W_{water} = (W_{wet(i-1)} - W_{wet(i)})/W_{water}$, fraction of the capillaries of $r_{(i)}$ in total pore volume.

Plot S_w versus P_c and r versus S_w. See Exercise 10.5 as an example.

EXERCISE 10.5

The capillary pressure data given in the following table was obtained from a porous plate measurement where air is displacing water. The core sample has a diameter of 3.77 cm, the length is 4.8 cm, and the porosity is 0.19. The water density is 1.0 g/cm^3 and the sample dry weight is 115.05 g. Calculate and plot the capillary pressure curve of the sample and indicate the irreducible water saturation and entry pressure. Plot the pore throat radius versus saturation fraction.

SOLUTION

$$V_p = \frac{\pi D^2}{4} \cdot L \cdot \phi = \frac{\pi \cdot 3.77^2}{4} \cdot 4.80 \cdot 0.19 = 10.18 \cdot cm^3$$

$$S_{wi} = 0.16 \text{ and } P_e < 0.1 \text{ bar}$$

The porous plate method is the most accurate measurement of capillary pressure in homogeneous and heterogeneous cores. Several plugs can be measured at a time. The limitation is that the capillary discontinuity may distort the results.

TABLE 10.4
Recorded Data

P_{air} (10^5 Pa)	Sample Weight (g)
1.1	123.04
1.7	120.33
2.1	117.50
2.9	116.65
4.6	116.29
6.6	116.20

TABLE 10.5
Calculated Data

Pair [bar]	Pw [bar]	Pc [bar]	Weight Sample (g)	dW (g)	Sw [-]	r [μm]	dSw: dW/W [-]
		0			1		
1.1	1	0.1	123.04	7.99	0.78	1.46	0.22
1.7	1	0.7	120.33	5.28	0.52	2.08	0.27
2.1	1	1.1	118.1	3.05	0.3	1.32	0.22
2.9	1	1.9	116.65	1.6	0.16	0.766	0.14
4.6	1	3.6	116.25	1.2	0.12	0.404	0.04
6.6	1	5.6	116.2	1.15	0.11	0.26	0

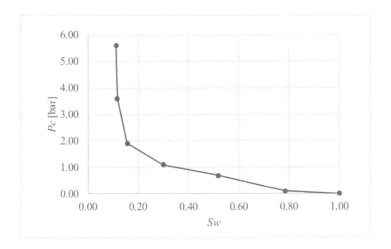

FIGURE 10.16 Capillary pressure versus water saturation, Exercise 10.5.

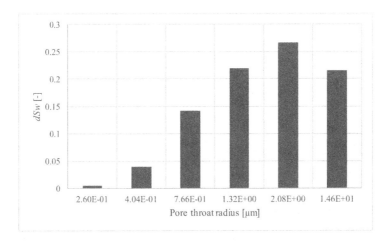

FIGURE 10.17 The pore throat radius versus saturation fraction, Exercise 10.5.

10.9.3 CENTRIFUGE

Capillary pressure can be measured by a high-speed centrifuge, which is the most common method applied in the laboratory. The method is based on the displacement of one fluid by the other under centrifugal force. For example, in primary drainage process, a 100% water-saturated sample is placed in the centrifuge bucket surrounded by a lighter non-wetting phase as seen in Figure 10.18a. Due to the centrifugal forces the lighter fluid tends to move inwards towards the axis of rotation, while the denser fluid from the core would want to move outwards, such that in the core the heavier fluid is displaced.

The opposite induces invasion of the heavier phase into the core and collection of the lighter phase production close to the axis, as in Figure 10.18b. Note that the volume displaced can be obtained in the neck volume. In case of a water wet core, using light oil and brine, scenario a. corresponds to a drainage process and b. to an imbibition process.

THEORY OF THE CAPILLARY PRESSURE MEASUREMENT BY CENTRIFUGE

Hassler and Brunner (1945) presented the basic concepts involved in the use of the centrifuge by relating the performance of a small core in a field of high acceleration.

If the cylindrical core of length L is subjected to an acceleration $a_c = -\omega^2 r$ where ω is angular velocity of the centrifuge and r is the distance from the axis of rotation, then we have:

$$\frac{\partial P_c}{\partial r} = \Delta \rho a_c \qquad (10.21)$$

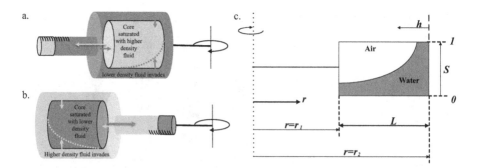

FIGURE 10.18 Mounting of the core and production collector dependent on the densities of initial saturation with the higher density fluid (a.) or lower density fluid (b.) compared to the surrounding fluid. c. Schematic diagram of a core in a centrifuge and its parameters.

Given the boundary conditions shown in Figure 10.18, the differential equation 10.21 can be solved by simple integration.

$$P_c(r) = \int_r^{r_2} \Delta \rho a_c \, dr \tag{10.22}$$

$$P_c(r) = -\int_r^{r_2} \Delta \rho \omega^2 r \, dr \tag{10.23}$$

The capillary pressure at the outer face of the core is zero, $P_c(r_2) = 0$, so:

$$P_c(r) = \frac{1}{2} \Delta \rho \omega^2 \left(r_2^2 - r^2 \right) \tag{10.24}$$

and for a continuous phase, the capillary pressure at the inner face of the core is:

$$P_{cL} = P_c(L) = \frac{1}{2} \Delta \rho \omega^2 \left(r_2^2 - r_1^2 \right) \tag{10.25}$$

Now, the main purpose is to relate the capillary pressure and saturation S for a given core, which gives the saturation in the core at equilibrium with the capillary pressure, $S = S(P_c)$.

The saturation at a distance h above the outer face of the core cannot be measured directly. However, the average saturation, which is the ratio of remaining liquid volume after production to pore volume, can be written as:

$$S_{avg} = \frac{1}{r_2 - r_1} \int_{r1}^{r2} S(r) \, dr \tag{10.26}$$

We will have a relationship of saturation as a function of capillary pressure, $S = S(P_c)$, so Eq. 10.26 can be expressed as follows by changing the integration variable.

$$P_c(r_2) = 0 \quad \text{and} \quad P_c(r_1) = P_{cL}$$

$$S_{avg} = \frac{1}{r_2 - r_1} \int_0^{P_{cL}} \frac{S(P_c) \, dP_c}{-\Delta \rho \omega^2 r} \tag{10.27}$$

An expression for r is obtained from Eq. 10.24.

$$r = r_2 \sqrt{1 - \frac{P_c}{\frac{1}{2} \Delta \rho \omega^2 r_2^2}} \tag{10.28}$$

and we obtain:

$$S_{avg} = \frac{1}{(r_2 - r_1)\Delta\rho\omega^2 r_2} \int_{P_{cL}}^{0} \frac{S(P_c)dP_c}{\sqrt{1 - \dfrac{P_c}{\dfrac{1}{2}\Delta\rho\omega^2 r_2^2}}} \tag{10.29}$$

and with mathematical manipulation it becomes

$$S_{avg}P_{cL} = \cos^2(\frac{\alpha}{2}) \int_{P_{cL}}^{0} \frac{S(P_c)dP_c}{\sqrt{1 - \dfrac{P_c}{P_{cL}}\sin^2\alpha}} \tag{10.30}$$

where

$$\cos\alpha = \frac{r_1}{r_2}$$

$$\cos^2(\alpha/2) = \frac{1}{2}(1 + \cos\alpha) = \frac{r_1 + r_2}{2r_2}$$

$$\sin^2\alpha = 1 - \cos^2\alpha = 1 - \frac{r_1^2}{r_2^2}$$

Eq. 10.30 cannot be solved so simply for the unknown function S. For small values of α (small core sample), the acceleration gradient along the core can be neglected. Assuming:

$$r_1/r_2 \approx 1$$

Then

$$\cos^2\left(\alpha/2\right) = 1 \quad \text{and} \quad \sin^2\alpha = 0$$

Eq. 10.30 is then reduced to:

$$S_{avg}P_{cL} = \int_{P_{cL}}^{0} S(P_c)dP_c \tag{10.31}$$

or in differentiation form:

$$S_L = \frac{d}{dP_{cL}}\left(S_{avg}P_{cL}\right) \tag{10.32}$$

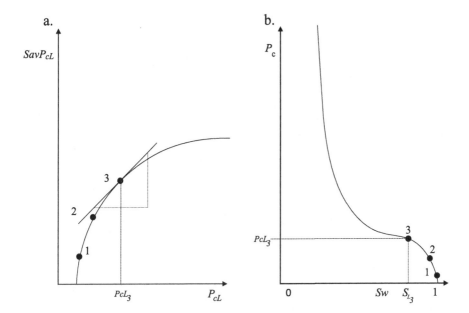

FIGURE 10.19 Graphical differentiation of $S_{av}P_{cL}$ - P_{cL} curve (a) to determine S-P_c curve (b).

In this method the cores are saturated with water (or oil) and rotated at increasing speed. The speed is increased in steps, and average fluid saturations at each speed are calculated from observation of the liquid produced. The liquid volume is read with a stroboscope while the centrifuge is in motion, and the speed of centrifuge is increased stepwise. When the run is over, the cores are removed and weighed.

The value of P_{cL} for each centrifuge speed is then computed from Equation 10.25, and the average saturation for each core is obtained from the dry and saturated weights and the corresponding pipette reading. From this data a smooth curve is prepared for each core. Figure 10.19a shows a typical $S_{avg}P_{cL}$ as a function of P_{cL} and points indicated on the curve are first, second, and third speed. The value of saturation that goes with each value of P_{cL}, which now represents the capillary pressure, is obtained from this curve by graphical differentiation according to Equation 10.32. A typical plot of P_c as a function of S is shown in Figure 10.19b. A complete capillary pressure curve by this method may be obtained in a few hours, and several samples are run simultaneously. The method is claimed to be accurate, will reach equilibrium rapidly, gives good reproducibility, and can produce high pressure differences between phases.

EXPERIMENTAL PROCEDURE FOR DRAINAGE CAPILLARY PRESSURE TEST

The objective of the experiment is to measure capillary pressure in a core sample which is 100% saturated with water (wetting phase) and plot the capillary pressure versus water saturation. The assumption behind this is water wetting of the sample, which means that water acts as wetting phase while oil acts as non-wetting phase.

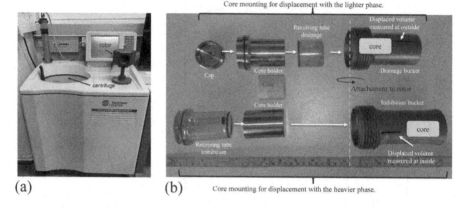

FIGURE 10.20 (a) Typical centrifuge including the rotor to attach the buckets; (b) accessories for drainage (top) and imbibition (bottom).

Therefore, the sample should be 100% saturated with water and displaced with oil to fulfil the drainage process. Figure 10.20a shows a typical centrifuge used in a core laboratory with a rotor that needs to be mounted in the centrifuge. Figure 10.20b shows the two different buckets with content to be mounted on the rotor.

Procedure for drainage:

Measure the bulk volume of a cleaned and dried core sample by measuring length and diameter. Weigh the dry core, W_{dry}.

1. Saturate the core sample with the brine/water with a known density of ρ_w. Weigh the saturated core, W_{sat}. Calculate the pore volume of the sample.
2. Place the core sample in a centrifuge core holder (Figure 10.21a–c) and connect to the receiving tube, which is filled with oil to the bottom of the core holder.
3. Put the core holder filled with the core and the receiving tube inside the drainage bucket and do the same for the other two samples and weigh all three buckets. All buckets should have the same weight within ±0.01 g. If there is a weight difference > 0.01 g among them, use small piece(s) of lead to make the buckets balance (see Figure 10.21d). This is necessary to balance the centrifuge while rotating. The centrifuge will stop in case of imbalance.
4. Open the top screw of the cap (see Figure 10.21e) and inject oil using a syringe so that the receiving tube and core holder are filled and the sample is surrounded with oil. Remove the air from the system by filling the void with the fluid using a syringe. Repeat the same procedure for the other two samples.
5. Note that if the experiment is supposed to be performed at an elevated temperature, then the centrifuge chamber, rotor, bucket that contains thesample, and the receiving tube need to be warmed up to reach the desired temperature before starting the experiment.

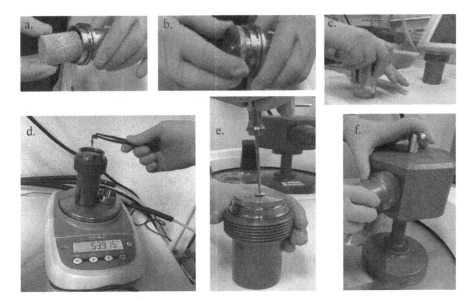

FIGURE 10.21 The mounting of the core in the centrifuge.

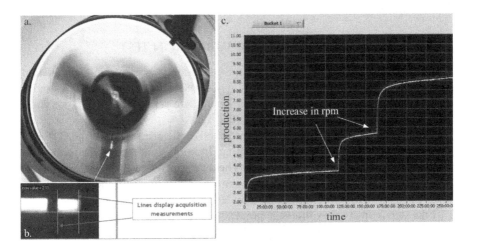

FIGURE 10.22 The centrifuge window (a) to determine the production (b) by monitoring the fluid level change over time. Knowing the container dimensions, this results in the production of the centrifuge versus time (c) Here for three increasing rotation speeds (rpm) (courtesy of Stratum).

6. Mount all three buckets to the rotor (see Figure 10.21f).
7. Place the rotor into the centrifuge chamber. Start the centrifuge to vacuum. After reaching a certain vacuum specified by the manufacturer, start to rotate at a speed as close to 500 revolutions per minute (rpm) as possible. It is important to start the experiment at the lowest possible rotation to be

able to determine the threshold pressure for pore entry. The choice of rpm is based on experience with the rock. Low permeability rock can start with rpm > 500; high permeability rocks shall start at rpm < 500. In case of an unknown rock, start at 100 and increase with steps of 100 rpm.

8. Mount the camera on top of the chamber and monitor the displaced fluid volume in the measuring tube for each sample while the centrifuge is in motion. An automated system with a camera connected to the computer allows better reading with higher accuracy.

9. The timing for each step depends on the permeability of the sample, which could vary from a few hours for high permeability samples to 1–2 days for low permeability samples. Go to the next step by increasing rotational speed if – *and only if* – the volume of collected fluid shows no further changes (see Figure 10.22). Record the collected volume (V_{coll}).

10. Repeat step 9. at higher rotational speeds until no further production occurs. This is the final stage of the experiment where the sample is saturated with oil and initial water (S_{wi}).

11. Take out the samples from the centrifuge and weigh each one. The volume of the produced fluid in the receiving bucket can also be read after pouring the produced fluid into a burette. These two values (weight of the sample and produced volume) can be used as quality check for the production obtained from step 10.

EXPERIMENTAL PROCEDURE FOR IMBIBITION TEST

The objective of the experiment is to measure capillary pressure in a core sample that is saturated with both phases – for example water as wetting phase at S_{wi} and oil as non-wetting phase – and plot the capillary pressure versus water saturation curve, with P_c reducing from zero to negative values.

The procedure for imbibition test is mostly like the primary drainage. The most important difference is the way of mounting. In imbibition, the arrangement is like the one shown in Figure 10.20b, bottom. The receiving tube will be placed on top of the core holder. This is the opposite to primary drainage where the receiving tube should be placed at the bottom of the core holder.

Like step 5. in primary drainage procedure, the top screw should be opened, but the receiving tube and core holder should be filled with water (not oil).

Procedure for imbibition:

1. Weigh the sample prior to the experiment.
2. Place the core sample in a centrifuge core holder.
3. Opposite to drainage process, the sample should be placed closest to the axis while the receiving bucket is placed furthest from the axis (Figure 10.20b).
4. Open the top screw and pour water using a syringe so that the receiving tube is filled and the sample is surrounded with water. Remove the air in the system.
5. Repeat the same procedure for the other two samples.
6. Put the core holder and the receiving tube inside the imbibition bucket and repeat the same for the other two samples and weigh all three buckets.
7. Follow steps 7 to 11 in the drainage process.

ERRORS

The main sources of error are the weight estimation and the determination of the production observing the interface in the window of the cups. In case of camera use, the resolution of the image will affect the estimation.

10.9.4 EXPERIMENT 10.2: DETERMINATION OF PRIMARY DRAINAGE CAPILLARY PRESSURE CURVE USING A CENTRIFUGE

Objective/description: Determine the drainage capillary pressure curve for a water wet core at room temperature using oil and brine. The oil density is lower than the brine density.

CALCULATIONS AND REPORT

1. Calculate pore volume of the core sample: $V_P = \dfrac{W_{sat} - W_{dry}}{\rho_w}$.

2. Convert rpm to rad/s: $\omega = \dfrac{2\pi\left(RPM\right)}{60}$.

3. Calculate capillary pressure at each step: $P_{cL} = \frac{1}{2}\rho_w\omega^2\left(r_2^2 - r_1^2\right)$,

Where:

       r_2 = the outer radius of the core holder, specific for the used centrifuge.
       $r_1 = r_2 - L$
       L = length of the core sample

4. Calculate average water saturation $S_{w,av}$ the core sample based on the volume of the water collected at corresponding capillary pressure: $S_{w,avg} = 1 - \dfrac{V_{coll}}{V_P}$.

5. Plot $S_{w,av}\,P_{cL}$ versus P_{cL}.

6. Draw tangents to the plot of the curve at each point and calculate the slope of each tangent. These slopes are the water saturation values S_w at the corresponding capillary pressure.

7. Plot the capillary pressures as a function of the water saturations.

DATA FORM

*Core No.: D: **cm** L: cm PV = **cm³***
See Exercise 10.6 as an example, where P_{cL} and $S_{w,av}$ are given.

TABLE 10.6
Capillary Pressure Measurement Using Centrifuge Method Calculation Table

Rpm	ω [rad/s]	P_{cL} [bar]	V_{coll} [cm³]	$S_{w,av}$ [-]	$S_{w,av}\,P_{cL}$ [bar]	S_w from Plot
x						
...						

EXERCISE 10.6

1. Capillary pressure as a function of saturation can be determined by centrifuging on a core sample. Make a sketch of a core in a centrifuge core holder and set important parameters. Briefly explain how a centrifuge measurement is performed.
2. Which equations and assumptions are used to derive the following equation (complete derivation is not required):

$$P_c(r) = \frac{1}{2}\Delta\rho\omega^2(r_2^2 - r^2)$$

The following data is measured in an oil-water drainage:

TABLE 10.7
Recorded Data

Capillary Pressure at the Top of the Core Sample (Closest to the Axis of Rotation) ($10^5\,Pa$)	Average Water Saturation in the Sample [-]
0.3	1.00
0.7	0.70
1.1	0.55
1.9	0.40
3.6	0.25
6.0	0.18

3. Plot the capillary pressure at the top of the sample versus the saturation at the top of the sample. Estimate the capillary entry pressure and irreducible water saturation.

$$\text{Hint: } S_L = \frac{d}{dP_{cL}}\left(\bar{S}P_{cL}\right)$$

SOLUTION

1. See Figure 10.18c for the sketch of the important parameters.

In a drainage capillary pressure curve, with air displacing water, a core saturated by water is surrounded by air. While centrifuging, the water drains out away from the rotational axis. Due to differences in centrifugal forces a saturation gradient is created with the higher water saturations farthest away from the axis of rotation.

2. $\dfrac{dP_c}{dx} = \Delta\rho a_c$

$a_c = \omega^2\pi$

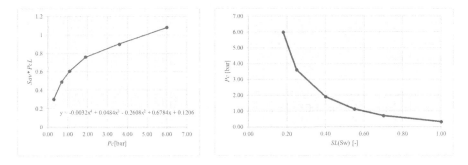

FIGURE 10.23 $S_{av}P_c$ versus P_c (left) and P_c versus S_L (right), with S_L obtained from the gradient of the left curve in each point of P_c.

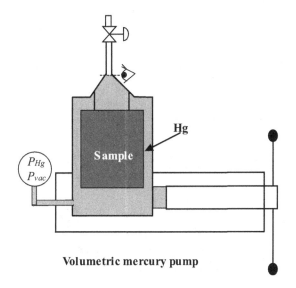

Volumetric mercury pump

FIGURE 10.24 Apparatus for mercury injection method.

From Figure 10.23 we see that the irreducible water saturation is $S_{wi} < 0.18$
The entry pressure is $P_e < 0.5$ bar

10.9.5 MERCURY INJECTION METHOD (MICP)

Capillary pressure measurement by mercury injection assumes that mercury is always non-wetting phase compared to air, which is a wetting phase. Therefore, capillary pressure measurement by mercury injection is a drainage process. The technique is based on stepwise injection of mercury into an air-saturated sample and determining the non-wetting phase saturation, dependent on the mercury pressure.

A mercury injection apparatus is schematically shown in Figure 10.24. The equipment consists basically of a mercury injection pump, a sample holder cell with a window for observing a constant mercury level, a pressure gauge, and a vacuum pump. Under vacuum a mercury pressure is set, holding its level constant. The change in invaded mercury can then be derived from the pump displacement.

In this method the mercury injected is calculated as a percentage of pore volume and related to pressure. A practical pressure limit on most equipment is about 15–25 MPa, but equipment for 400 MPa also exists (see Figure 10.25).

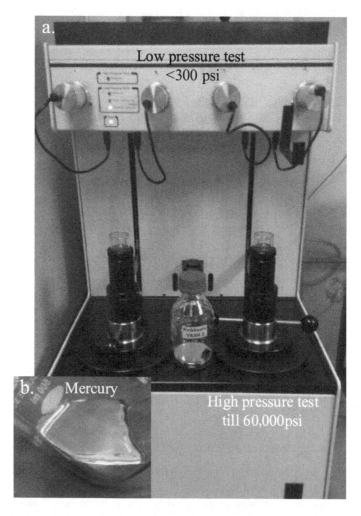

FIGURE 10.25 Commercial mercury porosimeter equipment, a. based on the principle of monitoring the change of the mercury; b. fluid level when pressed in steps into the core, in 2 stages, till 300 psi and 300–60,000 psi (courtesy of SINTEF Biotechnology and Nanomedicine).

EXPERIMENTAL PROCEDURE

1. Rotate the handle backward to withdraw the mercury, open the cap, and put the sample inside the chamber, which is partly filled with mercury.
2. Start to vacuum the sample until reaching a constant vacuum.
3. Rotate the handle forward to allow the mercury to be pumped into the chamber until the mercury appears at the Lucite window. Now the sample is totally immersed in mercury so that one may set the number on the ruler to zero, which will be the base for measurement of injected mercury in further steps.
4. Start to inject mercury by forward rotation of the handle so that mercury enters the sample. The volume of the injected mercury, read by the pump ruler, is noted for each pressure step.
5. Go to the next step by increasing the pressure *if and only if* reaching equilibrium for each pressure plateau.

Two important advantages are gained with the mercury injection method: a) the time for determining a complete curve is reduced to less than one hour, and b) the range of pressure is increased compared with the other methods. However, this method is suffering from the two most important disadvantages: a) the method is destructive since the sample is destroyed after the test, causing a permanent loss of the core sample, and b) it is difficult to transform the results to reservoir conditions because of the highly unrealistic fluid system and the uncertainty of wetting of mercury-solid. Additionally, the use of mercury involves significant HSE measures as its vapor is poisonous.

10.10   CONVERTING MERCURY CAPILLARY PRESSURE DATA TO OTHER FLUID SYSTEMS

Using mercury-air as the fluid pair, one will not obtain the irreducible saturation as when displacing water with air. Therefore, it is necessary to convert mercury-air data to water-air data. Purcell (Purcell 1949) was the first one who performed experiments and tried to establish a correlation between the air-mercury and air-brine system. He performed two different sets of capillary pressure measurement by mercury injection and porous plate method on six sandstone samples. He noticed that air-mercury capillary pressure data is approximately five times that of porous plate capillary pressure.

For air-mercury capillary pressure:

$$P_{cam} = 2 \sigma_{am} \cos \theta_{am}/r \qquad (10.33)$$

For air-water capillary pressure:

$$P_{caw} = 2 \sigma_{aw} \cos \theta_{aw}/r \qquad (10.34)$$

Conversion of Hg/air capillary pressure data to brine/air can be expressed as:

$$\left(P_c\right)_{water-air} = \frac{\sigma_{water-air}}{\sigma_{Hg-air}} \left| \frac{\cos\theta_{water-air}}{\cos\theta_{Hg-air}} \right| \left(P_c\right)_{Hg-air} \qquad (10.35)$$

where

$\left(P_c\right)_{water-air}$ = capillary pressure for brine-air system,

$\sigma_{water-air}$ = interfacial tension between water and air,

σ_{Hg-air} = interfacial tension between air and mercury,

$\theta_{water-air}$ = contact angle of brine and air,

θ_{Hg-air} = contact angle of mercury and air, and

$\left(P_c\right)_{Hg-air}$ = mercury-air capillary pressure.

Purcell assumed a value of 480 and 72 dynes/cm for mercury-air and water-air surface tensions and a value of 130° and 72° for contact angles of mercury and water
with rock sample. Substituting these values in the previous equation:

$$P_{c\ (gas\text{-}brine)} = P_{c\ (air\text{-}Hg)} \cdot 72\cos 0°/480\cos 130°$$

$$= 0.233\ P_{c\ (air\text{-}Hg)} \qquad (10.36)$$

$$\left(P_c\right)_{water-oil} = \frac{\sigma_{water-oil}}{\sigma_{Hg-air}} \left| \frac{\cos\theta_{water-oil}}{\cos\theta_{Hg-air}} \right| \left(P_c\right)_{Hg-air} \qquad (10.37)$$

$$P_{c\ (water\text{-}oil)} = P_{c\ (air\text{-}Hg)} \cdot 25\cos 30°/480\cos 130°$$

$$= 0.07\ P_{c\ (air\text{-}Hg)} \qquad (10.38)$$

DATA INTERPRETATION AND CALCULATION

Data form:

   $L =$ _____ cm, $D =$ _____ cm, Porosity = _____, $B.V. =$ _____cm³, $P.V. =$_____cm³

EXERCISE 10.7

Calculate the capillary pressure curve from the previous data for an oil-water system
as described in Equation 10.38.

SOLUTION

Use Equation 10.38 to convert to Pc for a water-oil system (RT). Calculate the cumulative injected mercury volume per injection pressure. The maximum volume is the
pore volume, which can be used to calculate the Hg and air saturation (= $1 - S_{Hg}$)
from the cumulative Hg volumes injected.

TABLE 10.8

Calculation Table and Example Data of a Mercury Injection Experiment

Measured Data		Calculated			
Pressure (bar)	Mercury Injected (cc)	Cumulative Injected Mercury [cc]	Pc Water-Oil [bar]	Hg Saturation [-]	Air Saturation [-]
0.0	0.0	0.0	0.00	0.00	1.00
2	0.0	0.0	0.14	0.00	1.00
3	0.1	0.1	0.21	0.01	0.99
5	0.3	0.4	0.35	0.03	0.97
10	0.8	1.2	0.70	0.10	0.90
15	1.2	2.4	1.05	0.20	0.80
20	1.5	3.9	1.40	0.32	0.68
25	1.8	5.7	1.75	0.47	0.53
30	2.0	7.7	2.11	0.64	0.36
35	2.2	9.9	2.46	0.82	0.18
40	2.2	12.1	2.81	1.00	0.00

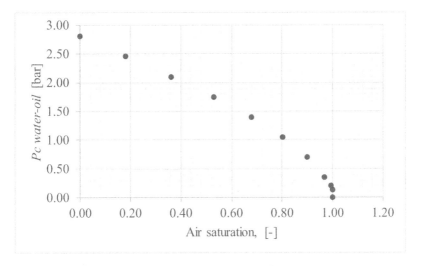

FIGURE 10.26 Capillary pressure curve for water oil derived from mercury injection test data given in Table 10.8.

Mercury injection is the best method for pore size characterization of the core sample (Swanson 1977). An example of the pore size distribution is given in Table 10.8 is given in Figure 10.27.

Pore size distribution in a core sample can also be expressed as pore size distribution index, λ. Brooks and Corey (1966) suggested a method that can be used to

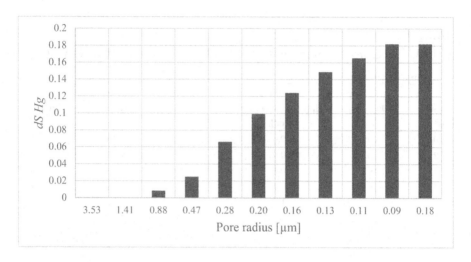

FIGURE 10.27 Mercury saturation versus pore throat radius, from Table 10.8.

identify the behavior of pore type in a core sample. They proposed the following relationship on experimental data as:

$$P_c = P_e (S_w^*)^{\frac{-1}{\lambda}}$$ (10.39)

Where P_c is drainage capillary pressure, P_{ce} the capillary entry pressure (displacement pressure), and S_w^* the normalized wetting phase saturation, which is defined as:

$$S_w^* = \frac{S_w - S_{wi}}{1 - S_{wi}}$$ (10.40)

And Equation (10.31) can be written in the form of:

$$\log P_c = \log P_e - \frac{1}{\lambda} \log S_w^*$$ (10.41)

The previous equation is a straight line on log P_c vs. log S^*_w, where the slope of the curve defines λ.

10.11 COMPARISON OF METHODS

In Table 10.9, the advantages and disadvantages of the methods are listed. The mercury injection method is primarily used for obtaining pore size distribution data, although it is a destructive method. Mercury-air capillary pressure curves have been found to be like water-air capillary pressure curves when the mercury-air pressure

TABLE 10.9
Comparison of the Methods

Method	Fluid Type	P_c Curve Type	Max. ΔP in Lab.	Test Time
Porous plate	Oil-water	Imbibition	2–5 bar	Weeks
	Gas-water	Drainage		
	Gas-oil			
Mercury injection	Hg-air	Drainage	100 bar	Minutes
Centrifuge	Gas-water	(Imbibition)	10 bar	Days
	Oil-water	Drainage		
	Gas-oil			

is divided by a constant. The constant can range from 5.8 to 7.5, depending on the nature of the rock.

The porous plate method is conceptually the simplest method, and it must be regarded as the standard method. Small and large samples can be used, and the choice of fluids is not restricted. A serious drawback is the limitation in pressure since most set-ups are limited to about 5 bar.

The centrifuge can reach capillary pressures higher than the porous plate method, and many samples can be measured in a relatively short time. A comprehensive presentation for professionals in the industry of the various methods for capillary pressure measurements are given in McPhee, Reed, and Zubizarreta (McPhee et al. 2015).

10.12 CONVERTING LABORATORY DATA TO RESERVOIR CONDITION

To use capillary pressure data in reservoir evaluation and numerical models, it is necessary to convert laboratory data to reservoir conditions. Laboratory data are obtained with a gas-water or an oil-water system, which does not normally have the same physical properties as the reservoir water, oil, and gas system.

As shown previously, by means of the capillary tube, the capillary pressure is expressed as

$$P_c = \frac{2\sigma\cos\theta}{r} \tag{10.42}$$

considering two specific cases wherein the laboratory (L) and reservoir (R) values are determined with gas-water and water-oil respectively. The capillary pressures become:

$$(P_c)_L = \frac{2\sigma_{wg}\cos\theta_{wg}}{r} \tag{10.43}$$

$$(P_c)_R = \frac{2\sigma_{wo}\cos\theta_{wo}}{r} \tag{10.44}$$

where

σ_{wg} = interfacial tension between water and gas used in laboratory test
θ_{wg} = contact angle between water and gas
σ_{wo} = interfacial tension between reservoir water and oil at reservoir conditions
θ_{wo} = contact angle among reservoir, water, and oil, and
r = radius of capillary.

Comparing the equations for laboratory and reservoir capillary pressure and assuming the radius of capillary constant with pressure and temperature, it is found that the reservoir capillary pressure is

$$(P_c)_R = \frac{\sigma_{wo} \cos\theta_{wo}}{\sigma_{wg} \cos\theta_{wg}}, \quad (P_c)_L = \frac{(\sigma\cos\theta)_R}{(\sigma\cos\theta)_L}(P_c)_L \qquad (10.45)$$

Note that it is difficult to determine the exact value of the contact angle for fluids in a porous matrix, and therefore the contact angles are often neglected. The equation becomes

$$(P_c)_R = \frac{\sigma_R}{\sigma_L}(P_c)_L \qquad (10.46)$$

EXERCISE 10.8

By centrifuge method, the water can be displaced in a core sample with air. The following capillary pressure data in Table 10.10 were determined for a core that may be assumed to be representative of a homogeneous oil reservoir:

TABLE 10.10
Recorded Data

S_w [%]	90	80	70	60	50	40	35	30	25
P_c [10^5 Pa]	0.070	0.090	0.123	0.160	0.225	0.345	0.460	0.800	1.800

Use this data to calculate the water saturation (S_w) in the oil reservoir at 8 m above the oil-water contact. The following data is provided:

The rock is completely water-wetting for all fluid systems.
Interfacial tension air/water: $\sigma^\circ_{aw} = 70. \ 10^{-3}$ N/m
Interfacial tension water/oil: $\sigma^\circ_{wo} = 35. \ 10^{-3}$ N/m
Water density: $\rho^\circ_w = 1010$ kg/m³
Oil density: $\rho^\circ_o = 700$ kg/m³
Gravitational acceleration: $g = 9.81$ m/s²

SOLUTION

Assuming: $\cos\theta^\circ{}_{lab} = \cos\theta^\circ{}_{res}$ and $r_{cap,lab} = r_{cap,res}$, Equation 10.46 can be used.

$$P_{c,res} = \frac{\sigma_{res}}{\sigma_{lab}} \cdot P_{c,lab}$$

$$P_{c,res} = \frac{35\cdot10^{-3}}{70\cdot10^{-3}} \cdot P_{c,lab} = 0.5\cdot P_{c\,lab}$$

$$P_c = \Delta\rho gh$$

$$h = 8\ m;\ g = 9.81\ m/s^2$$

$$P_{c,8m} = 310\cdot9.81\cdot8 = 24329\ Pa$$

Table 10.11 shows the calculated capillary pressure for the field conditions. According to Figure 10.28, water saturation can be read at 8 m above water-oil contact equal to 0.36.

TABLE 10.11

Calculation Table

$S_w\%$	90	80	70	60	50	40	35	30	25
$P_{c,lab}$ [10^5 Pa]	0.070	0.090	0.123	0.160	0.225	0.345	0.460	0.800	1.800
$P_{c,res}$ [10^5 Pa]	0.035	0.045	0.0615	0.080	0.113	0.175	0.230	0.40	0.90

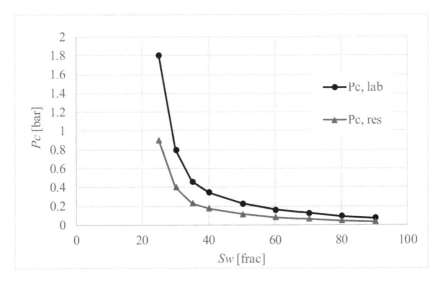

FIGURE 10.28 Capillary pressure versus water saturation for both lab and reservoir conditions.

EXERCISE 10.9

Based on capillary pressure curves determined in the laboratory, one can determine the water saturation in the reservoir. Figure 10.29a gives the capillary pressure curves for three different core samples (core samples 2, 5, and 9) from a given reservoir. The capillary pressure curves were measured in the laboratory for air displacing water at room conditions. Air-water interfacial tension is 70.10^{-3} N/m. For all core samples, water is assumed to be the wetting phase.

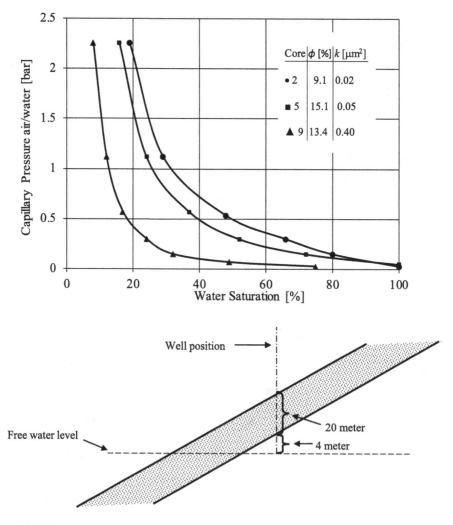

FIGURE 10.29 a. Capillary pressure curves for three different core samples. b. The cross-section of the reservoir.

The cross-section of the reservoir at a given well position is shown in Figure 10.29b. The lower and upper limit of the reservoir zone are as shown, respectively, 4 and 24 m above the theoretical oil water contact (free water level, $P_c = 0$). Other reservoir parameters at this well position are as follows:

$$\sigma_{o/w} = 30 \cdot 10^{-3}\,N\,/\,m \qquad \bar{\phi} = 15.1\%$$

$$\theta_{o/w} = 45^\circ \qquad \rho_o = 710\,kg\,/\,m^3$$

$$\bar{k} = 0.05\,\mu m^2 \qquad \rho_w = 1018\,kg\,/\,m^3$$

$\bar{k}$ and $\bar{\phi}$ are average values for the well position. Calculate the water saturation (S_w) at the upper and lower reservoir limits, at the specified well position.

SOLUTION

$$P_c = gh\,(\rho_w^\circ - \rho_o^\circ) \quad g = 9.81\ \text{m/s}^2, \quad \Delta\rho = (1{,}018 - 710)\ \text{kg/m}^3,$$

$$h_{upper} = 24\text{m} \qquad h_{lower} = 4\text{m}$$

$$P_{c\ upper} = 308 \cdot 9.81 \cdot 24 = 72{,}516\ \text{Pa}$$

$$Pc_{\ lower} = 308 \cdot 9.81 \cdot 4 = 12{,}086\ \text{Pa}$$

This must be converted into a laboratory condition:

$$P_{clab} = P_{cres} \cdot \frac{(\sigma cos\theta)_{lab}}{(\sigma cos\theta)_{res}} = P_{cres}\,\frac{70 \cdot 10^{-3}\,cos0}{30 \cdot 10^{-3}\,cos45} = 3.30 \cdot P_{cres}$$

The water saturation is determined from Figure 10.29a. Core 5 is used since this sample has the same φ and k as the mean values of the reservoir. Table 10.12 shows the final results. Alternatively, the J-curve can be calculated, from which S_w can be found; see next section.

TABLE 10.12
Calculation Table

	Upper	Lower
$P_{c\,res}$	0.725 bar	0.121 bar
$P_{c\,lab}$	2.393 bar	0.399 bar
S_w	17%	43%

10.13   LEVERETT J-FUNCTION. CONVERSION
TO FIELD CONDITIONS

Normally capillary pressure data measured in a reservoir laboratory is representative of a small part of the reservoir, which includes heterogeneity to some extent.
Therefore, it is necessary to combine all capillary pressure data from a given reservoir and convert to a single curve that will be representative of the capillary pressure
in the reservoir. Leverett (Leverett 1939) was a pioneer in this approach who studied
capillary pressure data from core samples with different porosity and permeability
values and found that the data could be expressed in a general form. He defined a
dimensionless function as follows, known now as Leverett J-function.

Where P_c is the capillary pressure, σ the interfacial tension, ϕ is porosity as
fraction, and k is the permeability.

$$J = \frac{P_c}{\sigma \cos\theta} \sqrt{\frac{k}{\phi}} \tag{10.47}$$

EXERCISE 10.9

The Leverett J-function plot (see Figure 10.30) represents an average dimensionless
capillary pressure curve for a heterogeneous reservoir. Suppose we have a reservoir with different layers of specific porosities, permeabilities, and fluid properties
$(\phi_i, k_i, \sigma_i, \cos\theta_i)$.

Briefly specify the procedure for calculating and plotting the Leverett J-function
curve.

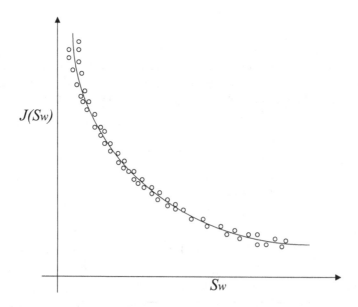

FIGURE 10.30   J-function versus water saturation.

TABLE 10.13

P_c for Different Saturations and the Corresponding Calculations

S_w	P_{c1}	P_{c2}	P_{c3}	$J = P_{c1}\dfrac{\sqrt{\dfrac{k_1}{\phi_1}}}{\sigma_1 \cos\theta 1}$	$J = P_{c2}\dfrac{\sqrt{\dfrac{k_2}{\phi_2}}}{\sigma_2 \cos\theta 2}$
Measured				Calculated	
20	–	–	–		
40	–	–	–		
60	–	–	–		
80	–	–	–		

SOLUTION

Procedure:

1. Calculate:

$$\frac{\sqrt{\dfrac{k_1}{\phi_1}}}{\sigma_1 \cos\theta_1} \quad \text{and} \quad \frac{\sqrt{\dfrac{k_2}{\phi_2}}}{\sigma_2 \cos\theta_2}$$

2. Choose different saturations and find P_c for each layer.
3. Calculate J for all these saturations (last two columns of Table 10.13).
4. Plot and find the best curve.

EXERCISE 10.10

In the following Table 10.14 are laboratory results from capillary pressure measurements of oil that displaces formation water for three cores:

TABLE 10.14
Recorded Data

Core No.	Permeability (md)	Porosity (%)	Pressure (bar)	0.034	0.068	0.102	0.136	0.204
			Saturation of formation water (% of pore volume)					
5	115	19.0		89.5	60.0	47.8	43.1	38.1
28	581	20.0		68.0	45.0	37.4	34.4	31.3
18	1640	27.0		65.0	36.4	31.0	31.0	28.3

Core No.			Pressure (bar)	0.272	0.340	0.408	0.476	0.544
			Saturation of formation water (% of pore volume)					
5				35.1	33.1	31.8	30.9	30.0
28				29.4	28.2	27.6	27.4	27.4
18				26.9	25.7	24.9	24.2	23.8

Interface tension for oil-formation water was measured to be 42×10^{-3} N/m. The contact angle is assumed to be $60°$. Calculate values for the Leverett $J(S_w)$ function for each point. Plot $\log J(S_w)$ against S_w and draw a curve that represents the points.

SOLUTION

Apply Equation 10.47 to the data gives the following data presented in Table 10.15 and 10.16, with the data of Table 10.16 plotted in Figure 10.31.

$$\sigma \cos \theta° = 42 \times 10^{-3} \cos 60° = 21 \times 10^{-3} \text{ N/m}$$

TABLE 10.15
Calculation Table

Core No.	k (10^{-12}m^2)	$\varnothing$	$\left(\dfrac{k}{\phi}\right)^{1/2}$ (10^{-6}m)
5	0.115	0.19	0.78
28	0.581	0.20	1.70
18	1.640	0.27	2.46

TABLE 10.16
Calculated Data

	Test No.					
	5		28		18	
P_c $(10^5$ Pa$)$	J	S_w	J	S_w	J	S_w
0.034	0.126	0.895	0.275	0.680	0.398	0.650
0.068	0.252	0.600	0.550	0.450	0.796	0.364
0.102	0.378	0.478	0.825	0.375	1.194	0.332
0.136	0.505	0.431	1.100	0.344	1.593	0.310
0.204	0.757	0.381	1.650	0.313	2.389	0.283
0.272	1.009	0.351	2.200	0.294	3.185	0.269
0.340	1.261	0.331	2.751	0.282	3.981	0.257
0.408	1.514	0.318	3.301	0.276	4.778	0.249
0.476	1.766	0.309	3.851	0.274	5.574	0.242
0.544	2.018	0.300	4.401	0.274	6.370	0.238

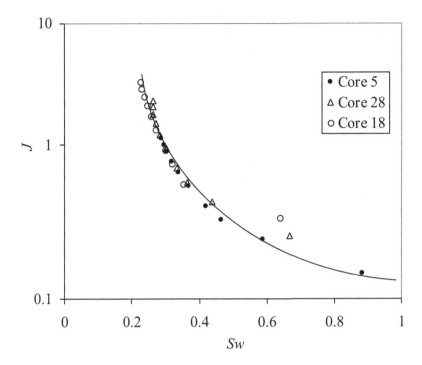

FIGURE 10.31   J-function versus water saturation.

REFERENCES

Amott, E. 1959. "Observations relating to the wettability of porous rock." *Transactions of the AIME 216*, 156–162.

Amyx, J.W., Bass, Jr. D.M., Whting, R.L. 1960. *Petroleum Reservoir Engineering.* Mc Graw-Hill.

Brooks, R.H., Corey, A.T. 1966. "Properties of porous media affecting fluid flow." *Journal of the Irrigation and Drainage Division 92*, 61–88.

Dake, L.P. 1978. *Fundamentals of Reservoir Engineering.* Elsevier.

Hassler, G.L., Brunner, E. 1945. "Measurement of capillary pressure in small core samples." *Transactions of the AIME 160*, 114.

Leverett, M.C. 1939. "Flow of oil-water mixtures through unconsolidated sands." *Transactions of the AIME 132*, 149–171.

McPhee, C., Reed, J., Zubizaretta, I. 2015. *Core Analysis. A Best Practice Guide in "Developments in Petroleum Science, Volume 64".* Elsevier.

Purcell, W.R. 1949. "Capillary pressures—their measurement using mercury and the calculation of permeability therefrom." *Transactions of the AIME.* Petroleum Transactions, AIME, Feb 1949: 39–48. T.P. 2544.

Swanson, B.F. 1977. "Visualizing pores non-wetting phase in porous rocks." In *Society of Petroleum Engineers, Annual Fall Technical Conference. SPE Paper 6857.* Society of Petroleum Engineers.

Zolotukhin, A.B., Ursin, J. -R. 2000. *Introduction to Petroleum Reservoir Engineering.* Høgskole Forlaget, Norwegian Academic Press.

11 Relative Permeability

11.1 INTRODUCTION

An important task of reservoir engineers is to predict the simultaneous flow of gases and liquids through reservoir rocks. Flow into or away from wells by production and injection and the fraction of liquid and gas that will be recovered during multiphase flow are two very important quantities that an engineer must predict. Similarly, for example, in storage of CO_2 in aquifers, the phases CO_2 and water are present, or if CO_2 is stored in an old hydrocarbon reservoir, the flow dynamics to predict storage potential might involve additionally the phases of oil and hydrocarbon gas. One means of making such predictions is by use of the adapted Darcy's equation for multiphase flow, which involves a parameter that quantifies the ratio of flow resistance of the various fluids. This parameter is called **relative permeability** and is a function of the fluid saturations in the porous rock. Relative permeability values are needed for multiphase flow description, applicable in both flow and recovery calculations. In many cases the relative permeability values have a significant influence on the solution compared to other variables in the calculations. It is therefore an important parameter to determine.

Relative permeability values can be obtained from laboratory flooding experiments but also from various sources like empirical correlations and capillary pressure relationships. Care needs to be taken that the data obtained represents in-situ reservoir behavior, since the rock used, fluids, and the flooding conditions may not be representative for the situation in the reservoir. Also, the concept of adaption of the single-phase Darcy equation is questioned (Darcy 1856). Still, research is ongoing to find a more accurate method for the description of multiphase flow in porous media (Mecke and Arns 2005; Hilfer 2006; Rücker et al. 2015; Herring et al. 2015; Armstrong et al. 2017; Khanamiri et al. 2018; Schlüter et al. 2017; McClure et al. 2018).

The subject of this chapter is to define relative permeability based on Darcy's law, discuss in detail the displacement theory that is the basis for obtaining the relative permeability from laboratory flooding, and finally the various laboratory techniques are presented.

11.2 DEFINITIONS

As mentioned before, absolute permeability refers to the ease of single-phase fluid flow, without any other phase present, meaning 100% saturation. In natural subsurface reservoirs, however, like petroleum reservoirs, the rocks are usually saturated with two or more fluids, such as water, oil, and/or gas. Therefore, it is necessary to generalize Darcy's law for multiphase flow by introducing the concept of effective permeability to describe the simultaneous flow of more than one fluid.

DOI: 10.1201/9781003382584-11

Effective permeability is the ability of the porous material to conduct a fluid when its saturation is less than 100% of the pore space, meaning when more than one fluid phase is present in the porous media. **Relative permeability** is the ratio of the effective permeability of a given phase, in presence of other phases (water and/or gas), to the absolute permeability k and can be written by the simple equation:

$$k_r = \frac{k_{eff}}{k_{abs}} \tag{11.1}$$

Where k_r is the relative permeability, k_{eff} the effective permeability, and k is the absolute permeability of the core sample. The relative permeability of each phase such as gas, oil, or water is mathematically expressed as:

$$k_{ro} = \frac{k_{eo}}{k} \tag{11.2}$$

$$k_{rg} = \frac{k_{eg}}{k} \tag{11.3}$$

$$k_{rw} = \frac{k_{ew}}{k} \tag{11.4}$$

and k_{rw} are the relative permeabilities of oil, gas, and water respectively, and k_{eo}, k_{eg}, and k_{ew} the effective permeability of oil, gas, and water phases, respectively. The relative permeability is dimensionless, and value ranges from zero to a maximum of one. Consequently, the sum of the relative permeability of each phase is always less than or equal to one:

$$k_{ro} + k_{rg} + k_{rw} \leqslant 1 \tag{11.5}$$

Therefore, when more than one fluid is present in the porous media, instead of the absolute permeability, the effective permeability, k_{eff} is to be used in the Darcy equation. This equation can be applied to all the phases present in the porous media. So, the generalization of the Darcy equation to multiphase flow can be written as:

$$q_n = -\frac{k_{eff\,n}\,A}{\mu_n}\left(\frac{\partial P_n}{\partial x} + \rho_n g \sin\alpha\right) \tag{11.6}$$

Where q is the flow rate in m³/s, k_{eff} the effective permeability in m², A the cross-sectional area in m², μ the fluid viscosity in N · s/m², $\frac{\partial P}{\partial x}$ the pressure gradient, ρ the fluid density in kg/m³, g the acceleration due to gravity, α the angle of inclination of dip, and n the fluid phase. Beside the assumptions valid for the single-phase Darcy equation, this equation assumes immiscible flow and continuity of the phases from inlet to outlet or from well to well. The latter assumption depends on the wettability, which is violated especially close to the end-point saturation of the non-wetting phase.

The relative permeability is mainly influenced by the following factors:

- Fluid saturations.
- Saturation history.
- Wettability.
- Temperature.
- Interfacial tension.
- Viscosity.

In this chapter we will focus on some important aspects of *two* phases of relative permeability. Three-phase systems—e.g., with three mobile phases like oil, gas, and water—are generally approached using correlations based on two relative permeability curves: Between gas and oil and between water and oil, like the Stone's models (Stone 1970, 1973).

Oil and water relative permeability curves are usually plotted as a function of water saturation as shown in Figure 11.1.

The directions of the curves point out the saturation histories, which are called drainage and imbibition. The *drainage* curve applies to processes where the wetting phase saturation is decreasing in magnitude and the *imbibition* curve applies to processes where the wetting phase saturation is increasing in magnitude; see Figure 11.2. Depending on the number of cycles, the curves can be labelled

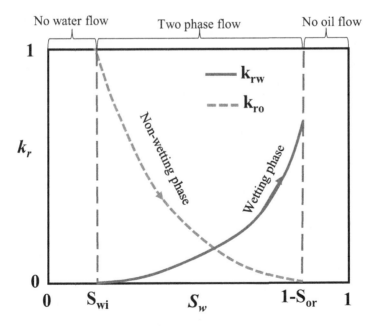

FIGURE 11.1   Typical water-oil relative permeability.

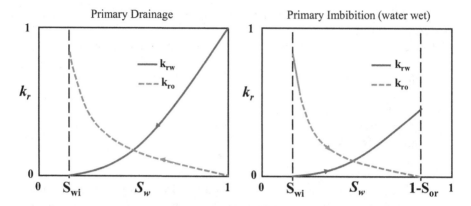

FIGURE 11.2 Typical water-oil relative permeability curves for drainage (left) and imbibition (right) for a water wet case.

as primary, secondary, tertiary, etc. It is worthwhile to notice that, in the primary drainage curve (see Figure 11.2 (left)), the saturation ranges from irreducible water (connate water) saturation (S_{wi}) to one (1) while in primary imbibition curve saturation ranges from irreducible water (connate water) to residual oil saturation (S_{or}) (see Figure 11.2 (right)).

In a petroleum reservoir, the drainage relative permeability curve can be applied in the process of oil migration from the source rock into a water wet reservoir. In this process, the reservoir is initially assumed to be 100% saturated with water. With the filling of the reservoir, the water saturation and its relative permeability will decrease while oil saturation and oil relative permeability will increase and reach the maximum value at maximum oil saturation. Similarly, the injection of gas in an aquifer can be described as a drainage process. Examples of imbibition are the injection of water in a water-wet oil reservoir or oil flow in an oil wet reservoir.

The relative permeability curves can be described by three main elements:

1. The end-point fluid saturations.
2. The end-point relative permeabilities.
3. The curvature of the relative permeability functions.

The saturation end points being the critical water saturation (or irreducible water or connate water saturation) S_{wc} and residual oil saturation S_{or} are two important parameters that should be taken into consideration. The end-point saturations shall be similar to the end-point saturations of the capillary pressure curve. It represents the capillary trapped, not the movable phase for the non-wetting phase and respectively the remaining film saturation for the wetting phase. Beyond S_{wc}, the water is not mobile and water relative permeability at water saturations below S_{wc} is zero. At the irreducible water saturation, the oil relative permeability, k_{ro}, equals 1 if effective permeability to oil is used as a reference. In the other words, at S_{wi}:

$$k_{ro} = \frac{kk_{ro} @ S_{wi}}{kk_{ro} @ S_{wi}} = 1 \tag{11.7}$$

k_{rw} is always equal to zero at this point because water is immobile.

The residual oil saturation or critical oil saturation S_{or} is the oil saturation below which the oil is immobile, and oil relative permeability is zero. However, at this saturation water relative permeability is at its maximum value because water is the only phase that is mobile. Therefore, boundary conditions in relative permeability curves can be expressed as:

At: S_{wi}, k_{ro} = 1; k_{rw} = 0, and

At: $S_w = 1-S_{or}$, k_{ro} = 0, and k_{rw} = maximum

The end-point saturations are important to determine the movable saturation range and are directly related to the ultimate amount of recoverable oil. For drainage this can define the storage potential of gas like H_2 or CO_2 within an aquifer being $1 - S_{wr}$, with S_{wr} being the remaining water saturation, or for imbibition of water in an oil field the final oil recovery can be defined as the ratio of all the recoverable oil in the rock sample, which is $(1-S_{wi} - S_{or})$ divided by available oil in the sample $(1-S_{wi})$.

$$RF_{ultimate\ HC} = \frac{1-S_{wi} - S_{or}}{1-S_{wi}} \tag{11.8}$$

The full curve can be approached by curve fitting, using different models available; this is further discussed in Section 11.4.

11.3 FLOW OF IMMISCIBLE FLUIDS IN POROUS MEDIA

The concept of relative permeability is fundamental to the study of the simultaneous flow of immiscible fluids through porous media. We consider first the case of linear displacement in a thin tube of porous material inclined at an angle α to the horizontal and with a cross section A and for this example with two phases present: water and oil. From Darcy's law, for the two phases, oil and water, we have:

$$q_w = -\frac{k_{ew}A}{\mu_w}\left(\frac{\partial P_w}{\partial x} + \rho_w g \sin\alpha\right) \tag{11.9}$$

$$q_o = -\frac{k_{eo}A}{\mu_o}\left(\frac{\partial P_o}{\partial x} + \rho_o g \sin\alpha\right) \tag{11.10}$$

The capillary pressure in the system assuming water as the wetting phase is written as:

$$P_o - P_w = P_c \tag{11.11}$$

The fluids are incompressible so that the continuity equation applies to each phase.

$$\frac{\partial q_w}{\partial x} = -\phi A \frac{\partial S_w}{\partial t} \tag{11.12}$$

$$\frac{\partial q_o}{\partial x} = -\phi A \frac{\partial S_o}{\partial t} \tag{11.13}$$

$$S_w + S_o = 1 \tag{11.14}$$

Adding Eqs. 11.12 and 11.13 to 11.14 yields Eq. 11.15:

$$\frac{\partial}{\partial x}(q_o + q_w) = 0 \tag{11.15}$$

such that the total flow rate $q_t = q_o + q_w$ is constant along the tube.

Now if we combine the Equations. 11.9, 11.10, and 11.11 to eliminate P_w and P_o and the following expression is obtained

$$q_o = -\frac{k_{eo} A}{\mu_o}\left(-\frac{\mu_w q_w}{k_{ew} A} + \frac{\partial P_c}{\partial x} - \Delta \rho g \sin \alpha\right), \tag{11.16}$$

we can define the fractional flow f_w of the flowing phases by:

$$f_w = \frac{q_w}{q_t} \tag{11.17}$$

$$f_o = \frac{q_o}{q_t} \quad \text{or} \quad f_o = 1 - f_w \tag{11.18}$$

The substitution of q_w and q_o in Eq. 11.16 yields:

$$f_w(S_w) = \frac{1 + \dfrac{k_{eo} A}{\mu_o q_t}\left(\dfrac{\partial P_c}{\partial x} - \Delta \rho g \sin \alpha\right)}{1 + \dfrac{k_{eo}(S_w)\,\mu_w}{k_{ew}(S_w)\,\mu_o}} \tag{11.19}$$

This is the fractional flow equation for the displacement of oil by water. For the displacement in a horizontal reservoir ($\alpha = 0$), and neglecting effect of capillary pressure gradient $\dfrac{\partial P_c}{\partial x}$, the fractional flow equation is reduced to:

$$f_w(S_w) = \frac{1}{1 + \dfrac{k_{ro}(S_w)\,\mu_w}{k_{rw}(S_w)\,\mu_o}} \tag{11.20}$$

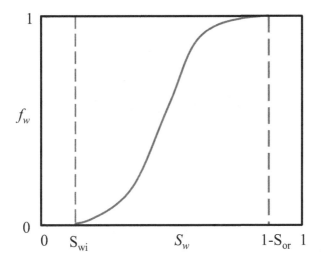

FIGURE 11.3 Typical fractional flow curve as a function of water saturation.

provided the oil displacement is strictly a function of water saturation, as related through the relative permeabilities. For a typical set of relative permeabilities, the fractional flow (Eq. 11.20) usually has the shape indicated in Figure 11.3, with saturation limit S_{wi} and $1 - S_{or}$, between which the fractional flow increases from zero to unity.

11.3.1 BUCKLEY-LEVERETT SOLUTION

Buckley and Leverett (1942) presented the basic equation for describing immiscible displacement in one dimension. For water displacing oil, the equation determines the velocity of a plane of constant water saturation travelling through a linear system. The Buckley-Leverett model discussed in this section is based on the following assumptions:

- Immiscible flow of two fluids in one dimension (no mass transfer between fluids).
- Diffuse, horizontal flow.
- Negligible capillary pressure.
- Incompressible fluids.
- Constant viscosity.
- Homogeneous rock (k and φ constant).
- Water is injected at $x = 0$ at constant rate q_w.

Note that on a core level it is generally assumed that the core is homogeneous such that the flow through a core can be described using the Buckley-Leverett solution.

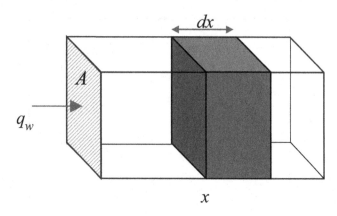

FIGURE 11.4   Mass flow rate of water through a linear volume element.

The conservation of mass of water flowing through the volume element shown in Figure 11.4 can be expressed as:

Mass flow rate in – mass flow rate out = rate of increase of mass in the volume element

$$q_w \rho_w \mid_x - q_w \rho_w \mid_{x+\Delta x} = A\varphi \cdot dx \frac{\partial}{\partial t}\left(\rho_w S_w\right) \tag{11.21}$$

or

$$q_w \rho_w \mid_x - \left(q_w \rho_w \mid_x + \frac{\partial}{\partial x}\left(q_w \rho_w\right)dx\right) = A\varphi \cdot dx \frac{\partial}{\partial t}\left(\rho_w S_w\right) \tag{11.22}$$

which can be reduced to:

$$\frac{\partial}{\partial x}\left(q_w \rho_w\right) = -A \frac{\partial}{\varphi \partial t}\left(\rho_w S_w\right) \tag{11.23}$$

and for the incompressible displacement where $\rho_w \approx$ constant:

$$\frac{\partial q_w}{\partial x}\mid_t = -A\varphi \frac{\partial S_w}{\partial t}\mid_x \tag{11.24}$$

The water saturation may be written as a full differential:

$$dS_w = \frac{\partial S_w}{\partial x}\mid_t .dx + \frac{\partial S_w}{\partial t}\mid_x .dt \tag{11.25}$$

and since it is the intention to study the movement of a plane of constant saturation, that is, $dS_w = 0$, then:

$$\frac{\partial S_w}{\partial t}\Big|_x = -\frac{\partial S_w}{\partial x}\Big|_t \cdot \frac{dx}{dt}\Big|_{S_w} \tag{11.26}$$

Furthermore,

$$\frac{\partial q_w}{\partial x}\Big|_t = \left(\frac{\partial q_w}{\partial S_w} \cdot \frac{\partial S_w}{\partial x}\right)_t \tag{11.27}$$

and substituting Eqs. 11.26 and 11.27 in Eq. 11.24 gives:

$$\frac{\partial q_w}{\partial S_w}\Big|_t = A\varphi \frac{dx}{dt}\Big|_{S_w} \tag{11.28}$$

Again, for incompressible displacement, q_t, is constant and, since $q_w = q_t f_w$, Eq. 11.28 can be expressed as:

$$\nu_{S_w} = \frac{dx}{dt}\Big|_{S_w} = \frac{q_t}{A\varphi} \frac{df_w}{dS_w}\Big|_{S_w} \tag{11.29}$$

This is the equation of Buckley-Leverett, which implies that, for a constant rate of water injection q_t, the velocity of a plane of constant water saturation, ν_{S_w}, is directly proportional to the derivative of the fractional flow equation evaluated for that saturation. With our assumptions, the fractional flow is strictly a function of water saturation, hence the use of the total differential of f_w in the Buckley-Leverett equation.

Integrating Equation 11.29 for the total time since the start of injection gives:

$$x_{S_w} = \frac{1}{A\varphi} \frac{df_w}{dS_w} \int_t^0 q_t \cdot dt \tag{11.30}$$

or

$$x_{S_w} = \frac{W_i}{A\varphi} \frac{df_w}{dS_w}\Big|_{S_w} \tag{11.31}$$

Where x_{S_w} is the position of plane of constant water saturation and W_i is the cumulative water injected with $W_i = q_w \cdot t$ [m³] if q_w remains constant, and it is assumed, as an initial condition, that $W_i = 0$ when $t = 0$. Therefore, at a given time after the start of injection (W_i constant), the position of different water saturation planes can be plotted, using Eq. 11.31, merely by determining the slope of the fractional flow curve for the value of each saturation. Considering the shape of the curve $x_{S_w} \sim S_w$ at one position x, two saturations can be present simultaneously. This is incorrect and therefore a correction is needed, which was provided by Welge (Welge 1952).

11.3.2 WELGE'S EXTENDED SOLUTION

Physically the solution can be explained as the slower moving saturations being overtaken and merging such that a liquid front is created. Over this liquid front, the saturation jumps from the initial saturation to a breakthrough saturation or front saturation. If e.g., water as the injection fluid in an oil reservoir at S_{wi} would be perfect in its displacement, the saturation before the front would be initial saturation S_{wi} and afterwards S_{or}. This is called piston-like behavior. Generally, this is not the case, and the saturation is not the residual saturation after the front. The front saturation can be derived by drawing a tangent along the fractional flow curve, with the initial saturation in the reservoir as a starting point (see Figure 11.5). The saturations between the initial saturation S_{wi} and front saturation S_{wf} do not exist. This so called **Welge method** solves the presence of having two saturations at similar positions. Mathematically the area at the frontal location, over the interval of front saturations, is zero.

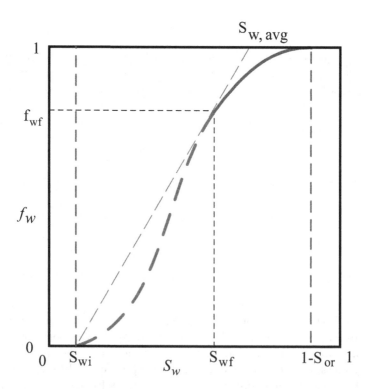

FIGURE 11.5 Tangent to the fractional flow curve, showing the part of the fractional flow curve that does not exist in the reservoir (dashed line, between S_{wi} and S_{wf}).

11.4   DISPLACEMENT EFFICIENCY

The Welge method can also be used elegantly to obtain average saturations behind the shock front, $S_{w,av}$. This consists of integrating the saturation distribution over the distance from the injection point to the front, thus obtaining the average water saturation behind the front, $S_{w,av}$, as shown in Figure 11.6a.

The situation depicted is at a fixed time, before water breakthrough in the producing outlet of the core, according to an amount of water injection W_i. At this time the maximum water saturation, $S_w = 1 - S_{or}$, has moved a distance $x1$, its velocity being proportional to the finite slope of the fractional flow curve at $f_w = 1$. Note that if the fractional flow curve asymptotically approaches $f_w = 1$, x_1 remains zero. The

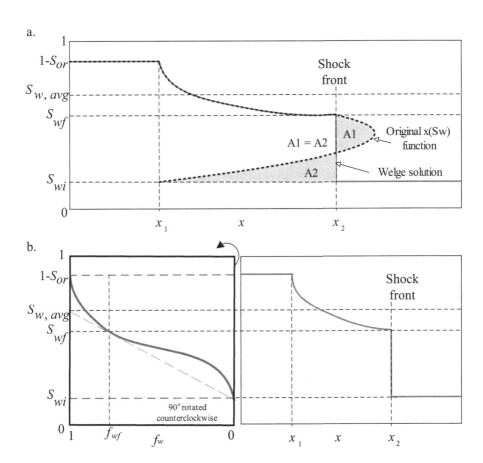

FIGURE 11.6   a. Eq. 11.31 plotted showing the water saturation distribution as a function of distance (dotted line) and the Welge's method determining the shock front where A1 = A2 (solid line). b. Shows how the parameters defining the profile of a. (repeated on the left side) correspond to the values derived from the fractional flow curve, which is tilted for this purpose.

flood front saturation S_{wf} is located at position x_2 measured from the injection point. Applying simple material balance:

$$W_i = Q \cdot dt = x_2 \cdot A \cdot \varphi \cdot \left(S_{w,avg} - S_{wi} \right) \tag{11.32}$$

or

$$S_{w,avg} - S_{wi} = \frac{W_i}{x_2 A \varphi} \tag{11.33}$$

and using Eq. 11.31, which is applicable up to the flood front at x_2, then:

$$S_{w,avg} - S_{wi} = \frac{W_i}{x_2 A \varphi} = \frac{1}{\frac{df_w}{dS_w}|_{S_{wf}}} \tag{11.34}$$

The average water saturation behind the front, $S_{w,avg}$, can be determined graphically by drawing a tangent to the curve $f(S_w)$, starting from the initial point (S_{wi}) as is illustrated in Figure 11.6.

To satisfy Equation 11.34, the tangent to the fractional flow curve, from the point $S_w = S_{wi}, f_w = 0$, must have a point of tangency with coordinates $S_w = S_{wf}$ and $f_w = f_{wf}$, and the extrapolated tangent must intercept the line $f_w = 1$ at the point $S_w = S_{w,avg}$. Figure 11.5 illustrates this approach. This method of determining S_{wf}, f_{wf} and $S_{w,avg}$ requires that the fractional flow curve must be plotted for the entire water saturation range: $S_{wi} < S_w < 1 - S_{or}$.

As noted earlier, the use of these equations ignores the effect of the capillary pressure gradient, $\partial P_c / \partial x$, which is only present behind the flood front for $S_{wf} < S_w < 1 - S_{or}$ and for a large flow system or when low rates of displacement are used. Also, capillary effects can be observed at the front, which is less sharp due to capillary flow, the so-called smearing out the front, leading to an earlier water breakthrough as predicted by this method.

The part of the fractional flow curve for saturations less than S_{wf} is, therefore, virtual, and the first real point on the curve has the coordinates S_{wf} and f_{wf}, corresponding to the shock front.

The displacement efficiency can be derived from the $S_{w,avg}$ behind the front. The higher the $S_{w,avg}$ at breakthrough, the more efficient the water, meaning f_{wf} (S_{wf}) needs to be closest to 1. Considering Eq. 11.20, this indicates that the fraction $\frac{k_{ro}(S)}{k_{rw}(S)} \frac{\mu_w}{\mu_o}$ needs to remain larger with increasing S_w, such that f_w remains low at higher water saturations. So k_{ro} shall remain high or k_{rw} low. Wettability or interfacial tension can affect this, as will be discussed later. A high viscosity ratio $\frac{\mu_w}{\mu_o}$ also helps, where μ_w shall be larger than μ_o. The ratio $\frac{k_{ro}(S)}{\mu_o}$ or $\frac{k_{rw}(S)}{\mu_w}$ is called mobility, λ. And the ratio of mobilities of the wetting phase and the non-wetting phase is called mobility ratio, generally defined as

$$M = \frac{\lambda_w}{\lambda_o} \quad \text{and} \quad \frac{k_{ro}(S)}{k_{rw}(S)}\frac{\mu_w}{\mu_o} = \frac{1}{M}. \tag{11.35}$$

From Eq. 11.31 the breakthrough time of the injection fluid with a fractional flow of f_{wf} at S_{wf} can be derived, knowing the length of the core L and area of flow (A).

$$t(S_{wf}) = \frac{LA\varphi}{q_T \frac{df_w}{dS_w}|_{Swf}} \tag{11.36}$$

11.5 PRODUCTION PROFILES

From the Buckley Leverett analysis production profiles also can be derived, knowing the relative permeability curve. Before breakthrough, the injected fluid is not produced yet, so only the fluid inside the core is produced. $q_{inj} = q_{production}$. The RF versus PV injected is then linear. At breakthrough both water and oil are produced till the ultimate recovery factor is reached where the oil becomes immobile, reaching S_{or}. At breakthrough, the RF curve will start deviating from the linear trend based on the mass balance; q_{total} is constant but q_o is not constant anymore and depends on $f_w > f_{wf}$. The breakthrough time depends on the steepness of the tangent to obtain the $S_{w,av}$ and S_{wf}. The steeper, the lower the $S_{w,avg}$, the quicker the BT. The $S_{w,avg}$ will give the RF (based on the original oil in place) at and after breakthrough.

$$RF\ after\ BT,\ or\ BT = \frac{S_{w,avg} - S_{wi}}{1 - S_{wi}} \tag{11.37}$$

Before breakthrough, the RF can be determined using the duration and flow rate of injection, t, and q_i, respectively and the volume of oil present in the pore space.

$$RF\ before\ BT, or\ BT = \frac{q_i \cdot t}{PV(1 - S_{wi})} \tag{11.38}$$

Figure 11.7 displays a typical graph for recovery factor versus injected pore volume. Breakthrough is easily noticeable when the production curve begins to bend and reaches a plateau.

In summary, the Buckley-Leverett theory can be used directly to analyze flooding performance dependent on relative permeabilities by e.g.:

- Calculation of saturation profile $S_w(x, t)$
  - Saturations at core outlet.
  - Average core saturations.
- Generation of production profiles and recovery factors (t).
- Determination of breakthrough time and accompanying fractional flow and average saturation.

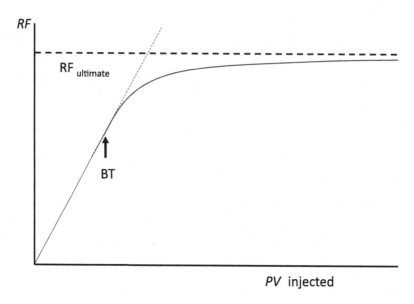

FIGURE 11.7   Production profile with indicated breakthrough and the ultimate recovery ($RF_{ultimate}$).

The Buckley-Leverett theory requires as input relative permeability, in case this shall be measured, the theory is to be applied inversely for the analysis of the phase production, average core saturations (t), pressure drops (t), and injection rates (t).

EXERCISE 11.1

Sketch the cumulative liquid injection volume, the cumulative oil, and the brine production volumes as a function of time corrected for dead volume (no values needed, relative trends only). The injection flow rate remains constant. Mark the breakthrough time of the injection fluid at the core outlet.

SOLUTION

Figure 11.8 illustrates the relationship between water and oil production rate over time. Initially, oil production follows a linear increase until it reaches the breakthrough point. At breakthrough time, water starts to be produced, causing the oil production to deviate from its linear pattern, ultimately reaching the final volume of recovery. Upon reaching this point, the rate of injection equals the water production rate.

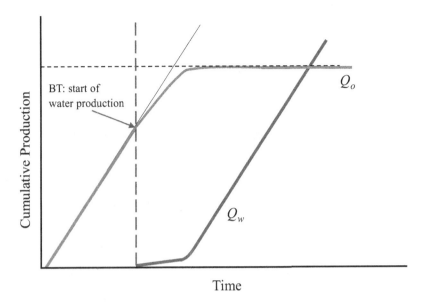

FIGURE 11.8 Cumulative water and oil production versus time in an unsteady-state experiment.

11.6 RELATIVE PERMEABILITY MODELS

Several models have been developed to describe the shape of relative permeability curve. The two main models are the Corey correlations (Corey 1954; Brooks and Corey 1966; Corey and Rathjens 1956) and LET models (L, E, and T are matching parameters in the models) (Lomeland 2018). Here, only the two-phase Corey correlation is presented. The data from the experiments is generally matched to these models such that, for the purpose of flow simulation, relative permeability functions can be used to describe the relative permeability for each saturation needed. Beside the description of the end points, $k_{ro}(S_{wi})$, $k_{rw}(1 - S_{or})$, the curvature shall be described with Corey exponents, for both the wetting and non-wetting phase, for water and oil respectively, c_w and c_o. The Corey correlation, with S_w as a variable, is:

$$k_{ro}\left(S_w\right) = k_{ro(Swi)}\left[\frac{S_{w,max} - S_w - S_{orw}}{S_{w,max} - S_{wi} - S_{orw}}\right]^{c_o} \qquad (11.39)$$

$$k_{rw}\left(S_w\right) = k_{rw(Sorw)}\left[\frac{S_w - S_{wcr}}{S_{w,max} - S_{wcr} - S_{orw}}\right]^{c_w} \qquad (11.40)$$

For the correlation, the end-point saturations are to be defined as illustrated in Figure 11.9 with S_{wmin} as minimum water saturation, S_{wcr} as critical water saturation,

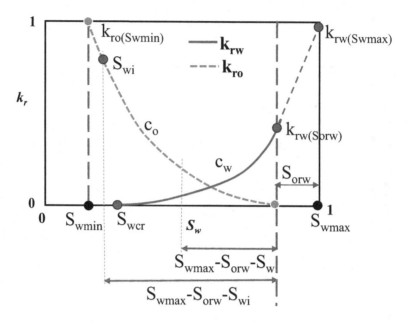

FIGURE 11.9 Water-oil relative permeability model using Corey exponent, with S_w as variable. The marked saturation intervals correlate to the nominator and denominator of Equation 11.39.

TABLE 11.1

Typical Oil and Water Wet Corey Exponents, c_o and c_w

Wettability	c_o	c_w
Water-wet	2–4	5–8
Intermediate-wet	4–6	3–5
Oil wet	6–8	2–3

S_{wi} as initial water saturation, and S_{orw} as residual oil saturation to water, as well as the relative permeabilities corresponding to the end-point saturations: $k_{rw(Sorw)}$ as water relative permeability at residual oil saturation, $k_{rw(Swmax)}$ as water relative permeability at max water saturation, and $k_{ro(Swmin)}$ as oil relative permeability at minimum water saturation. Generally, $S_{wcr} = S_{wi}$.

Typical Corey exponents can be found in Table 11.1.

11.7 PARAMETERS AFFECTING THE RELATIVE PERMEABILITY CURVES

11.7.1 WETTABILITY

Depending on the wettability, the flow path of the wetting and non-wetting fluids will differ. Simply illustrated, in a water wet case, water will form a film over the grain surface. During the imbibition process, with an increase in water saturation, the water film swells along the walls and the oil phase moves into the center of the pores. In general, it can be said that along the walls more friction is present in comparison to a flow through the pore center. Because of this, with increasing water saturation, oil flows relatively better in a water wet case based on relative permeability in comparison to water. The opposite is valid in an oil wet case.

The end point from the $k_{rw}(S_{or})$ in an oil wet case is generally higher than for a water wet system. This is caused by the morphology of the trapped oil. In a water wet case, the remaining oil is capillary trapped in front of the pore throats as droplets, blocking the water flow and causing the k_{rw} to be low. In an oil wet case, the oil trapped is adsorbed onto the rock grains, not obstructing the water from flowing resulting in a higher $k_{rw}(S_{or})$.

Additionally, based on a capillary bundle model, the pores initially filled with water are the smaller pores, with the higher resistance to flow, leading to a lower k_{rw} for a water wet case. Oil wet pores are the larger pores after aging, so then the oil flows in the larger pores, having lower resistance to flow.

Craig (Craig 1971) presented several rules of thumb indicating the effect of wettability on relative permeability, as shown in Table 11.2.

Figure 11.10 presents typical relative permeability curves under two different conditions of water wet and oil wet.

Anderson (Anderson 1987) discussed the effect of wettability on relative permeability and stated that the differences in relative permeabilities measured in strongly water wet and oil wet systems are caused by the differences in fluid distribution. A comparison of displacement efficiency dependent on wettability can be made based on the fractional flow curve; see Figure 11.11 for an example. Based on the relatively high k_{ro} and low k_{rw} for the water wet case, the fractional flow curve remains lower than for the oil wet case. Therefore the $S_{w,av}$ at breakthrough is much higher for the water wet case. However, the time of breakthrough is later, as the tangent is smaller. This is because more oil is displaced so more pore volume is flooded, leading to a later breakthrough.

The general shapes of relative permeability and fractional flow curves for strongly water wet and strongly oil wet rocks are illustrated in Figure 11.12 and Figure 11.13, respectively. Examining the fractional flow curve reveals that the strongly water-wet sample has a significantly limited two-phase flow region compared to the strongly oil wet sample. Consequently, this leads to a very small two-phase saturation range and transition zone for the strongly water-wet sample, making extrapolation of the

---

TABLE 11.2
Craig's Rules of Thumb Relating Wettability and Relative Permeability

Characteristics	Water Wet	Oil Wet
Initial water saturation, S_{wi}	Greater than 20–25%	Less than 15%
S_w at $k_{rw}=k_{ro}$	Greater than 50%	Less than 50%
k_{rw} at S_{or}	Less than 30%	Greater than 50%

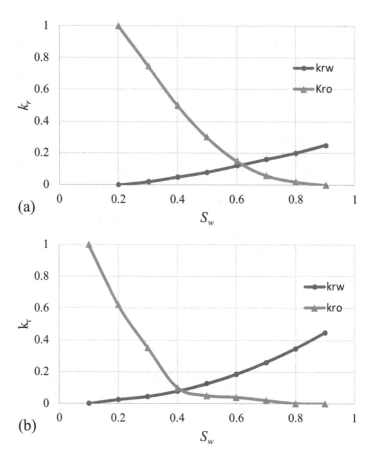

FIGURE 11.10   Schematic of relative permeability curves for (a) water wet and (b) oil wet cases.

relative permeability curves challenging. In contrast, the strongly oil wet sample shows broader two-phase flow region, resulting in wider saturation ranges and making the extrapolation of relative permeability much easier.

A sketch of recovery factor (RF) versus pore volume injected (PV_{inj}), based on the fractional flow curves for an oil water system of different wettability is given in Figure 11.14. The residual oil saturation (S_{or}) is equal for both cases, therefore, they will ultimately reach the same final recovery factor asymptotically.

The breakthrough time depends on the steepness of the tangent to obtain $S_{w,av}$ and S_{wf}. The steeper the tangent, the lower the average water saturation and the quicker breakthrough occurs. Therefore, in case A, breakthrough will happen earlier at a lower recovery factor, while in case B, breakthrough will occur at a higher recovery factor. Some production will continue after breakthrough, especially for the oil wet case. In this example, the amount of residual oil capillary trapped is similar, so both

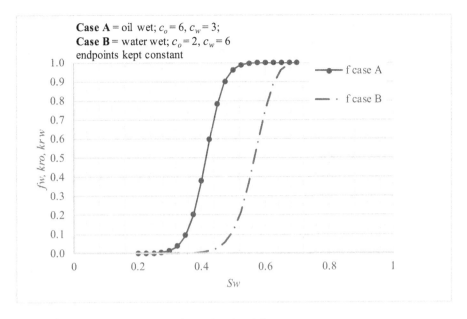

FIGURE 11.11   Effect of wettability on fractional flow curve.

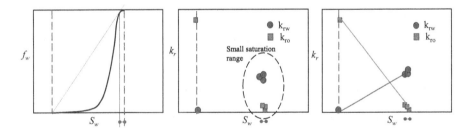

FIGURE 11.12   Typical relative permeability and fractional flow curves for water wet sample.

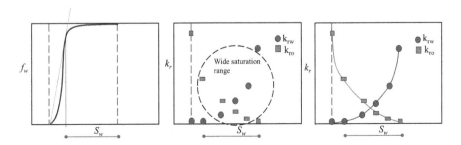

FIGURE 11.13   Typical relative permeability and fractional flow curves for oil wet sample.

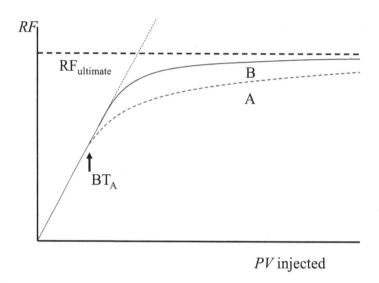

FIGURE 11.14 Dependence of the recovery on wettability with A being more oil wet than B.

cases are expected to reach the same ultimate recovery, although this might not always be the case.

Note that in comparing core flooding experiments the breakthrough time, the amount produced, and the level of ultimate recovery can help to classify the cores according to their wettability. But be reminded that there is a limit in PV injected during a multiphase flow experiment based on time and budget available. This makes it especially challenging to reach the actual residual oil in oil wet cores with an extensive period of two-phase flow after breakthrough. In oil wet cores, the oil phase wets the grains, forming a surface film, which can be produced to quite low levels as long as flooding occurs (Anderson 1987). The amount of remaining liquid phase then depends on the duration of the flood. Bump rates, increasing the viscous forces by increasing the injection flow rates, might give some acceleration of production to reach the plateau value. It is a judgement of the experimentalist as to when this is reached.

11.7.2 Interfacial Tension

Figure 11.15 shows the generalized effect of interfacial tension (IFT) for high, low, and zero IFT on the relative permeability curve. When IFT is reduced starting at S_{wi}, first the amount of capillary trapped oil is reduced, so the end point $1 - S_{or}$ increases towards 1. Additionally, capillary pressure reduces or becomes negligible, so there is no preference anymore for the fluid to flow in bigger or smaller pores based on capillary pressure.

So, with reducing IFT, the multiphase flow will become ideal with no friction present, such that the relative permeability curves approach straight lines.

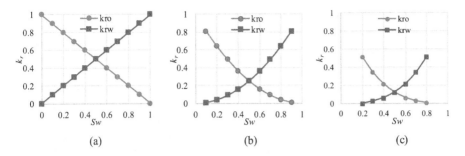

FIGURE 11.15 Effect of IFT on relative permeability: (a) High IFT, (b) low IFT, and (c) zero IFT.

Consequently, the fractional flow curve becomes a straight line from S_{wi} to $(1 - S_{or})$ such that the flow becomes piston-like displacement. Note that S_{wi} also can approach zero at ultralow IFT in a drainage experiment.

As the measurement of relative permeability at low IFT is a very difficult task, relative permeability correlations may be used instead. It is, however, essential to measure a single set of relative permeability at high IFT as a base curve. Several correlations have been developed to account for the effect of IFT on relative permeability (Coats 1980; Khazam et al. 1980; Amaeful and Handy 1982; Karimaie and Torsæter, 2008).

11.7.3 SATURATION HISTORY

Multiple sequences of fluid injection can occur in the reservoir either in imbibition or drainage modes. For example, during CO_2 injection, drying out occurs, depositing salt around the well bore, which can reduce the permeability. As solution, water sequentially is injected in slugs to dissolve the salt. This leads to multiple cycles of drainage and imbibition. Another example is water alternative gas (WAG) injection where water injection and gas injection are carried out alternately for periods of time to provide better sweep efficiency and reduced gas channeling.

Secondary drainage and imbibition are not similar to primary curves. The dependency on the history is called relative permeability hysteresis. The primary drainage starts from 100% wetting phase saturation, while the secondary drainage starts from the residual saturation of the non-wetting phase. The non-wetting phase then could find different paths to move through depending on the trapped non-wetting phase already present. This will reduce the ease of flowing of both phases. This might change the location of the residual wetting phase such that the initial state for the secondary imbibition will similarly be reduced. The history dependence of the relative permeability curves indicates that in the model of relative permeability other parameters might be included to create a history-independent relative flow description.

11.8 MANIPULATION/ADAPTION OF RELATIVE PERMEABILITY

Efforts are made to improve the microscopic sweep efficiency as the Buckley-Leverett theory describes. Based on the parameters that affect the relative permeability curve, concerning both the end points as well as the curvature, wettability and interfacial tension are parameters that can be manipulated in the field via the chosen water injection strategy. This is part of the measures of enhanced oil recovery (EOR) methods or in general the enhancement of storage and production. In Table 11.3 an overview is given of the most common methods applied in oil fields to improve recovery during water/gas injection. At gas injection, the injected phase is a non-wetting phase, so the *drainage* process needs to be optimized like water injection in an oil wet reservoir or CO_2 or hydrogen in an aquifer. In a water wet reservoir, water injection optimization focuses on optimizing the imbibition relative permeability curve. Generally, chemicals are used for optimization, like in surfactant or alkali flooding, so-called chemical EOR. Out of environmental concerns, green chemicals are considered or compositional changes of fluids in place, like low salinity.

Several techniques are developed to reduce interfacial tension (IFT), discussed in Section 11.5.2. Surfactants are surface active molecules that can reduce the surface energy at the liquid-gas or liquid-liquid interface dependent; see Section 8.2. This enables phases, including capillary trapped oil droplets, to deform more easily. Therefore, they can be mobilized more easily, leading to a reduced residual oil saturation. Capillary forces are also reduced during flow, increasing the relative permeability of both the wetting and non-wetting phase. Alkali, like sodium hydroxide and sodium carbonate, can form with natural acidic oil components natural surfactants, depending on brine salinity, rock mineralogy, and temperature. This reduces the IFT. Additional effects can be pH change, emulsification, and wettability change, all parameters that can affect the relative permeability. Surfactant (S) and alkali (A) flooding are often combined as AS flooding.

Note that the presented methods focus on microscopic sweep efficiency, the optimization of recovery at the pore scale. This requires that the fluids get to the pores.

TABLE 11.3
EOR Measures Targeting Improved Microscopic Sweep

IFT Reduction/Miscibility

Surfactants	Chemicals functioning at the liquid-liquid or liquid-gas interface to reduce the interfacial tension to values < 10^{-3}mN/m
Alkali e.g. NaOH	Reaction with oil components to form natural surfactants inside the reservoir
Miscible gas injection	At full miscibility an IFT does not exist. The gas needs to be miscible like e.g. CO_2

Wettability Changes from Oil- to Water Wet

Low salinity water flooding	Change of original brine composition to stimulate desorption of oil components initially adsorbed to the rock surface
Surfactants	Surfactants adsorbing at the rock surface changing wettability or the adsorption of oil components

A good sweep in the reservoir, such that all pores can be reached, the so-called volumetric sweep, depends on low heterogeneities in permeability and porosity on a field scale level and mobility ratio of the fluids. For a good mobility ratio and therefore volumetric sweep, the injection of specifically surfactant and alkali is often combined with polymer (P), increasing the viscosity of the injection fluid, (see Section 4.5.2), called ASP flooding.

Completely removing the interfacial tension can be obtained when the two phases are fully miscible, like ethanol and water in a miscible gas injection. Dependent on the temperature and pressure CO_2 can be on first contact miscible with oil. During miscible CO_2 flooding the mixture then flows as a single phase and no residual phase remains, so oil becomes completely mobile where the CO_2 arrives in the pores. See Jarrell (2002) for more details.

A different kind of surfactant can be used to change the wettability towards more water wet conditions by adsorption to the rock or by stimulation of oil desorption from the rock surface. This can be of interest when changing an oil wet rock matrix in a fractured carbonate reservoir. Low salinity is a green alternative, tuning the salinity in composition and quantity, stimulating a wettability change. Some more details are addressed in Chapter 9.6.

EXERCISE 11.2

Figure 11.16 a and b represent two models of how water displaces oil in a reservoir. The x-axis represents distance (from, for example, an injection well or inlet of a core plug). The y-axis represents water saturation. The figures can be considered as a cross-section of the reservoir.

- Figure 11.16a shows a displacement model in which the water moves like a piston and pushes the oil in front of it (piston displacement). Only immobile oil (S_{or}) remains in the reservoir behind the waterfront. In front only oil flows.
- Figure 11.16b shows a model where we have a waterfront where the water saturation in the reservoir suddenly increases with a leap and then there is a gradual increase in the water saturation behind the front (Buckley-Leverett displacement). In fact, both oil and water flow behind the front.
- Figure 11.17 shows the relative permeability curves for the reservoir we consider. Note that for piston-like displacement we only consider the end points. In this case, suppose that water is in the wetting phase.

1. Define what we mean by an imbibition process and a drainage process. What kind of displacement process do we have in this case?
2. We assume that the model in Figure 11.16a can be used.
   What is the relative permeability of the water at $d = 40$ m?
   What is the relative permeability of the oil at $d = 80$ m?
3. We assume that the model in Figure 11.16b gives a better picture of the displacement mechanism. What is the relative permeability of the oil and water at $d = 40$ m in this case?

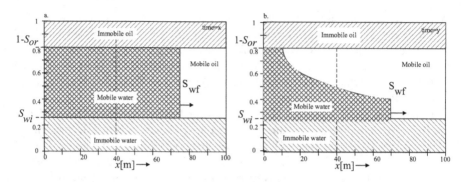

FIGURE 11.16   a. Piston displacement, b. Buckley-Leverett displacement.

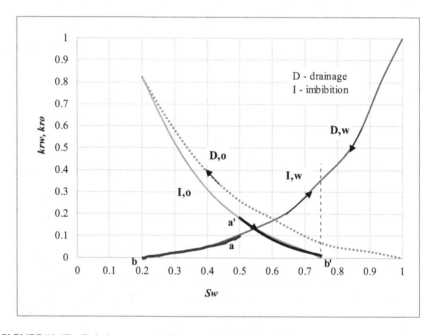

FIGURE 11.17   Relative permeability curve for Exercise 11.3.

SOLUTION

1. Since water is the wetting phase, the imbibition curve represents conditions where water saturation is increasing, and drainage curve represents conditions where oil saturation is increasing.

2. Use the model in Figure 11.16a (piston displacement) and the end points of the relative permeability curve of Figure 11.17. Behind the front, the water saturation is $S_w = 1 - S_{or} = 0.75$. Read the relative permeability for water from Figure 11.17. At $S_w = 0.75$ we have $k_{rw} = 0.35$. The relative permeability

of oil at $d = 80$ m (in front of the waterfront): The water saturation is $S_{wi} = 0.20$ and relative permeability for oil at $S_{wi} = 0.20$ is $k_{ro} = 0.82$

3. Using the model in Figure 11.16b (Buckley-Leverett displacement) we find that $S_w = 0.50$ when $d = 40$ m. Since water displaces oil and water is the wetting phase, we have a system where the wetting phase increases in saturation and the imbibition curve must be used. From the curves we read that $S_w = 0.50$ is giving $k_{rw} = 0.11$ and $k_r = 0.18$.

The conditions behind the front are represented by the line segment {a b} on the relative permeability curve for water and {a' b'} on the relative permeability curve for oil, shown in Figure 11.17. Points a and a' show relative permeability at the front, at $S_w = 0.50$.

EXERCISE 11.3

We are studying the displacement of oil with water in a laboratory model. The model consists of a cylindrical porous medium, 1 m long with a diameter of 10 cm. The flow in the model takes place along the axis of the cylinder, which is horizontal. The model has an oil and water saturation of 60% and 40% respectively before starting the injection. It is assumed that the water displaces only half of the oil in the model, so the average oil saturation is therefore 30% behind the injected water and 60% in the front. The water injection pressure is kept constant at 2 atm and we produce at atmospheric pressure of 1 atm. The model's porosity and permeability are 20% and 1 Darcy, respectively. The viscosity of oil and water is 2 and 1 cp,

---

TABLE 11.4

Systematic Representation of Data Placed in the Schematics of the Core

Δp_1	Δp_2
$S_o = 0.3$	$S_o = 0.6$
$k_{ro} = 0.0$	$k_{ro} = 0.8$
$S_w = 0.7$	$S_w = 0.4$
$k_{rw} = 0.6$	$k_{rw} = 0.1$
x	L-x

TABLE 11.5

Oil-Water Relative Permeability Measurements

Water Saturation	Water Relative Permeability	Oil Relative Permeability
40%	0.1	0.8
70%	0.6	0.0

respectively. Table 11.4 summarizes the data just described. Relative permeability for oil and water is given in Table 11.5 for the relevant saturations.

1. What is the water injection rate at the start of the injection and how much oil and water are produced simultaneously on the outlet (production rates)?
2. Calculate the cumulative amount of oil produced and the oil production rate at the time the waterfront passes the center of the model.

SOLUTION

1. Assume capillary pressure and fluid compressibility can be neglected.

Use the Darcy eq. $q_i = A \dfrac{kk_{ri}\Delta p}{\mu \Delta L}$, with $I = o$ or w

$$q_o = \frac{\pi \cdot 10^2}{4} \frac{1 \cdot 0.8}{2} \frac{1}{100} = 0.314 \frac{cm^3}{s}$$

$$q_w = \frac{\pi \cdot 10^2}{4} \frac{1 \cdot 0.1}{1} \frac{1}{100} = 0.079 \frac{cm^3}{s}$$

$$q_T = q_o + q_w = 0.393 \frac{cm^3}{s}$$

2. $N_p = V_p \cdot \Delta S_o = \dfrac{\pi \cdot 10^2}{4} \cdot 50 \cdot 0.2 \cdot (0.6 - 0.3) = 235.6 \ cm^3$

$$q_T = A \frac{kk_{ro1}}{\mu_o} \frac{\Delta p_1}{\dfrac{L}{2}} + A \frac{kk_{rw1}}{\mu_w} \frac{\Delta p_1}{\dfrac{L}{2}} = A \frac{kk_{ro2}}{\mu_o} \frac{\Delta p_2}{\dfrac{L}{2}} + A \frac{kk_{rw2}}{\mu_w} \frac{\Delta p_2}{\dfrac{L}{2}}$$

$$\frac{k_{rw1}}{\mu_w} \Delta p_1 = \left(\frac{k_{ro2}}{\mu_o} + \frac{k_{rw2}}{\mu_w} \right) \Delta p_2$$

$$\frac{0.6}{1} \Delta p_1 = \left(\frac{0.8}{2} + \frac{0.1}{1} \right) \Delta p_2$$

$0.6\Delta p_1 = 0.5\Delta p_2$

Also: $\Delta p_1 + \Delta p_2 = 1$

That gives: $\Delta p_1 + 1.2\Delta p_1 = 1$

$\Delta p_1 = 0.455 atm$

$\Delta p_2 = 0.545 atm$

$$q_o = \frac{\pi \cdot 10^2}{4} \frac{1 \cdot 0.8}{2} \frac{0.545}{50} = 0.342 \frac{cm^3}{s}$$

Note that the differential pressure over the cylindrical sample remains constant, which means that when a lower viscous fluid is injected, the total flow rate at the outlet increases during the displacement.

11.9 RELATIVE PERMEABILITY MEASUREMENT METHODS

Usually in core laboratories, relative permeability can be measured by three different methods:

1. Unsteady state (USS).
2. Steady state (SS).
3. Centrifuge.

The measurements of absolute permeability and relative permeability for oil and water are some of the most important tasks in core laboratories. Generally, the tested sample plug is saturated initially with a wetting phase using a vacuum pump and the absolute permeability for the wetting phase is measured (see Chapter 7). Then the relative permeability measurements are conducted under two-phase flow, steady or unsteady method. The steady-state method is based on injection of two phases simultaneously, resulting in constant pressure drops and flow rates in the core. The unsteady-state method is performed with injection of a single-phase injection fluid displacing at constant injection flow rate or injection pressure at the second phase in the core, leading to respectively changing pressures or flow rates during the experiment and two-phase flow after the injection fluid reaches the outlet. Based on the data collected in the two measurements, the absolute and relative permeability are calculated. The centrifuge is a variation on to the unsteady-state test using centrifugal forces to displace the mobile fluid in the core. Test equipment for USS and SS tests can vary from a bench-top at ambient conditions using a synthetic oil and brine till advanced tests for simulating high pressure, high temperature reservoir condition using reservoir fluids.

11.9.1 CRITICAL EXPERIMENTAL CONSIDERATIONS

Decide on:
- Confining pressure, to prevent the flow of injected fluid through the annular space between the core plug and the sleeve during experiment. It can serve as additional pressure to supply a compressing force on the core.
- Pressure drop or flow rate controlled.
- Experimental conditions.
  - Temperature.
  - Absolute pressure (back pressure).
- Oil to be used.
  - Dead, live, composite, or synthetic oil.
  - Viscosity match of the oil with the reservoir oil.
- Brine composition.
- Core sample.

- Length; composite.
- Core wettability, is aging needed?
- Setting initial water saturation.
- As preparation data is needed from other experiments, Table 11.6 shows an overview and details where additional information can be found.

11.9.2 UNSTEADY-STATE METHOD

The unsteady-state method (USS) is based on the theory of Buckley and Leverett (Buckley and Leverett 1942), which explains mechanism of fluid displacement in porous media (see Section 11.3). Figure 11.18 shows a standard set-up for relative permeability measurements used in the laboratory at room temperature. An injection pump is supplying two vessels of injection fluids, mainly oil and water, while the fluid for confining pressure is supplied by a separate pump. Differential pressure

---

TABLE 11.6

The Data Needed to Analyze and Prepare for the Relative Permeability Experiment

Core bulk volume	Chapter 5
Pore volume	Chapter 5
Absolute permeability	Chapter 7
Liquid density	Chapter 4
Liquid dynamic viscosity	Chapter 4

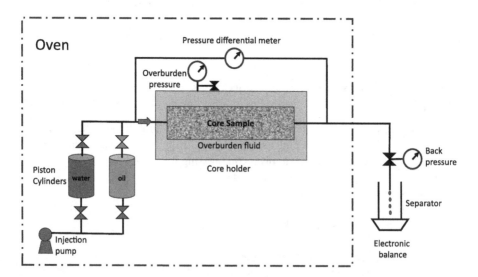

FIGURE 11.18 Schematics of relative permeability apparatus for USS. (Note for simplification no bypass line is drawn).

is measured over the core. A back pressure regulator is installed in the outlet to control the outlet pressure and enable experiments at higher pressure than atmospheric pressure.

The procedure for performing an unsteady-state test is relatively simple and fast. In the following, a water-oil test is described. The procedure is similar for gas-oil or water-gas systems. The confining fluid can be a gas or liquid. Here Isopar-L is used as an example.

The concept of unsteady-state relative permeability measurement is illustrated in Figure 11.19. First the core is saturated with 100% water and then the sample is desaturated by injecting oil until no more production of water is obtained. Water

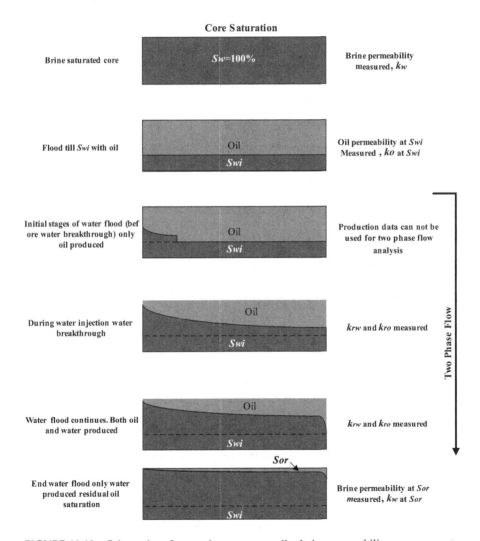

FIGURE 11.19 Schematics of unsteady-state water-oil relative permeability measurement.

production is recorded and S_{wi} calculated. The effective oil permeability is then measured at S_{wi}. Afterwards, oil is displaced by constant rate of water injection and over
time the pressure drop across the core and the production rates of oil and water will
be recorded. The experiment ends when no more oil is produced. At this point the
effective water permeability can be measured at S_{or}.

11.10   EXPERIMENTAL PROCEDURE

The procedure for performing an unsteady-state test is relatively simple and fast compared to the other two methods. In the following a water-oil test is described, but the
procedure in principle is the same for gas-oil or water-gas. This procedure is only part
of the documentation. As described in Chapter 2 (working in the lab), the procedure is
based on the measures identified in the risk assessment that is to be made; see Table 11.7.

General PPE to be used: Lab coat, safety glasses, safety shoes, gloves (in case cores
surface needs protection).

11.10.1   Pre-flooding Preparation Prior to
Experiment (Water-Oil Imbibition)

The preparations are similar to the preparations for the liquid permeability described
in Chapter 7. A summary of the preparations includes:

- Saturate a plug with brine (see Chapter 5, porosity).
- Sample loading in core holder (see Section 7.3, absolute permeability).

Additional for multiphase flow:

- Apply confining pressure followed by back pressure.
  - The fluid needed for providing confining pressure can be either a gas
    (such as nitrogen) or a liquid (such as Isopar-L).
  - Confining pressure can be provided through a pump for liquids or directly
    from a gas source. It is advisable to maintain the confining pressure at a
    level of approximately 35 bar or 500 psi higher than the pore pressure at
    all times to be able to seal the rubber sleeve around the core sample if the
    equipment pressure limit allows for this. At any time of the experiment,
    the following relation should be true:

Confining pressure > Pore pressure > Back pressure

  - After applying the confining pressure, allow a minimum of 30 minutes
    for pressure stabilization.
  - Once the confining pressure is set, increase the back pressure slowly to
    the preferred level.
- Leakage test of sleeve and core holder (see Section 7.3, absolute
  permeability).

TABLE 11.7

The Risk Assessment to the Core Flooding with Brine and Oil at Room Temperature and Low Pressures; H Stands for Health, M for Material, E for Environment, and R for Reputation; P Represents the Probability of Occurrence

RISK ANALYSIS Multiphase Flow/Relative Permeability

Unit/Institute: Date:

Responsible line manager (name):   Revised: 08.2021 AN

Responsible for activities being risk assessed (name):

Participants in the risk assessment (names):

Description of the activity, process, area, etc.:

The relative permeability is measured performing drainage, imbibition, and EOR flooding at room temperature. An initially salt water filled core is place in a core holder, after which, sequentially, oil, salt, water, and a surfactant solution are flooded through the core. Flow rates are set using a pump and the pressure differences over the core are measured as well as oil and brine effluent volumes per time interval. The inlet pressure is measured with a pressure transducer; the outlet pressure is atmospheric. The pump pumps oil into a closed vessel, bringing a piston in motion, which presses the flooding fluid of the vessel into the flow system. The effluent is collected in set time intervals in different containers. To avoid liquid bypassing the core, a sleeve is pressed around the core by use of gas (which gas) supply from a gas bottle. The sleeve is a closed system.

Activity/process	Unwanted incident	Existing risk-reducing measures	(P)	Consequence (C) H	M	E	R	Risk value (P ×C)	Risk reducing measures – suggestions	Residual risk (Pnew* Cnew)	Changes
Mounting the core in the core holder with vacuum from flowing water	Wrong mounting of the core: Cell leakage	Maximum volume gas release; unlimited air from 7 bar flow line	1	2	1	1	1	5	Update procedure: After sleeve check	3	3 (S = 1)
Mounting the core in the core holder with vacuum from flowing water	Wrong mounting of the core: Cell leakage	Maximum system overpressure = flow lines; black 100 bar or teflon 35 bar, core holder XX bar; pressure transducers max 10 bar. Maximum pressure for the meter is 70 bar.	1	0	4	0	0	4	Check maximum pressure coreholder, maximum pressure pressure gauge		
Sleeve mounting	Damaged sleeve or wrong mounting: Over pressurizing the system by supply of the gas bottle	Use of a valve reducing the maximum gas supply to 10 bar	1	2	1	1	2	6	Mount extra pressure transducer to monitor the pressure. Update procedure: Check for sleeve leakage	6	6
Use of liquid container for fluid supply to the pump	Damage/spill oil	Close bottle and place bottle on a spill tray/ bucket	2	2	1	1	1	10		4	4 (P = 1, S H = 1)

(Continued)

TABLE 11.7 (Continued)

The Risk Assessment to the Core Flooding with Brine and Oil at Room Temperature and Low Pressures; H Stands for Health, M for Material, E for Environment, and R for Reputation; P Represents the Probability of Occurrence

Activity/process	Unwanted incident	Existing risk-reducing measures	Consequence (C)					Risk value (P×C)	Risk reducing measures – suggestions	Residual risk (Pnew* Cnew)	Changes
			(P)	H	M	E	R				
Sucking up pumping fluid	Dirty oil/particles in oil that can plug the flow lines of the pump	Use of filter at entrance of pump inlet flow line. Cover the pump fluid bottle; only hole for flowline	1	1	4	1	0	6			
Setting the sleeve pressure	Damaged sleeve or wrong mounting; Over pressurizing the system by supply of the gas bottle	Use of a valve reducing the maximum gas supply to 10 bar. Procedure: Only set sleeve pressure once core is mounted.	1	3	1	0	1	5			
Filling the inlet tubing	Spill	Accepted brine/oil spill	5	1	0	0	0	5	Consider use of vacuum. With oil: fumehood use; reduce exposure to evaporate, use of spill tray	0	0 (P = 0) 0 (H = 0)
Setting the pressure/flow rate	Not enough pump fluid	Maximum pressure: 400 bar, 10ml/min. What if pump is pumping air? = Setting P0	2	3	2	1	1	14	Procedure update: Check volumes needed, have enough oil supply. Check setting electric power	7	7 (P = 1)
Flow	Too high pressure from pump	Maximum pressure: 400 bar, 10ml/min. Setting in pump: maximal < 35bar and minimal pressure, above zero	0,5	2	4	0	0	3			
Flow	leakage: Spill	Procedure: leakage test before the experiment. No overnight, unattended experiment. Max spill without stopping is content reservoir + pumping fluid volume. 2–3 liters	1	2	2	1	0	5			
Flow	Fines release; flow line blocking – pressure build up	Weakest point = flow lines, max 35 bar. Pump max pressure below that. Experiment only with physical attendance	1	2	4	0	1	6	Consider filter, but that causes pressure drop.		
Volume of reservoir is used up: Oil	Oil reservoir: Pump fluid is similar oil as in reservoir so reservoir is "re"filled with the pump oil	None	2	1	1	0	1	6			

Item	Description	Consequence / Procedure						RPN	Action		Result
Volume of reservoir is used up: brine	Brine reservoir: Pump oil displaces the heavier brine being injected from the top	Oil will flow in flowline; flow lines are transparent, oil is colored, so influx of oil into flowline is visible. Procedure: Stop pump if so. Oil in the flow lines is not detrimental for the system.	2	1	1	0		4	Update procedure; check fluid contents in pump system before starting the experiment; if it occurs shut pump	2	2 ($P=1$)
Volume of reservoir is used up: EOR	EOR fluid; pump oil displaces piston, which presses the EOR fluid in the system. If piston is lowered till the bottom, the pump builds up pressure in the system.	The maximum pump pressure is set in the pump. Max pump pressure = 400 bar. Limit is set to < 35bar.	2	2	1	0		10	Clear procedure, risk assessment pump	5	5 ($P=1$)
Core dismounting	No pressure release from sleeve before opening the coreholder; sudden gas pressure release. Will core shoot out/ gas accumulation at inlet? If not, opening of vacuum inlet.	Volume = max XX cm³; 20 bar = 2 liters. Gas is nitrogen. Suffocation danger is limited if gas flow line from gas bottle is closed.	2	3	1	0	2	12	Use manometer to check pressure is released. Update procedure.	6	6 ($P=1$)
								0			
Set up design	Long loss flow lines Long distance between gas bottle and sleeve Missing clear overview over set-up.							0	Reconsider set-up		
Pump risk assessment		See separate risk assessment						0			
Use of three-way valves	General: valve turned the wrong way: Core inlet, ruine core experiment Core outlet, flow in bypass/blocking outlet Valve gas bottle outlet		2	1	2	0	0	6	Reduce amount of three-way valves	3	3 ($P=1$)
Use of three-way valves	Valve turned the wrong way: Dore inlet; extreme: Ruined core experiment e.g. air in tubes (more common) or flood in bypass		2	1	2	0	0	6	Update procedure, instruct on use of three-way valve	3	0 ($P=0$); ($P=1;3$)
Use of three-way valves	Valve turned the wrong way: Core outlet, flow in bypass/blocking outlet	Pump shut off at high pressure, pressure transducer shows high pressures	2	2	2	0	1	10	Update procedure, instruct on use of three-way valve. Avoid use of three-way valve at outlet	5	0 ($P=0$); ($P=1;5$)
Use of three-way valves	Valve turned the wrong way: Valve gas bottle outlet		5	2	2	0	1	25	Update procedure, instruct on use of three-way valve – remove! Add outlet valve.	5	0 ($P=0$); ($P=1;5$)

11.10.2 MAIN EXPERIMENTAL PROCEDURE

The multiphase flow experiment can start with a single-phase liquid permeability; see Section 7.3 for the details. We start the experiment with a core saturated with brine.

11.10.2.1 Drainage (Oil is the Displacing Fluid)

1. Inject multiple pore volume of brine into the core sample to ensure full saturation and then assess the liquid permeability following instructions outlined in Chapter 7.3.
2. Change the inlet pump fluid to oil.
3. Make sure that the pressure regulator at inlet shows zero pressure and flow rate shall also be zero. Check that the outlet fluids are to be captured in the container, potentially monitoring its volumes.
4. Set the pump to the initial flow rate 1 ml/min, and a flow will start.
   Air bubbles can appear in the outlet flow due to air captured at the outlet end cap.
   The pressure across the core must be large enough to make capillary end effects and gravity effects negligible.
5. Per time step, take a reading of the pressure, water production, and oil production and calculate how much water is left in the core.
   a. Measure water volume change and total volume change, preferably using a burette.
   b. Alternatively use 5 or 10 ml tubes; use a cap or wax paper to close off the tubes when filled.
   c. Or the total production can be measured by monitoring the change in weight using a balance. Based on the density of the two phases the separated production volumes can be derived.
6. Continue the flooding experiment until no more water is coming out of the core ~8 PV.

11.10.2.2 Imbibition (Brine is the Displacing Fluid)

7. Change the injection fluid to brine.
8. Record the cumulative water and oil production at time zero of the imbibition if possible.
9. Set the pump to the initial flow rate, and a flow will start. Time = 0.
10. Note the time when the liquid enters the core holder, as control value for dead volume.
11. Per time step (the smaller the better but consider the measurement accuracy), take a reading of the pressure, water production, and oil production, and record the time of initial water production.
12. Record the brine breakthrough time at the core holder outlet (in the flow line) and the time when it flows into the glass tube. After breakthrough,

both water and oil phases are produced, which is the starting point for two-phase flow and relative permeability calculation (imbibition).

13. Perform the flooding experiment until no more oil is coming out of the core. ~8 PV.
14. Increase the flow rate (bump rate).
15. Take a reading again per time step of the water production and oil production.
16. Perform the flooding experiment until no more oil is coming out of the core. ~4 PV.
17. Stop the pump.

11.10.3 POST FLOODING PROCEDURE AT THE END OF THE EXPERIMENT

The following steps should be followed once the experiment is completed:

18. Stop injection pressure.
19. Gradually reduce the back pressure until it is fully released.
20. Gradually lower the inlet pressure by opening the valve.
21. Slowly reduce the confining pressure until it is entirely released.
22. Take out the sample from core-holder.
23. Weigh the core and calculate the residual water saturation, $S_{wr(end)}$ and compare it with the $S_{wr(prd)}$, calculated from production data.
24. Clean all lines using hexane or ethanol.

11.10.4 ADVANTAGES AND DISADVANTAGES

Advantages of performing the unsteady-state method are:

1. Substantially quicker than the steady-state method.
2. The process better resembles the mechanisms taking place in the reservoir and gives better end-point data.
3. Simpler experimentally and better adaptable to reservoir condition applications.
4. Smaller amounts of fluids required.

The main disadvantages are:

1. Relative permeability data will not be available over the entire saturation range depending on brine efficiency till breakthrough and might in the extreme case be restricted to end-point data only.
2. Discontinuities in capillary pressure at the core ends may lead to distortion of the pressure data and recovery measured, the so-called capillary end effect, further discussed in section 11.8.
3. Substantially more advanced data analysis is necessary.

11.11 DATA ANALYSIS

With the recording of cumulative water injection, pressure drop, and produced oil volume, it is possible to calculate relative permeability by a theory developed by Welge (Welge 1952), which is discussed in Section 11.3.2. Note that the theory was explained knowing the relative permeability curves and deriving production profiles from that. Now the problem is inverse. Having obtained the production data and deriving the relative permeability data. here an analytical solution will be discussed, neglecting capillary end effects and use of the pressure data. Numerically, simulations can be applied using the pressure and production data to derive capillary pressure and relative permeability curves simultaneously.

11.11.1 DATA PREPARATION

Before the data is used for either simulation or analytical analysis, we need to correct for the fact that our pressure data shall represent the pressure at the time when the injected fluid enters the core and production data shall represent production (time and volume) from the core outlet. Pressure data shall be corrected for starting time, where production data shall be corrected for time and volumes produced from the volumes other than the pore volume.

11.11.2 DEAD VOLUME DETERMINATION AND CORRECTION

The **dead volume** is the amount of produced liquid that comes from the flow lines and all volumes that are not the core sample. There is dead volume before the inlet of the core as well as between the core outlet and the separator. Recording the production data from the time the pump started, the produced volume needs to be corrected for dead volume production both for time and volume, as for the data analysis only production data from the core is to be considered.

Similarly, pressure data is needed starting at the time the injection fluid enters the core. To obtain the pressure measurement when the injection fluid enters the core inlet, a *time* correction is needed corresponding to $t' = q/V_1$, the time needed for the injection fluid to flow through V_1, the inlet dead volume, with the flow rate q. The new time for the pressure readings is t^*. For the production to match with production of the core outlet at $t^* = 0$, the production data needs a time correction $q/(V_1 + V_2)$ and the production volume needs a correction by reduction of "$V_1 + V_2$." Figure 11.20 shows the different steps for the correction.

The measurement of V_1 and V_2 can be done once the set-up is mounted and can be based on geometrical calculations of the set-up (e.g. flow line volume measuring length of the line times inner area of the line, normally 1/16 inch (1.59 mm) inner diameter line is used in all core flooding apparatus, plus valve connection volume and entrance volume of core holder). The uncertainty of this calculation influences the accuracy of the production volumes. Therefore, it is generally advised to have the

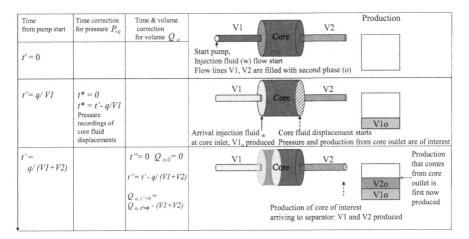

FIGURE 11.20 Illustration of the dead volume correction where phase o is in the flow lines at the start of injection of phase w. The start t^* of the pressure readings needs to be corrected for the production time of V_1, $t^* = t' - q/V_1$. The production volume needs to be corrected for $V_1 + V_2$, where time t'' is then corrected as $t' - q/(V_1 + V_2)$.

core volume several times larger than the dead volume and reduce the dead volume as much as possible.

Exercise 11.4   Dead Volume Correction of Pressure Data

At the imbibition start, injecting water, the flow lines are filled with oil. The dead volume in front of the core V_1 is 0.9 cm³. Correct the pressure data for the presence of dead volume.

Solution

With a flow rate of 1 cm³/minute it takes the waterfront $q/V_1 = 54.07$ seconds to arrive at the core inlet. This gives a time correction for the pressure data of 54 seconds (Table 11.7, Figure 11.21).

For this data set then the $t = 0$ needs to be interpolated. In this case the pressure response appears to have a delay; it does not show zero and remains constant longer than expected. If the pressure drop in the flow lines can be assumed to be negligible, then only when the fluid enters the core will the pressure start to change.

Exercise 11.5   Dead Volume Correction on the Cumulative Volumes of an Imbibition Experiment

At the imbibition start, injecting water, the flow lines are filled with oil; see Figure 11.20. A burette was used to monitor the oil and water production, with the oil lighter than

TABLE 11.7

Pressure Data to n USS Imbibition Experiment Core A

Dead volume in front, V_1	901.24	mm³
q	1.00	ml/min
Time correction: q / V_1	54.07	s
$t_{recorded}$ [s]	$t_{corrected}$ [s]	dP [bar]
0	−54	0.04
80	−27	0.04
150	96	0.04
180	126	0.04
240	186	0.05
300	246	0.08
360	306	0.12
420	366	0.14
480	426	0.18

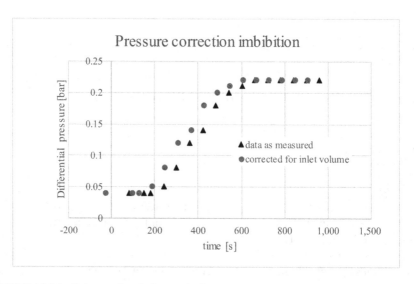

FIGURE 11.21 Pressure data before and after time correction of 54 seconds presented in Table 11.7.

the water. The top level is here called oil-gas contact (GOC), the oil-water contact WOC.

Make a correction to the cumulative production volumes for the presence of dead volume of V_1 and V_2.

Solution

The difference between WOC and GOC is the oil volume; the change of WOC is the production of brine. The flow lines are filled with oil before the start of the experiment, so the oil volume produced needs a correction for the volumes in the flowline, which are $V_1 + V_2$. The corresponding time shift is $q/(V_1 + V_2)$. The volume of only water needs a time shift for the total dead volume produced $q/(V_1 + V_2)$, but no production correction is needed as the lines were filled with oil initially. See Table 11.8 for the data and the corresponding corrected data, only for the cumulative oil and water volumes. Note that the cumulative volume data for oil and water plotted in Figure 11.22 shows the trends as expected, seen in Figure 11.8. Figure 11.23 shows the corrected data of Figure 11.22.

TABLE 11.8
Cumulative Volume Data for a USS Imbibition Experiment (cor = corrected), Core A

V_1				901	mm³			
V_2				884	mm³			
q				1.00	ml/min			
Time correction			$q / (V_1 + V_2)$	107.1	s			
Volume correction oil			$V_1 + V_2$	1.78	ml			
t [s]	t_{cor} [s]	OWC [height in cm³]	OGC [height in cm³]	$V_{oil\ cum}$ [cm³]		$V_{oil\ cum\ cor}$ [cm³]	$V_{w\ cum}$ [cm³]	$V_{tot\ cum\ cor}$ [cm³]
0	−107	49.7 (lowest point)	49.7	0		−1.78	0	−1.8
60	−47	49.7	49.6	0.1		−1.68	0	−1.7
120	13	49.7	48.5	1.2		−0.58	0	−0.6
180	73	49.7	46.5	3.2		1.42	0	1.4
240	133	49.7	45.5	4.2		2.42	0	2.4
300	193	49.7	44.8	4.9		3.12	0	3.1
360	253	49.7	44.1	5.6		3.82	0	3.8
420	313	49.7	43.3	6.4		4.62	0	4.6
480	373	49.7	42.5	7.2		5.42	0	5.4
540	433	48.9	41.6	7.3		5.52	0.8	6.3
600	493	48.9	40.5	8.4		6.62	0.8	7.4
660	553	48.2	39.5	8.7		6.92	1.5	8.4
720	613	47.3	38.5	8.8		7.02	2.4	9.4
780	673	46.2	37.7	8.5		6.72	3.5	10.2
840	733	45.7	36.5	9.2		7.42	4.0	11.4
900	793	44.0	35.4	8.6		6.82	5.7	12.5
960	853	43.0	34.5	8.5		6.72	6.7	13.4

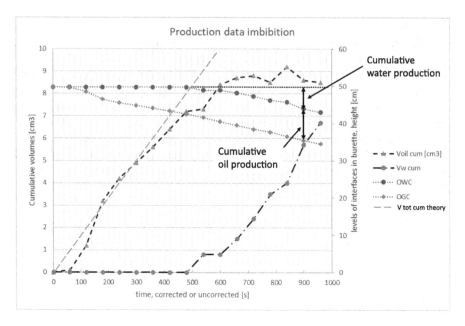

FIGURE 11.22   The recorded production data of an USS imbibition experiment where water was injected in an oil saturated core at S_{wi}.

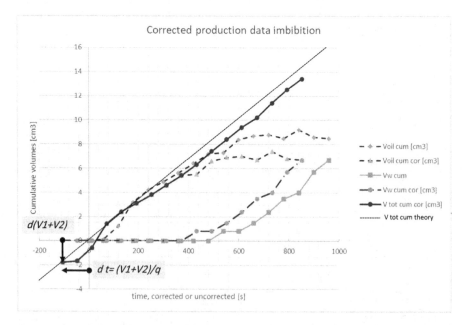

FIGURE 11.23   The cumulative oil volume ($V_{oil\ cum}$) corrected in time and dead volume ($V_{oil\ cum\ cor}$). Cumulative water volumes ($V_{w\ cum}$) are corrected for time only.

Note that the mass balance shows errors in the WOC and GOC readings as the total volume produced is not equal to what was injected, which means that the corrected volumetric data does not go through (0.0). The errors arise from emulsion building at interphase, production along the wall of the burette, and low production changes that are not visible with the grading or the burette.

You can consider the correction as an axis transform for both time and cumulative production volumes having the new origin:

$$t_{corrected} = t_{measured} - q/(V_1 + V_2) \text{ and } V_{cum\ corrected} = V_{cum} - (V_1 + V_2).$$

11.11.3   END-POINT DETERMINATION

The end points of the relative permeability curves can be derived easily applying Darcy's law using the data at the end of the drainage respective to the imbibition experiment where only oil or water is flooding. The following equations are to be used:

$$k_{ro}(S_{wi}) = \frac{Q_o \mu L}{k_{abs} A \Delta P_0} \text{ and } k_{rw}(S_{or}) = \frac{Q_w \mu L}{k_{abs} A \Delta P_w} \quad (11.41)$$

The corresponding end-point saturations, saturations reached at the end of the drainage respective to imbibition experiments, need to be determined based on the mass balance, or if there is a possibility to weight the core, the following correlation can be applied:

$$M_{rf} = M_r + S_{w1} \cdot PV \cdot \rho_w + (1 - S_{w1})PV \cdot \rho_o \quad (11.42)$$

$$S_{w1} = \frac{M_{rf} - M_r - PV \cdot \rho_o}{PV(\rho_w - \rho_o)} \quad (11.43)$$

Where S_{w1} is the end point water saturation, M_{rf} is the mass of rock and fluids inside, M_r is the mass of dry rock, PV is the pore volume, ρ_o is the oil density, and ρ_w is the water density.

Exercise 11.6

After a drainage experiment and aging, S_{wi} was established, and a water flooding experiment was performed. Data available is shown in Table 11.9 and Table 11.10.

1. Which method can be used to determine the S_{wi} in the original core?
2. What are the additional assumptions when the concept of relative permeability factor is introduced in the two-phase flow Darcy's equation?
3. Calculate the end points of the relative permeability curve and make a sketch.

Solution

1. Dean Stark cleaning.
2. Immiscible flow, two continuous phases.
3. The end-point saturations for the drainage and imbibition of core 5 are shown in Table 11.11. In Table 11.12 the calculated relative permeability values are presented for the corresponding end point saturations given in Table 11.11. Table 11.13 shows the summarizing end point data for the imbibition curve for Core 5.

11.11.4 MULTIPHASE PRODUCTION

The following data analysis is the application of the so-called JBN method, with JBN standing for the grounders of the work, Johnson, Bossler, and Naumann (Johnson et al. 1959). It can be considered the opposite of the Buckley-Leverett theory, now having production data to derive the relative permeability data.

TABLE 11.9
End-Point Data from the Drainage and Imbibition Experiment

Core 5.	q [ml/min]	ΔP [bar]
End imbibition experiment (only water flow)	4	0.416
	2	0.208
End drainage experiment (only oil flow)	4	0.184
	1	0.046

TABLE 11.10
Core Data for Experimental Data Given in Table 11.9

Core 5.	Standard Core Data	
Core length	4.57	cm
Core diameter	3.82	
Total core weight volume	52.26	cm^3
Dry weight	106.24	g
Wet weight	117.76	g
1D = 9.869$10^{-13}$ m^2		
exsol d 60 viscosity	1.0335	cp
exsol d 60 density	0.785	g/cm^3
Brine viscosity	1.16	cp
Brine density	1.019	g/cm^3
Dead volume before the core	4	cm^3
Dead volume after the core	2.0	cm^3
Drainage result; total volume of water produced	15.23	cm^3
Imbibition result; total volume of oil produced	13.153	cm^3

TABLE 11.11
End-Point Saturation Calculation Using Data from Table 11.9 and Table 11.10

Drainage core 5	dV_w total displaced, cm^3		15.228
	Oil volume – dead volme, cm^3		9.228
	S_o,[-]	9.228/11.30 =	0.817
	S_w remaining in core, [-]	$1 - S_o = S_{wi}$	0.183
	V_w remaining in core, cm^3	11.30 – 9.228 =	2.074
Imbibition core 5			
	V_o produced total, cm^3		13.153
	Dead volume, cm^3		6.000
	dV_o total displaced, cm^3	13.153 – 6 =	7.153
	V_o remaining in core, cm^3	9.228 – 7.153 =	2.075
	V_w total, cm^3	2.074 + 7.153 =	9.226
	S_w final, [-]	9.226 / 11.30 =	0.816
	$S_{or,}$ [-]	1 – 0.816 =	0.184

TABLE 11.12
End-Point Calculation of Relative Permeabilities, Corresponding to the Saturations Found in Table 11.11

Core 5	q [ml/min]	dp [bar]	q [m^3/s]	$dP \cdot A / (\mu dL)$ [m/s]	k [m^2]	k_{eff} [-]	$k_r =$ k_{eff} / k_{abs}	$k_r =$ k_{eff} / k_{effo}
End imbibition	4	0.340	$6.67 \cdot 10^{-8}$	734,056.97	$9.08 \cdot 10^{-14}$	$9.20 \cdot 10^{+1}$	0.45	0.49
	2	0.170	$3.33 \cdot 10^{-8}$	367,028.48	$9.08 \cdot 10^{-14}$			
End drainage	4	0.150	$6.67 \cdot 10^{-8}$	362,995.20	$1.84 \cdot 10^{-13}$	$1.86 \cdot 10^{+2}$	0.91	1.00
	1	0.037	$1.67 \cdot 10^{-8}$	90,748.80	$1.84 \cdot 10^{-13}$			

TABLE 11.13
The End-Point Relative Permeability Saturations from Table 11.11 and Relative Permeabilities from Table 11.12 Combined

Core 5		k_{ro}	k_{rw}
S_{wi}	0.183	0.91	0
S_{or}	0.816	0	0.45

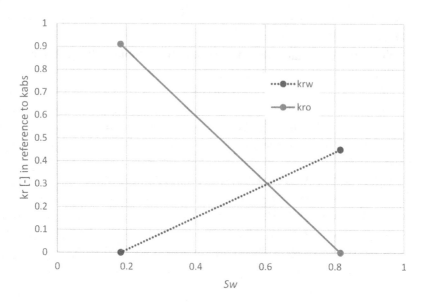

FIGURE 11.24   The relative permeability curve for the data of Exercise 11.6.

The saturations obtained are average saturations over the core, where for the derivation of fractional flow, which we monitor at the core outlet, saturations need to be converted to the saturations at the core outlet. Welge's extension of the Buckley-Leverett method states that:

$$S_{o,2} = S_{o,avg} - W_{iPV}(f_w)_2.$$

$$S_{w,2} = 1 - S_{o,2} \qquad (11.45)$$

where subscript 2 denotes the outlet end of the core.

With W_{iPV} as cumulative injection in pore volumes $= \dfrac{Q*t}{ALx\phi}$ and $(f_w)_2$ as the water fraction flowing at outlet face of the core. $S_{o,avg}$ is the average oil saturation in core and $S_{w,2}$ and $S_{o,2}$ are respectively the water and oil saturation at outlet face.

$$S_{o,avg} = \frac{(1 - S_{wi})*V_p - N_p}{V_p} \qquad (11.46)$$

where N_p is cumulative oil produced during imbibition and V_p is pore volume.

Since Q_o and $S_{o,\,avg}$ can be measured experimentally, f_{w2} can be determined from the plot of W_{iPV} as a function of $S_{o,avg}$:

$$(f_w)_2 = \frac{dS_{o,avg}}{dW_{iPV}} \qquad (11.47)$$

and the oil saturation at the outlet $(S_{o,2})$ may be calculated using Equation 11.44.

Since viscosities are known, the relative permeability ratio k_{ro}/k_{rw} can be determined from Eq. 11.20, with $f_w = f_{w,2}$:

$$\frac{k_{ro}}{k_{rw}} = \frac{\mu_o}{\mu_w}\left(\frac{1}{f_{w,2}} - 1\right) \text{ and } k_{ro} = k_{rw} * \frac{\mu_o}{\mu_w}\left(\frac{1}{f_{w,2}} - 1\right) \quad (11.48)$$

The relative permeability for water can be determined using the well-known Johnson–Bossler–Naumann (JBN) method (Johnson et al. 1959) using the following equations:

$$k_{rw} = (f_w)_2 \frac{d\left(\dfrac{1}{W_{iPV}}\right)}{d\left(\dfrac{1}{W_{iPV}I_r}\right)} \quad (11.49)$$

$$k_{rw(i)} = (f_w)_{2(i)} \frac{\left(\dfrac{1}{W_{iPV(i)}}\right) - \left(\dfrac{1}{W_{iPV(i-1)}}\right)}{\left(\dfrac{1}{W_{iPV(i)}I_{r(i)}}\right) - \left(\dfrac{1}{W_{iPV(i-1)}I_{r(i-1)}}\right)} \quad (11.50)$$

With W_{iPV} as cumulative injection in pore volumes and I_r as the relative injectivity defined as:

$$I_r = Relative\,injectivity = \frac{I_{2-phase}}{I_{abs}} = \frac{u/\Delta P}{u/\Delta P_i} \quad (11.51)$$

I_{abs} is the absolute (water) injectivity defined as:

$$I_{abs} = \frac{V_w}{\Delta t \cdot A \cdot \Delta P_i} = \frac{u}{\Delta P_i} \quad (11.52)$$

With A = cross-section area of the core, u = velocity, and ΔP = pressure difference. Δt is the time step. If the experiment is performed under constant flow rate u then

$$I_r = Relative\ injectivity = \frac{I_{2-phase}}{I_{abs}} = \frac{\Delta P_i}{\Delta P} \quad (11.53)$$

If the experiment is performed under constant pressure,

$$I_r = \frac{\Delta u_t}{\Delta u_i} = \frac{\Delta V_t}{\Delta V_i} \quad (11.54)$$

with ΔV as the total volume of water plus oil produced in each time step and ΔV_w as the volume of water produced in each time step Δt. By knowing k_{rw}, k_{ro} can be found by:

$$k_{ro}k_{rw} = \frac{\mu_o}{\mu_w}\left(\frac{1}{f_{w,2}}-1\right). \text{ such that } k_{ro} = k_{rw}\cdot\frac{\mu_o}{\mu_w}\left(\frac{1}{f_{w,2}}-1\right)$$

11.12   DATA QUALITY CONTROL

Data quality control is a crucial aspect of data interpretation. Figure 11.25 illustrates the typical trend of cumulative oil production and differential pressure. Generally, the laboratory data should be analyzed with respect to general trend, monotonicity, and smoothing.

Check for a linear trend till water breakthrough for oil production. Depending on the mobility of the injection fluid, the pressure will increase simultaneously (lower mobility, higher pressure) or opposite. During this phase, the volume of water injected is equal to the volume of oil produced ($PV_{inj} = N_p$). However, at a certain point, deviation from this linearity occurs, defining the point of breakthrough. Subsequently, a gradual decline in oil production rate becomes noticeable until eventually no oil is produced anymore. The differential pressure shows a rising trend, reaching its maximum at the breakthrough point and subsequently decreasing until it stabilizes when no oil is produced anymore. It is imperative to ensure that the maximum differential pressure point aligns with the breakthrough point as a part of quality-control measures.

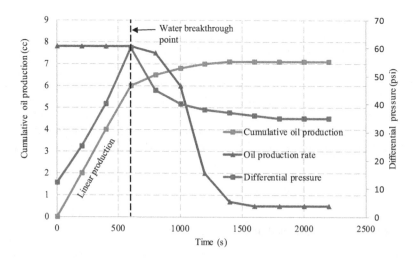

FIGURE 11.25 Typical trend of differential pressure, oil production rate, and cumulative production.

11.12.1 EXPERIMENT 11.1: UNSTEADY-STATE OIL-WATER RELATIVE PERMEABILITY MEASUREMENT

Description: Determination of the relative permeability of oil and brine in a sandstone core at room temperature using an unsteady-state experimental set-up. Note that for this experiment the core and liquid characteristics need to be available (see Table 11.6).

CALCULATIONS AND REPORT

Data needed as preparation for the experiment:

Liquid data: $\rho_w =$ _____, $\mu_w =$ _____, $\rho_o =$ _____, $\mu_o =$ _____
$A =$ ____cm^2, $V_{pore} =$ ____cm^3, $V_b =$ ____cm^3
$L =$ ____cm, $k_{abs} =$ ____m^2, dry core weight
$W_{dry} =$ ____g; saturatured core weight $W_{sat} =$ _____g,
Weight of core after test, $W_{isat} =$ _____g.

Formulate your expectations/hypothesis. Consider how much volume can be expected to be produced from the core and check your results during the experiment. Check whether data are realistic. Similar for pressure data, compare it to the water pressure measured during the absolute permeability experiment for the flow that is to be used in the USS experiment. At residual oil saturation the pressure drop at equal flow rate shall be higher. Depending on whether the oil viscosity is higher or lower than the brine, the pressure drop over the core during drainage shall respectively increase or decrease.

Table 11.7 gives an example on the recordings of the pressurre data. Table 11.8 gives an example on the recordings of the production data.

Based on the data given in Table 11.7 and Table 11.8 and data discussed in Exercise 11.4, 11.5, and 11.6 and additionally the experimental data of Core A presented in Table 11.14, an example is given for the end-point calculation and the history match using numerical simulation.

TABLE 11.14
Core Data Belonging to Data Presented in Table 11.7 and Table 11.8

	Core A			Core A
Length [cm]	5.7	at end of drainage	q [cm^3/min]	1
Diameter [cm]	3.7		dP [bar]	0.08
Dry core weight [g]	124.75		dVw total displaced [cm^3]	11.23
Wet core weight [g]	138.90		end imbibition core weight [g]	137.98
Porosity [-]	0.2265		Q [cm^3/min]	1
k [mD]	246		ΔP [bar]	0.18
Exxsol D60 density [g/cm^3]	0.785		dV_o total displaced [cm^3]	7.15
Exxsol D60 viscosity [cp]	1.0335			
Brine density [g/cm^3]	1.019			
Brine viscosity [cp]	0.96			

TABLE 11.15

End-Point Calculation Data for Core A, Table 11.14

	q [cm³/ min]	q [m³/s]	dp [bar]	$dP \cdot A$ / $(\mu \, dL)$ [1/m²]	k_{eff} [D]	k_{wr} [-]	k_{ro} [-]
End drainage (oil flow)	1	$1.67 \cdot 10^{-8}$	0.08	146,015.17	0.114		0.495
End imbibition (brine flow)	1	$1.67 \cdot 10^{-8}$	0.18	1,807,736.26	0.0369	0.148	
						0.32	1

Based on the Darcy equation the k_{rw} and k_{ro} can be determined, knowing the equilibrium pressures at respectively the end of drainage (at S_{wi}) and imbibition (at S_{or}), as shown in Table 11.15.

The corresponding mass balance starts at the end of drainage. The PV is $\pi r^2 \cdot$ length $\cdot$ porosity $= \pi \cdot (1.85)^2 \cdot 5.7 \cdot 0.22 = 13.87$ cm³.

11.22 cm³ of water was produced, displaced by the oil. The oil saturation is therefore $S_o = 11.22/13.87 = 0.81$ and $S_{wi} = 1 - 0.81 = 0.19$. Then $13.87 - 11.22 = 2.63$ cm³ of water remains in the core.

At the end of imbibition, 6.7 cm³ was produced (corrected for dead volume). An additional 0.4 cm³ was produced after increasing the flow rate (bump rates), so 7.2 cm³ oil was produced in total.

From the drainage experiment it is known that 11.22 cm³ oil was initially in the core, so 11.22-7.2 – 4.02 cm³ oil remains. With a PV of 13.87 cm³, the amount of water is $13.87 - 4.02 = 9.85$ cm³, giving $S_w = 0.71$.

From the weight of the core $S_{w1} = \dfrac{137.98 - 124.75 - 13.87 \cdot 0.785}{13.87(1.019 - 0.785)} = 0.72$. This is within the accepted error range. The corresponding relative permeabilities based on the end point values given in Table 11.16 are shown in Figure 11.26.

With the use of 1D numerical simulation and history matching the production and pressure data, the following relative permeability was derived. The data is not matched perfectly as the comparison of the data points and simulated results show, especially the pressure data (Figure 11.27).

11.12.2 STEADY-STATE METHOD

Leverett (Leverett 1939) was the first who suggested steady-state measurement of relative permeability in a water-oil system. In the steady-state method, two fluids are injected simultaneously at a fixed volumetric ratio, commencing at a low water/ oil ratio, until the produced ratio is equal to the injected ratio and stabilization of the pressure drop across the core is achieved. Core saturations must be measured at each

TABLE 11.16

Input Data

S_{wi}	0.46	c_o	2.81
$k_{ro}(S_{wi})$	0.5	c_w	1
S_{or}	0.19		
$k_{rw}(S_{or})$	0.20		

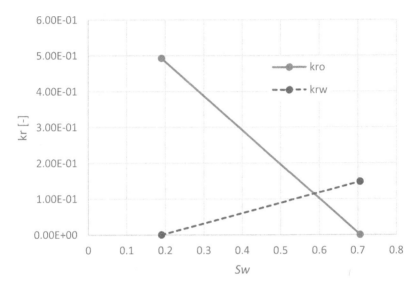

FIGURE 11.26 The relative permeability curve from the USS imbibition experiment presented previously (note the end points are not normalized to $k_{ro}(S_{wi})$).

equilibrium by performing a mass-balance calculation for each phase. This can be done based on the production or, if the set-up allows, the weight change of the core can be measured. This is repeated through changing the ratio of the injection phases by increasing the water fraction and eventually terminating the oil injection at 100% water flow (see Figure 11.28).

The equipment for conducting a steady-state test is like the unsteady-state apparatus and is given in Figure 11.30 with a modification on the inlet side of the endcap of the core holder to provide injection of two fluids simultaneously. The endcap has a pattern so that the two fluids can be mixed, also called a mixer head. Additionally, the pressure difference measured shall be from both phases separately. This can be obtained by using phase-selective membranes in front of the differential pressure transducer, such that the fluid in contact with the transducer is selective. It is assumed here that the two phases are continuous. This

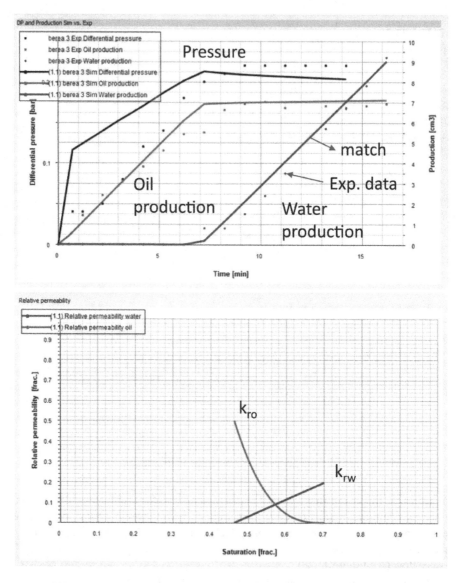

FIGURE 11.27 History matching. The top image shows the match of the simulated data (lines) with the data points of oil, water production, and the pressure difference. The bottom image shows the corresponding relative permeability.

assumption can be violated during imbibition, close to the end point where the saturation of the displaced, non-wetting phase loses continuity. From the mass balance of injected and produced fluids the changes in saturation can be derived, but based on the large quantities this is considered not to be very accurate.

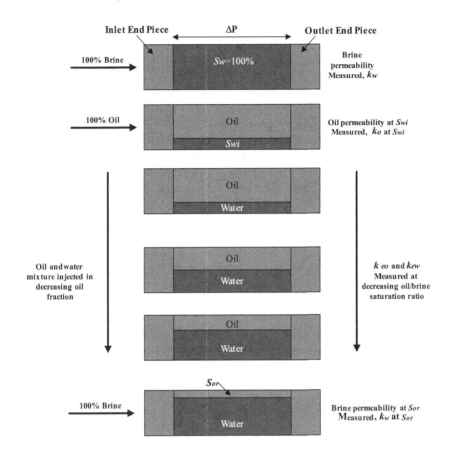

FIGURE 11.28   Schematic of steady-state water-oil relative permeability measurement.

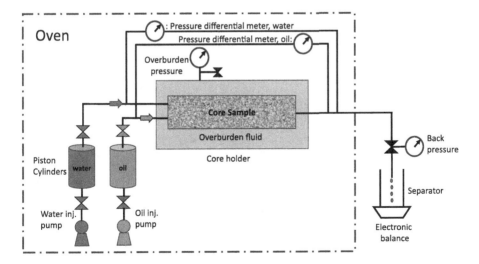

FIGURE 11.29   General schematic of relative permeability apparatus for SS experiment.

Therefore, additionally the core set-up is standardly weighted to obtain the saturation at the end of each flow rate ratio. Alternatives are resistivity measurements or gamma ray scanning, illustrated in Figure 11.30. The set-up shall be adapted accordingly.

The experimental procedure can be summarized by the following points:

1. Follow steps 1 to 15 in the unsteady-state method given in Section 11.7.1.
2. In the next step, two fluids, water and oil, at a fixed volumetric ratio (e.g., 95% oil and 5% water), are simultaneously injected into the core. The volume of produced fluids and pressure drop should be recorded. The injection should be continued until the volumetric ratio at the production face becomes the same as the volumetric ratio at the injection face. Then the system is in a steady-state condition. Record the oil and water pressure drop across the core and produced volumes of oil and water.
3. Increase the volumetric flow rate ratio so that the water phase ratio increases (e.g., 85% oil and 15% water). Like the previous step, the process should continue until reaching the same phase ratio at the outlet face of the core. Record pressure drop across the core and produced volumes of oil and water.
4. Continue the process in several steps by increasing the ratio of water to oil, and in the final step, only water is injected until residual oil saturation is reached.
5. Weigh the core to extract an independent estimation of the final saturations in the core.

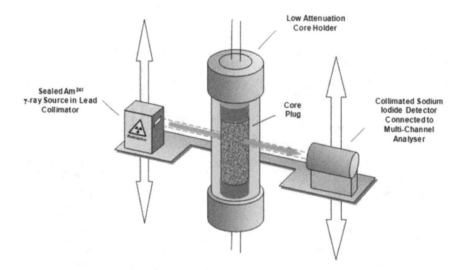

FIGURE 11.30 An example of the use of gamma ray scanning on a coreholder (courtesy of Stratum).

11.13 CALCULATION

The phase saturation and individual effective permeability of each phase can be calculated from the following procedure:

1. Using material balance, one can write:

$$M_{rf} = M_r + S_{w1}.PV.\rho_w + (1 - S_{w1})PV.\rho_o$$

$$S_{w1} = \frac{M_{rf} - M_r - PV.\rho_o}{PV(\rho_w - \rho_o)}$$

Where S_{w1} is the new water saturation at first injection rate, M_{rf} is the measured mass of rock and fluids inside, M_r is the mass of dry rock, PV is the pore volume, r_o is the oil density, and r_w is the water density.

Effective water and oil permeability at S_{w1} can be obtained from:

$$k_{eo}(S_{w1}) = \frac{q_o \mu_o L}{A \Delta P_o} \tag{11.55}$$

$$k_{ew}(S_{w1}) = \frac{q_w \mu_w L}{A \Delta P_w} \tag{11.56}$$

End point relative permeabilities at S_{wi} : $k_{ro} = 1$ and $k_{rw} = 0$

1. For saturation point (S_{w1}):

$$k_{ro} = \frac{k_{eo} @ S_{w1}}{k_{eo} @ S_{wi}} \tag{11.57}$$

$$k_{rw} = \frac{k_{ew} @ S_{w1}}{k_{eo} @ S_{wi}} \tag{11.58}$$

2. The calculation in step 1. may be repeated for other saturation points (S_{w2}, S_{w3}, S_{wn}).
3. The last step will be the end-point relative permeability at S_{or}; $k_{ro} = 0$

$$k_{rw} = \frac{k_{ew} @ S_{or}}{k_{eo} @ S_{wi}} \tag{11.59}$$

Calculated values can be plotted as water saturation vs. water and oil relative permeabilities.

The serious experimental problem with the steady-state method is that saturations in the core must be measured or calculated from material balance. This can be inaccurate due to the large amount of pore volumes flooded through the core. There are several other methods to measure saturations in the core such as electrical resistance and capacitance methods; physical and chemical methods; core imaging techniques like nuclear magnetic imaging (NMRI) and gamma ray imaging (GR); or electromagnetic radiation absorption methods, weighing, etc. These methods can replace the material balance or can be used to support the direct volumetric measurements. Another problem is capillary end effect, which also occur in the USS experiment further discussed in Section 11.8. Advantages of the steady-state method are that it is conceptually straightforward and gives relative permeability data for the whole saturation range.

11.13.1   EXPERIMENT 11.2: STEADY-STATE RELATIVE PERMEABILITY

Description: Determination of the relative permeability of oil and brine in a sandstone core at room temperature using the steady-state method. Note that for this experiment the core and liquid characteristics need to be available (see Table 11.6).

CALCULATIONS AND REPORT

Liquid data: $\rho_w = $ —————, $\mu_w = $ —————, $\rho_o = $—————, $\mu_o = $ —————
$A = $____cm^2, $V_{pore} = $____cm^3, $V_b = $ ____cm^3
$L = $____cm, k_{abs}
$k_{ai} = $ ____ m^2, dry core weight $W_{dry} = $ ____ g; saturated core weight $W_{sat} = $ ____ g
Weight of core after test, $W_{isat} = $ ————— g.

See Exercise 11.7 for an example on the data analysis.

EXERCISE 11.7

Calculate and plot relative permeabilities for oil and water based on the laboratory data from a relative permeability test at a constant pressure drop over the core given in Table 11.17.

TABLE 11.17
Recorded Data, Exercise 11.7

Data from the Steady-State Permeability Test			
Saturation [%]		Flow Rate [cm^3/s]	
Water	Oil	Water	Oil
100	0	0.55	0.00
90	10	0.33	0.00
80	20	0.17	0.01
60	40	0.04	0.08
40	60	0.01	0.23
30	70	0.00	0.35

Core data:

Length = 3.65 cm
Diameter = 2.40 cm
Water viscosity = 0.001 Pa.s
Oil viscosity = 0.0015 Pa.s
Outlet pressure = 100,000 Pa
Pressure drop across the core = 100,000 Pa

What is the absolute permeability of the core sample?

In the interval of the reservoir from which the core sample is taken, the water saturation is 65%. To be profitable to produce from this interval, the water-oil ratio produced under reservoir conditions must be less than 0.9. Should this interval be produced?

SOLUTION

Effective permeabilities are found by calculating water and oil flow separately as:

$$k_w = q_w \mu_w \cdot \frac{\Delta L}{A \Delta p} \quad \text{and} \quad k_o = q_o \mu_o \cdot \frac{\Delta L}{A \Delta p}$$

Relative permeability is then found by dividing by absolute permeability.

We find absolute permeability when the core is 100% saturated with a fluid. By inserting data from the test in previous equations we find relative permeability as a function of water saturation. The calculations are given in Table 11.18.

Absolute permeability is calculated by using equations above with data for $S_w = 100\%$.

$$k = 0.55 \cdot 1 \cdot \frac{3.65}{\pi \left(\frac{2.40}{2} \right)^2 \cdot 1} = 0.444 D$$

Relative permeability as a function of water saturation is plotted in Figure 11.31. To determine whether the interval should be produced, Darcy's law is used for both the water phase and the oil phase. Remember that Darcy's law applies under reservoir condition:

$$\frac{q_w}{q_o} = \frac{k_w}{\mu_w} \cdot \frac{\mu_o}{k_o}$$

TABLE 11.18

Calculated Relative Permeabilities

S_w	k_w (md)	k_o (md)	k_{rw}	k_{ro}
100	444	0	1.00	0.00
90	266	0	0.60	0.00
80	137	12	0.31	0.03
60	32	97	0.07	0.22
40	8	278	0.02	0.63
30	0	423	0.00	0.96

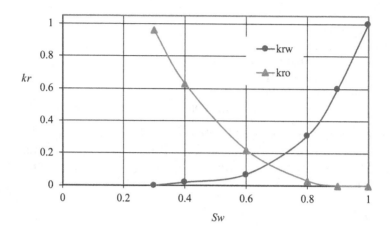

FIGURE 11.31 Relative permeability as a function of water saturation.

From the plot in Figure 11.31 we can read $S_w = 0.65$:

$$k_{rw} = 0.11 \rightarrow k_w = 49 \text{ mD}$$
$$k_{ro} = 0.16 \rightarrow k_o = 71 \text{ mD}$$

When this is inserted in the previous equation it gives $\dfrac{q_w}{q_o} = 1.0$
So, the interval should not be produced.

11.13.2 CENTRIFUGE TECHNIQUE

The centrifuge method is based on enabling a saturated sample to drain by a second phase through rotating the complete system at a constant speed and measuring liquid production vs. time. The centrifuge method offers an advantage over steady- and unsteady-state methods, by substituting viscous forces with centrifugal forces and being much faster. Consequently, results from centrifuge tests are less susceptible to the influence of capillary end effects. Moreover, this method allows the measurement of very low values of the only displaced phase, although it doesn't provide data on the displacing phase, because only the displaced phase production can be measured in this method. Therefore, to determine oil relative permeability, one should conduct an experiment where water or gas displaces oil. Conversely, to measure drainage water relative permeability, an experiment involving oil or gas displacing water should be performed (in a different saturation history).

Hagoort (Hagoort 1980) pioneered the method for determining relative permeability by centrifuge in a gas-oil gravity drainage process. He reported several measured recovery vs. time curves for different core samples using a centrifuge and proposed to use this data for calculating oil relative permeability by applying a gravity drainage process.

Determining relative permeability from a centrifuge is based on the analysis of fluid flow, where the Bond number, commonly referenced as N_b or B_o, signifies the relationship between gravitational forces and surface or interfacial forces. The Bond number equation (Hagoort 1980) is given by:

$$N_b = \frac{k \cdot \Delta \rho_{og} \cdot r \cdot \omega^2}{\sigma} \tag{11.60}$$

Where k is permeability (m²), $\Delta \rho$ is the density difference between the fluid phases (kg/m³), ω is angular velocity ($2\pi \cdot$ revolutions/s), r is the distance from the center of the centrifuge to the center of the plug (m), and σ is the IFT of the fluid phases (N/m). When $N_b \ll 1$, it means that the flow is minimally influenced by gravitational forces, while $N_b \gg 1$ indicates that gravitational forces overcome the interfacial forces.

It is assumed that oil relative permeability is a function of oil saturation as given in the equation:

$$k_{ro} = k_{ro}^o S_o^{*n} \tag{11.61}$$

Where k_{ro}^o is a dimensionless curve-fitting parameter that may not have any significance with respect to the end point of the oil relative permeability curves. To calculate relative permeability, the following parameters are needed: Reduced porosity, normalized oil saturation, and dimensionless time, which can be expressed as:

$$\phi^* = \phi\left(1 - S_{org} - S_{iw}\right) \tag{11.62}$$

$$S_o^* = \frac{S_o - S_{org}}{1 - S_{org} - S_{iw}} \tag{11.63}$$

$$t_D = \frac{\Delta \rho_{og} r \cdot \omega^2 k}{\mu_o \phi^* L} t \tag{11.64}$$

Measurements during the centrifuge displacement give production N_p vs. dimensionless time. t_D. N_p is cumulative oil production after breakthrough (fraction of movable pore volume). Thus, at a particular t_D, N_p and dN_p / dt_D are known. A plot of $\log\left(1 - N_p\right)$ vs. $\log t_D$ results in a straight line as seen in Figure 11.32.

From the slope and intercept with the $t_D = 1$ ordinate, the two parameters k_{ro} and n are obtained. The slope gives n and Equation (11.65) gives k_{ro}^o. See complete derivation in Hagoort (Hagoort 1980).

$$k_{ro}^o = \frac{1}{n}\left[\frac{n-1}{n\left(1 - N_p\right)}\right]^{n-1} \tag{11.65}$$

k_{ro} is obtained from Equation 11.61.

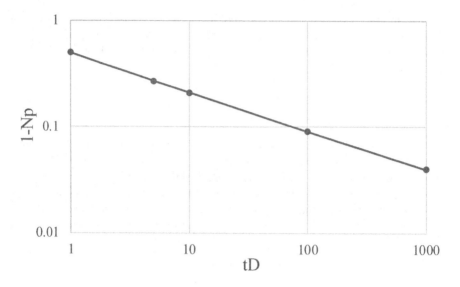

FIGURE 11.32   Cumulative production vs. dimensionless time (data adapted from Hagoort (1980)).

11.14   CAPILLARY END EFFECT

In addition to practical challenges encountered with field-rate equivalent injection in the laboratory, low injection rates are also creating some laboratory-related issues such as the capillary end effect.

In multiphase core flooding the sudden change of capillary pressure over the inlet and outlet of the cores can lead to capillary end effects, an accumulation of the wetting phase at the core in/outlets. Inside the porous media there is a capillary pressure, while outside of the core no capillary pressure is present and the $P_c = 0$. So, depending on the magnitude of capillary pressure, there will be a saturation drop over the core sample. In other words, because of the capillary pressure discontinuity, the wetting fluid is trapped at the outlet end of the sample.

Leverett and Lewis (Leverett and Lewis 1941) highlighted that as the wetting phase exits the core, it experiences resistance due to an abrupt change in capillary forces. This resistance leads to a change in wetting phase saturation and a subsequent reduction in the permeability of the non-wetting phase near the outlet (Gupta and Maloney 2016; Moghaddam and Jamiolahmady 2019). Figure 11.33 illustrates the schematics of capillary end effect in core flooding experiment depending on wettability. Water and gas are injected into the system at a known rate and the bold line in the figure shows the saturation distribution along the core. The value of $S_{w,avg}$ represents the average saturation resulting from the known injection rate. In the vicinity near the outlet end, the saturation distribution undergoes changes, which can be

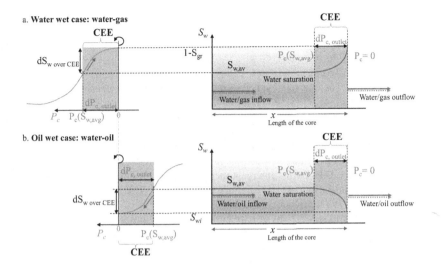

FIGURE 11.33 Illustration of capillary end effect (CEE) for a. a water wet core co-injected with water and gas and b. an oil wet core co-injected with water and oil.

attributed to the jump in capillary pressure, visualized by the accompanying flipped capillary pressure curve. The P_c belonging to $S_{w,avg}$ changes reduces to zero over the outlet. In the absence of this end effect, the bold line would maintain a constant linear trend.

In capillary end effect, capillary forces are dominant over viscous forces. One method to reduce the capillary end effects is therefore to increase flow rates and thereby viscous forces (Masalmeh et al. 2014). The flow rates in the laboratory are therefore generally higher than the rates to be expected in the field.

Other ways to increase viscous forces is increasing the fluid viscosities, but this might change the representability of the fluids for the reservoir conditions.

Smaller capillary forces can be reached by reduced interfacial tensions, but this changes the properties of the fluids. Obviously lower capillary forces depend on the capillary pressure curve of the rock, so low pore entry pressures (larger pore sizes) reduce the capillary end effects as the capillary forces are smaller. Permeability and porosity are directly related. There are several correlations that describe the mentioned measures to minimize the capillary end effects (Rapaport and Leas 1953; Mohanty and Miller 1991).

REFERENCES

Amaeful, J.O., Handy, D.G. 1982. "The effect of interfacial tension on relative oil/water permeabilities of consolidated porous media." *SPE Journal* 371–381.

Anderson, W.G. 1987. "Wettability litterature survey—part 6: The effects of wettability on water flooding." *Journal of Petroleum Technology* 391605–1622.

Armstrong, R.T., McClure, J.E., Berill, M.A., Rücker, M., Schlüter, S., Berg, S. 2017. "Flow regimes during immiscible displacement." *Petrophysics* 58 (1) 10–18.

Brooks, R.H., Corey, A.T. 1966. "Properties of porous media affecting fluid flow." *Journal of the Irrigation and Drainage Division 92* (61–88) 61–88.

Buckley, S.E., Leverett, M.C. 1942. "Mechaisms of fluid displacement in sands." *Transactions of the AIME 146* 107–116.

Coats, K.H. 1980. "An equation of state compositional model." *SPEJ* 363–376.

Corey, A.T. 1954. "The interrelation between gas and oil relative permeabilities." *Producers Monthly 19* 38–41.

Corey, A.T., Rathjens, C.H. 1956. "Effect of stratification on relative permeability." *Journal of Petroleum Technology 8* (12) 69–71.

Craig, F.F. 1971. "The reservoir engineering aspects of waterflooding." *SPE Monograph Series 3.*

Darcy, H. 1856. "Les fontaines publiques de la ville de dijon." *Dalmont.* Paris.

Gupta, R., Maloney, D.R. 2016. "Intercept method-a novel technique to correct steady-state relative permeability data for capillary end effectrs." *SPE Reservoir Evaluation and Engineering 19* (2) 316–330.

Hagoort, J. 1980. "Oil recovery by gravity drainage." *SPE Journal 20* (03) 139–150.

Herring, A.L., Andersson, L., Schlüter, S., Sheppard, A., Wildenschild, D. 2015. "Efficiently engineering pore-scale processes: The role of force dominance and topology during nonwetting phase trapping in porous media." *Advances in Water Resources 79* 91–102.

Hilfer, R. 2006. "Macroscopic capillarity and hysteresis for flow in porous media." *Physical Review E 73* (1).

Johnson, E.F., Bossler, D.P., Naumannn, V.O. 1959. "Calculation of relative permeability from displacement experiments." *Transactions of the AIME 216.*

Karimaie, H., Torsæter, O. October 2008. "Effect of interfacial tension on relative permeability curves in gas-oil gravity draiange." *Society of Core Analysis (SCA).* Abu Dhabi.

Khanamiri, H.H., Berg, C.F., Slotte, P.A., Schlüter, S., Torsæter, O. 2018. "Description of free energy for immiscible two-fluid flow in porous media by integral geometry and thermodynamics." *Water Resources Research 54* (11) 9045–9059.

Khazam, M., Danesh, A., Tehrani, D.H., Todd, A.C. 1980. "Dynamic validation of phase behavior models for reservoir studies of gas injection schemes." *SPE Reservoir Engineering* 391–401.

Leverett, M.C. 1939. "Flow of oil-water mixtures through unconsolidated sands." *Transactions of the AIME 132* (01) 149–171.

Leverett, M.C. and Lewis, W.B. 1941. "Steady flow of gas-oil mixture through unconsolidated sands." *Transactions of the AIME 142.*

Lomeland, F. 2018. "Overview of the family of versatile correlations for flow functions." In *International Symposium of the Society of Core Analysts, SCA2018-56.* Society of Core Analysts.

Masalmeh, S.K., Sorop, T.G., Suijkerbuijk, B.M., Vermolen, E.C., Douma, S., Van Del Linde, H., Pieterse, S. 2014. "Low salinity flooding: Experimental evaluation and numerical interpretation." *International Petroleum Technology Conference (IPTC).*International Petroleum Technology Conference (IPTC).

McClure, J.E., Armstrong, R.T., Berill, M.A., Schlüter, S., Berg, S., Gray, W.G., Miller, C.T. 2018. "Geometric state function for two-fluid flow in porous media." *Physical Review Fluids 3* 084306.

Mecke, K., Arns, C.H. 2005. "Fluids in porous media: A morphometric approach." *Journal of Physics: Condensed Matter 17* (9) 503–534.

Moghaddam, N.R., Jamiolahmady, M. 2019. "Steady-state relative permeability measuremnents of tight sand shale oil rocks considering capillary end effect." *Transport in Porous Media 128* (1) 75–96.

Mohanty, K.K., Miller, A.E. 1991. "Factors influencing unsteady relative permeability of a mixed-wet reservoir rock." *SPE Formation Evaluation 6* (3) 349–358.

Rapaport, L.A., Leas, W.J. 1953. "Properties of linear waterfloods." *Journal of Petyroleum Technology 5* (5) 139–148.

Rücker, M., Berg, S., Armstrong, R.T., Georgiadis, A., Ott, H., Schwing, A. 2015. "From connected pathway flow to ganglion dynamics." *Geophysical Research Letters 42* 3888–3894.

Schlüter, S., Berg, S., Li, T., Vogel, H.-J., Wildenschild, D. 2017. "Time scales of relaxation dynamics during transient conditions in two-phase flow." *Water Resources Research 53* (6) 4709–4724.

Stone, H.L. 1970. "Probability model for estimating three-phase relative permeability." *JPT*, February.

Stone, H.L. 1973. "Estimation of three-phase relative permeability and residual oil data." *Journal of Canadian Petroleum Technology*, October–December 53–61.

Welge, H.J. 1952. "Simplified method for computing oil recovery by gas or water drive." *Transactions of the AIME 195* 91–98.

12 2D and 3D Flow Visualization

12.1 INTRODUCTION

In the methods presented to characterize the rock, so far, the porous media has been considered as a black box or a 1D model. In experiments like the porosity and absolute and relative permeability measurements, the exact structure of the rock is not a necessary parameter to evaluate. By applying Kozeny-Carman (see Chapter 7.2), the porosity is correlated to permeability using the capillary bundle model, where the distribution of the size of the capillaries represents the pore throats of the porous media. The reality is much more complicated.

Traditionally, cross sections of cores and thin sections are studied to obtain information on grain size, cementation, and porosity and pore structure. This, however, provides two-dimensional information and no dynamics are considered. For information on the multiphase flow behavior in porous media and how porous media has an effect on the flow, micromodels in 2D are constructed to visualize optically the fluid flow in a thin cell with a limited height. Experiments can be done more quickly, as less volume is present. Micromodels are commonly used for EOR screening studies, analyzing the remaining oil saturations after flooding with different injection fluids, like surfactants at different concentrations, polymers, or nanoparticles.

With the development of 3D observation tools initially for medical applications, like computed tomography scanning (CT scanning) or neutron magnetic resonance imaging (NMRI), in the field of geoscience the possibility also became available to visualize the porous media in 3D (Ketcham and Carlson 2001). The possibilities are broad, varying from obtaining morphological characterization, to localizing which minerals are where and what pore structure they form, to dynamic flow studies to derive information on fluid distribution during flow or identifying flow path with tracers (Anbari et al. 2018; Jahanbakhsh et al. 2020; Karadimitriou and Hassannizadeh 2012; Taylor et al. 2013). Modelling the fluid flow is part of the understanding of flow in porous media and there the knowledge of the porous media itself is crucial. Porous media knowledge is important both for continuum models where the porous media is represented by grid cells larger than the representative elementary volume (REV) or pore scale modelling where the pore structure needs to be described. For knowledge on the appropriate averaging or direct pore level modelling, a precise description of the pore structure is important, which can best be obtained from 3D imaging.

Different equipment has specific areas of application with partial overlap. The potential is still explored and expanded with continuous developments in spacial and time resolution and configuration. The focus in this chapter lies on 2D micromodels

DOI: 10.1201/9781003382584-12

(12.2) and CT scanning (12.3.1). In Section 12.3.2 nuclear magnetic resonance imaging and confocal microscopy will be briefly introduced.

12.2 2D FLOW VISUALIZATION

The main purpose of the use of 2D flow cells in a flooding experiment is the visualization of fluid flow within the pore space using a light-transmissive set-up using visible fluorescent light. Hele-Shaw used parallel plates in 1889 to observe velocity streamlines along different single geometries using tiny traceable air bubbles within a single-phase fluid system (Hele-Shaw 1898a, b). The currently used parallel plate system is named after Hele-Shaw as the Hele-Shaw cell.

As an alternative to represent the pore space, Chaternever presented observations of multiphase flow experiments in a Hele-Shaw cell filled with a single layer of beads in 1952 (0.007 inch diameter) (Chaternever 1952).

Nowadays 2D cells are created with tiny pillars between two flat surfaces, which form patterns representing a pore structure. In this manner a 2D model is created such that light can shine through and fluid flow in porous media can be observed. Based on the difference in colors or differences in reflective indexes at the fluid boundaries, the fluids can be visualized with a camera or microscope, dependent on the scale. To enable transmission of light through the cell and not to have overlapping fluid streams, the cells have a small depth in comparison to their width and length, generally in the range of 2–3 mm on a large scale to one-tenth of a μm between 20–50 μm on microscopic scales. The model can therefore be considered 2D because the depth of the model is limited.

It is critical that capillary forces (F_{cap}) created between the top and bottom plates are high enough such that in horizontal set-ups the effect of gravitational forces (F_g) in the depth can be neglected, with symmetrical interfaces at both the top and bottom plate; see Figure 12.1 (right). If not, then interfaces on the top and bottom plates are not equal and can cause effects on flow and give more challenging conditions for interface detection. This is not an issue in vertical cells where the gravity is parallel to the flow direction; see Figure 12.1 (left)

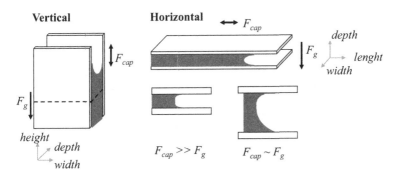

FIGURE 12.1 The effect of gravity (F_g) and capillary forces (F_{cap}) on the fluid in the micromodel.

The Bond number (N_b) describes the ratio between gravitational and capillary forces:

$$N_b = \frac{gravitational\ forces}{capillary\ forces} = \frac{hg\Delta\rho}{\sigma/h} \tag{12.1}$$

with σ = interfacial tension [mN/m], $\Delta\rho$ = the density difference between the two phases [kg/m^3], g = the gravitational acceleration [m/s^2], and h [m] = the distance between the plates in this case. Derived from this, the model depth h shall be much smaller than the capillary length l, which defines the limiting height where capillary forces can be assumed dominant over the gravitational forces in a horizontal model (Eq: 12.2).

$$l = \sqrt{\frac{\sigma}{\Delta\rho g}} \tag{12.2}$$

If gravitational forces start to play a role, there might be some difference in the geometry of the interface in the horizontally oriented flow cells; see Figure 12.2, bottom right. Figure 12.2 shows the effect on the optical pattern of the contact angles, both with the top and bottom plates and the pillar structure. Dependent on the wetting conditions, the fluids will create curved interfaces in contact with the top and bottom plate. This creates optically a transition zone at the phase boundaries instead of a sharp boundary.

Depending on the purpose of the experiment the dimensions of the cells in length and width can vary from a couple of cm in length to tenths of cm; see Figure 12.3 for two examples.

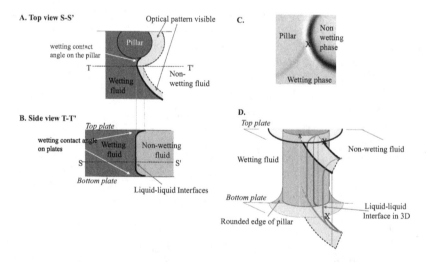

FIGURE 12.2 The effect on the optical pattern of the contact angles, both with the top and bottom plates and the pillar structure. In C and D, the X shows the location measured, where the white x shows the actual location to measure contact angles.

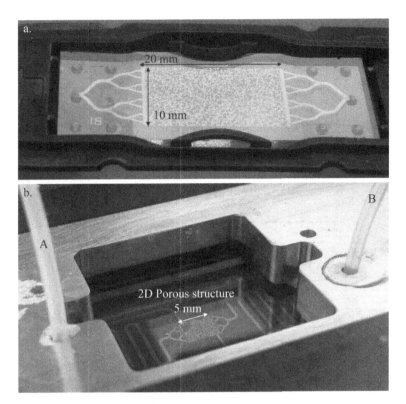

FIGURE 12.3 a. Example of a 2D micromodel representing a porous media from Micronite (EOR PR 20.2), b. a different flow cell connected with an inlet and outlet flow line (A, B).

The flooding equipment used to set up flow in the cells is quite similar to a core flooding set-up, including pumps to regulate flow rate or pressure and pressure/flow rate measurements for single and multiphase flow experiments (Figure 12.4). The microscope is placed over the horizontal set-up. Alternatively, a prism is used to direct the light to cameras at different positions.

12.2.1 FLOODING CELL GEOMETRY AND PATTERN

Depending on the purpose of the experiment, the cell geometry and pattern need to be decided on. There are different kinds of cells developed over the years to study multiphase flow in porous media.

As mentioned before, the simplest form is a Hele-Shaw cell where the distance between the plates is uniform (Hele-Shaw 1898). This can be used to study, for example, convective flow of denser CO_2-saturated brine formed by diffusion of CO_2 from a gas cap into undersaturated brine underneath; see Figure 12.5 for an example. This is a very small cell to study convective flow of 5 cm high and 1.5 cm width. Sizes can vary, generally being several decimeters in width and height (Taheri et al. 2017), but also cells are built in meter scale (Eikehaug et al. 2024).

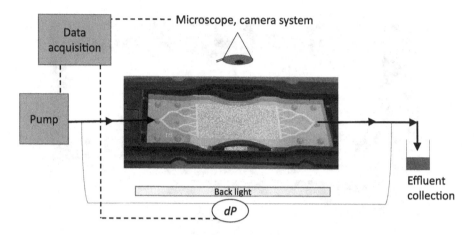

FIGURE 12.4 Schematics of a micromodel set-up.

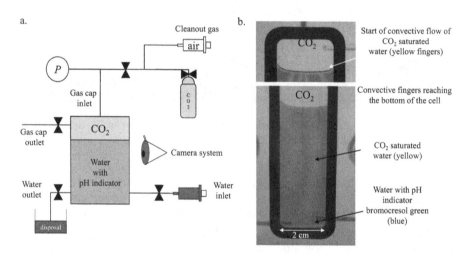

FIGURE 12.5 Example of a Hele-Shaw cell to study natural convective flow of CO_2. a. Shows a general set-up, with pressure-controlled gas cap, without mass monitoring. b. With the pH indicator, CO_2 saturated water with a pH ~4 (yellow) can be distinguished from water with a pH ~5.6 (blue).

Also flow around objects or bulk emulsion or foam behavior can be studied (Hele-Shaw 1898). Packing the spacing between the parallel plates with beads, a semi-porous system can be made to study flow in porous media. Due to the spherical symmetry of the beads, the fluid distribution along the beads is 3D. This effect is reduced by use of pillar structures instead, formed e.g., by etching or lithography (Karadimitriou and Hassannizadeh 2012; Anbari et al. 2018). This enables creation of patterns in any form (see for more information Section 12.2.3).

12.2.2 Creating Flooding Cell Geometry and Pattern

There are several methods to create the patterns dependent on the materials the micro-model will be made of. Glass, silicon, or polymers such as polymethyl(methacrylate) (PMMA) or poly(dimethylsiloxane) (PDMS) are commonly used (Anbari et al. 2018). The choice of material is influenced by the costs and availability of the manu-facturing method but mainly mechanical strength, temperature stability, chemical stability, flexibility for contact angle manipulation, and transparency are criteria for the material selection. Table 12.1 gives an overview of currently available materials and its important characteristics (Anbari et al. 2018; Jahanbakhsh et al. 2020).

There are different methods available to create models with unique patterns, each with its advantages and disadvantages.

Lithography is the general description of methods using a mask or stamp princi-ple to form a flooding pattern on a basis, called a **substrate,** with a coating material. After coating the full substrate, the coating material can be removed on the loca-tions where the mask is present or alternatively not present dependent on the method applied. For example, **curing** of an initially applied coating can be done e.g. by ultraviolet (UV) light as performed in optical lithography. The UV-sensitive coatings solidify when exposed to the UV light. The mask represents then the pore structure of the model, enabling the pillars to be exposed to the UV light. After the local curing, the nonreacted coating can be removed. This can also include removal of the original substrate from a deeper pattern. The latter is called **etching**. Etching can also be used as standalone technique, where the mask pattern is the pillar structure, enabling the pore structure to be etched away in the coating. Etching is the direct removal of mate-rial of the substrate by chemical interaction either by liquid (*wet* etching) or gasses or plasma (*dry* etching).

TABLE 12.1
Summary of Different Possible Micromodel Materials and the Characteristic Properties, Adapted from Anbari et al. (2018)

Material	Operational Temperature Range Compared to Reservoir Conditions (Max 40°C) [°C]	Not Compatible with	Wetting	Wetting Treatment
Glass	All reservoir T	-	Water wet	Silanization (Vukovic et al. 2023)
Silicon	All reservoir T	-	Oil wet	Silanization
PDMS Poly(dimethylsiloxane)	Max ~50 (Karadimitriou et al. 2013)	Non-polar solvents	Oil wet	
PMMA Poly(methyl methacrylate)	Max ~85 (SAS 2021)	Alcohols, acetone, benzol Hydrocarbons (Specialties 2024)	Water wet Neutral wet	Polymer coating

A **mold** can be made using optical lithography, the inverse of the pillar pattern, where the pores are lifted out of the surface. The mold is filled with a monomer solution or non-crosslinked polymer solution, which is then polymerized in the mold to obtain the pattern of interest, like making sand cookies with a plastic mold. This method is called soft lithography. Also, in hot embossing, the mold is used but then is pressed under high-temperature or high-pressure conditions patterns into thermoplastic polymers, which are deformable depending on pressure and temperature conditions.

More advanced methods include laser applications in the form of laser etching or stereo lithography where a laser is used to cure resins layer by layer, comparable to 3D-printing techniques (Karadimitriou and Hassannizadeh 2012).

Note that after the pattern generation a top plate needs to be attached to finalize the manufacturing of the 2D model. Hereby it is important that the plate is sealed and contacts all structures such that no fluid can flow over the structure. Also, the plate surface properties and pillar surfaces shall be similar with regard to, for example, contact angles, roughness, and surface reactivity.

A new development is 2.5D models where a varying depth is created on purpose (Xu et al. 2017) to study snap-off better.

For more details on the different techniques and advantages and disadvantaged see Karadimitriou and Hassannizadeh (2012) and Jahanbakhsh et al. (2020).

12.2.3   2D Representation of Rock Structure

Within the limitations of using a 2D models system, representation is to be considered. Pattern choice and wettability are important design criteria.

Pattern Choice

Note that 2D is not 3D, and it is an important consideration as far as what pattern to work with if the model is to be used to study flow in porous media. Questions arise as to how the pattern can influence the results and whether the results obtained in the 2D model are representative for 3D flow. Thin section slices cannot directly be used to recreate the porous media in 2D. Not all pores will be connected on the 2D slice, as they are in 3D. With the thin section as a mask, the pore throats can be created artificially, making additional connections between the pores. Alternatively, any other pattern can be designed. Commonly, homogeneous regular patterns are used like hexagonal packed or face cubic centered. Pillars forming the grains can be a random shape depending on the grain cross section but also can be circular or square. Figure 12.6 shows some examples. Figure 12.6A. shows an analogy to a nodal network of capillary tubes with equal cross sectional areas. B and C are derived from face-centered cubic (fcc) or body-centered cubic (bcc) structures resulting in a pillar structure, with regular pore throats and pore bodies. D shows a random distribution and shapes of pillars (lighter shaded) based on a thin section. Features like fractures can be added. The pattern choice has a direct influence on the porosity and permeability of the model.

Flow modelling can be used to support the choice of pattern and cell dimensions. Lattice Boltzmann (Chen and Doolen 1998) or computational fluid dynamics (CFD)

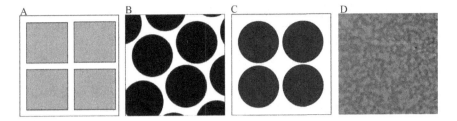

FIGURE 12.6   Examples of 2D micromodel patterns.

(Gidaspow and Jiradilok 2009) are examples of pore scale modelling tools, based on the solutions to the Navier-Stokes equation. Reservoir simulators like Eclipse, T-navigator, or CMG are examples of larger-scale (cm scale and above) models based on Darcy's equation; mass balance and energy and flow conservation laws using grid cells are representative elementary volumes (REV), continuum scale modelling. The grid model gets assigned a distribution of averaged rock properties dependent on the geological data (Mattax and Dalton 1990).

Wettability

Fluid flow in porous media is significantly influenced by rock-fluid interaction, described by the parameter of wettability (Chapter 9). The wettability of the micromodel is therefore an important parameter to consider. The original material of the model will determine the initial wettability in reference to the liquids used. Glass and silicon are generally water wet and the polymers PDMS and PMMA are oil wet in water-air systems. This wettability can be altered e.g., by the application of a surface coating or a chemical coating respectively or by chemical cleaning of the surface. In flooding experiments, it is important that the applied method of surface modification is stable and not changing in time. Coating stability is critical to test. The silanization reaction is a common coating applied in micromodels (Vukovic et al. 2023) and used here for discussion on kinds of wettability modification.

For a coating that chemically reacts with the surface, the wettability can be set homogeneously over the model, performing a single-phase flood of coating solution through the model. For chemical coating the exposure to the surface concerning duration and/or concentration determines the coating density, which allows tuning of the contact angle and wettability state of the surface. Fractional wetting can, for example, be obtained by introducing the reactive coating with a remaining saturation of a non-reactive liquid.

12.2.4   IMAGE PROCESSING

To analyze the experimental data of a flooding experiment or for characterization of the micromodel, images are made of the model. They need to be processed and analyzed. Depending on the objective of the experiment, different analyses can be performed, for example saturation determination, contact angle measurements between phases, or simply determination of cell porosity. One of the main aspects needed in the image processing is therefore the identification of the different phases.

Images

Images are a matrix of pixels, where each pixel obtains a numerical value based on the brightness of the light captured representing the greyscale of the image in case of a greyscale images (8-bit images). In this case the scale varies from 0 = full black to 255 = full white. Other scales are color images (red, green, blue (RBG)), which, as the abbreviation suggests, have three scales of red, blue, and green to cover the full color pallet. Binary images have just 0 and 1 as numerical values. For simplification we will focus here only on greyscale images to explain the fundamentals of image processing and analysis.

Segmentation

Before performing an analysis on the different phases in the image, first these need to be identified. Based on difference in greyscale a **segmentation** can be performed where the different scales of grey are grouped or binned. This is based on the fact that each phase, being oil, gas, brine, and solid, will have another distribution of greyscales; see Figure 12.7. Ideally, they differ such that a division can be made based on values of grey. Depending on the number of phases to differentiate, different phases can be segmented directly using cut off values, if they differ significantly. But in practice if greyscales overlap for the different phases the phase identification becomes more challenging.

This situation can occur, for example, if brine is not distinguishable from the solid, in an image with brine, oil, and solid phase based on color scale. An additional

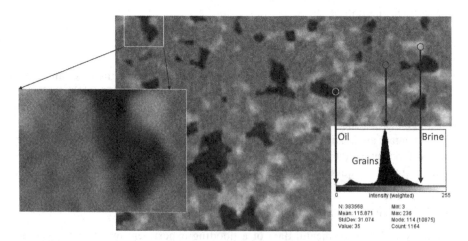

FIGURE 12.7 Micro-CT scan image of a sandstone filled with oil (black) and brine (white) and the histogram of the greyscale values of the image. The zoom shows the pixels of the image, representing a grey value.

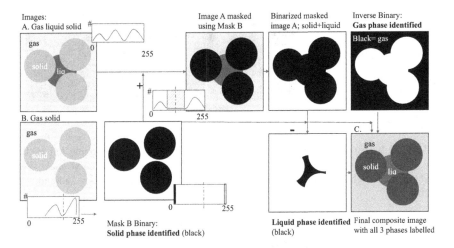

FIGURE 12.8 The segmentation of the image A., being image C., illustrated by the images and corresponding histograms. Image B is used to identify the solid phase using the histogram (inlay) and used as a mask for image A such that the solid + liquid can be identified. This can be used to segment the gas phase, and its difference with the mask B will give the liquid phase.

operation on the images is then necessary, which is illustrated in Figure 12.8. First, another way is to be found to identify one of the phases, in this case the solid phase, which does not change during the flooding. This can be used as **mask** later. The solid phase mask can be obtained, for example, if in an image with only the solid and gas phase; the solid can be identified using the histogram of the greyscale distribution. Therefore, ideally an additional image shall be made from a cell filled with only air, such that the solid is identifiable, image B in Figure 12.8. Using this segmented image as a **mask** enables the isolation of region outside the solid and can be used to analyze that leftover area for liquid and gas present. Applying the mask sets the pixels to zero in the output image. All other grey values belong then to either the liquid or gas phase. Binarizing this image such that the liquid phase becomes zero (black too) will give a directly inverted gas phase and comparing it to the mask gives the liquid phase. Hereby subtraction or addition operations are performed on different segmentations to obtain the information of interest, resulting in a final composite image.

By having identified the phases, analysis can be done, e.g., how much area of the pore space is filled with brine or oil, what the shape of the fluid is, how many droplets there are, and how it changes in time or with pressure. In Figure 12.9 an example is given on the processing of an image of a microchip after waterflooding in an oil-filled microchip.

In the images, **shading** or **noise** will have the result that the segmentation is not perfect and some error will be present in the analysis. For example, some pixels might belong to the brine phase, but they are segmented as an oil phase. Improvement of lightning conditions, color contrast between the fluids, camera exposure settings, or material cleanness or roughness might be of help to improve the image quality.

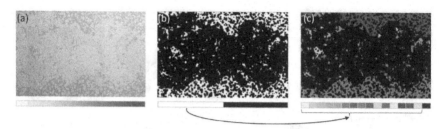

FIGURE 12.9 (a) Pre-processing image of a microchip after water flooding (oil = gold, glass, and brine transparent). (b) Segmented, binary image to enable quantitative analysis (oil in white). (c) Oil clusters colored individually for qualitative analysis (from Aadland et al. (2020)).

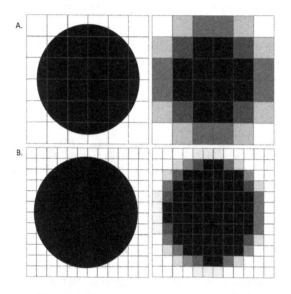

FIGURE 12.10 The effect of resolution on the greyscale and boundaries. Left are the actual boundaries of a spherical object identical in A and B and right for A the low resolution image and in B a higher resolution image with 4× smaller pixel size. The consequences for the smearing of the boundaries is much larger at low resolution A.

The **image resolution,** which is the areal size of the pixels where the value is the averaged signal from that area, plays a role in the error as well. In Figure 12.10 an example is given as to how a lower resolution results in a less sharp object edge detection, which then will have an influence on the accuracy of determining area or shape of the object.

Image Resolution

There are multiple image processing tools available to try to modify images to make segmentation specifically and image analysis in general more accurate based on the

Image processing: **Close routine**

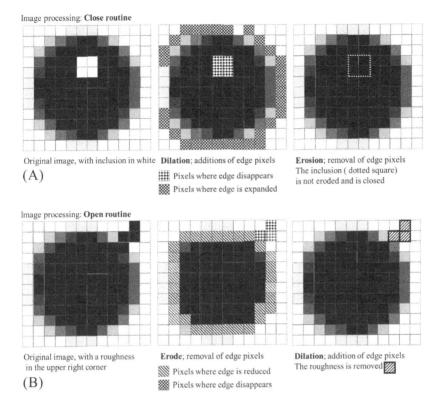

Original image, with inclusion in white   **Dilation**; additions of edge pixels

▦ Pixels where edge disappears

▨ Pixels where edge is expanded

Erosion; removal of edge pixels
The inclusion (dotted square)
is not eroded and is closed

(A)

Image processing: **Open routine**

Original image, with a roughness
in the upper right corner

(B)

Erode; removal of edge pixels

▧ Pixels where edge is reduced

▨ Pixels where edge disappears

Dilation; addition of edge pixels
The roughness is removed ▨

FIGURE 12.11   The close and open operations explained. "Close" first takes away a number of layers of pixels at the edges (A). "Open" removes them (B).

images available, for instance, filters for noise reduction or contrast enhancement. Worth mentioning for object identification are **erosion** and **dilation** or, if sequentially applied, **close** or reverse **open** are possible to smoothen and remove rough or isolated regions (see Figure 12.11A and B for a visualization).

For more details on image processing and analysis we refer to the literature (Solomon and Breckon 2011; ImageJ 2023).

12.2.5   2D Representation of Rock Structure; Characterizing the Micromodel

Like rocks, the micromodel needs to be characterized. Initially they can be designed including the parameters of structure dimensions, porosity, absolute permeability, and pore volume.

Porosity Determination

In general, a template is made where the pore space can be directly tuned by the ratio of the flooding area and total area of the model (only the porous media, without inlet and outlet geometries). Similarly, the porosity can be derived directly from a model image, which is segmented for pore space and pore grains. The relative area of the pores compared to the total model area (only considering the porous media region of the model) is the porosity.

$$\varphi = \frac{Pore\ area}{Total\ porous\ media\ area} \tag{12.3}$$

Pore volume (*PV*) can be derived from the total volume of the model V_{model}

$$PV = Pore\ area \cdot model\ height \tag{12.4}$$

$$PV = Total\ porous\ media\ area \cdot modelheight \cdot \varphi \tag{12.5}$$

Note that, based on the smearing of the edges of the objects in the images and the possibility of insufficient contrast, the area selected as pore space includes an uncertainty. Alternative methods such as helium porosity (Chapter 5) could be used, but due to the small volumes this can lead to inaccurate data.

Permeability Estimation

Capillary bundle model or simplified Hagen-Poiseuille channels cannot be used directly to predict and design a model, especially with irregular structures. Instead, permeability can be designed by pore scale modelling using, for example, Lattice Boltzmann- or Navier-Stokes-based models (Gidaspow and Jiradilok 2009; Blunt 2013).

Besides permeability prediction, permeability estimations can be made by measuring directly on the model according to Darcy's law, as described in Chapter 7. Based on the model size, liquid permeability is recommended to obtain a large enough pressure drop to be measured.

Wettability Measurement

Fluid flow in porous media is significantly influenced by rock-fluid interaction, described by the parameter wettability (Chapter 9). The wettability of the micromodel is therefore something to consider; see Figure 12.12 for two examples of hydrophilic and hydrophobic conditions. As mentioned, the original material of the model will determine the initial wettability in reference to the liquids used. The wettability of the model can be tested by measurement of contact angle in situ in the model. Similar to the 2D derivation of contact angle using droplets or bubbles on a surface, the three-phase contact point and interfaces need to be identifiable. The interface pattern is influenced by contact angles formed also on the top and bottom plate (Figure 12.2) (van der Net et al. 2007) and possible non-uniform shaping of the pillar structures. Figure 12.2 shows in A an actual three-phase contact point that is visible and standardly used to measure the contact angle of interest, with B the contact angles formed are theoretically visible in a cross-section along the depth of model. C is an actual image made, where the three-phase contact point that is

Hydrophylic   Hydrophobic

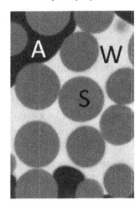
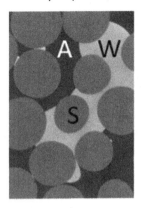

FIGURE 12.12 Two examples of hydrophilic and hydrophobic conditions in a micromodel based on the contact angles observed (A = air, W = water, S = solid) (courtesy of Tomislav Vukovic).

derivable is represented by a black X. This corresponds to the actual three-phase contact point under the assumption that the patterns are symmetrical, only possible using perfect parallel light.

Challenges found for contact angles measurements using droplets on plates like line pinning due to surface roughness can occur similarly in micromodels. Generally, the determination of contact angles determined in 2D micromodels gives a larger spread than with single droplets placed on surfaces (Vukovic et al. 2023). Additional uncertainties in image analysis are determination of the three-phase contact point, but consequently finding the correct gradient in the image along the interface due to boundary smearing due to the limitations in resolution also increases the error.

EXERCISE 12.1

1. Based on which consideration can an actual 3D model be considered 2D?
2. What are the advantages of using a 2D model for studying multiphase flow in porous media and what are the disadvantages? Name one for each.
3. Which parameters of a 2D micromodel can be tuned independent of the manufacturing process? Name two.
4. Describe image segmentation of a three-phase system.
5. How can wettability be determined in a 2D micromodel? Additionally, name two sources of error.

SOLUTION

1. When the depth of the model is smaller than the capillary length, the model can be considered 2D. Additionally, with no differences in the z-direction considering pillar diameter, the model can be considered 2D.

2. Advantage: With a transparent model the model can be visualized with optical light directly.
   Disadvantage: 2D models do not cover all effects that might occur in 3D.
3. Pillar structure, wettability.
4. Image segmentation is the assignment of pixels (greyscales) to a specific phase in the image. The image consists of pixels containing values of the greyscale. The segmentation can be done by cut-off assigning greyscales to a phase directly, but if phases overlap in greyscale then image subtraction can be performed. Then e.g., an image of two phases (rock-gas) can be subtracted from the three-phase image such that rock can be identified. Then, with a cutoff, the two phases (no rock) can be identified.
5. By introducing the two phases in the system of which the wettability in combination with the model is to be determined. Then the contact angles is determined by the standard imaging of the phases in the model, finding the three-phase contact line and gradient along the interfaces from that point. Sources of error are a) equilibrium force has not reached balance in the model by e.g., pinning and b) optical effects; image smearing that makes the interfaces not sharp. Or the three-phase contact point is hard to find (Figure 12.2 C).

12.3   3D FLOW VISUALIZATION: COMPUTED TOMOGRAPHY SCAN

The use of 2D models to visualize flow gives valuable information and possibilities to study flow in porous media, but the lack of the third dimension cannot be always discarded. There might be effects that can only occur in 3D flow like snap-off (Wardlaw 1982) or Haines jumps (Haines 1930; Berg et al. 2013; Sun 2019). In addition, the grains (pillars) are not connected in the 2D micromodel. This will have effects on relative permeability with a change in flow path or distribution of wettability, which can lead to other flow dynamics, preventing an extrapolation of the 2D data to 3D. Therefore, it is also important to study and visualize flow in three dimensions, where then actual rocks could be used. Here we will discuss the method of computed tomography scanning (CT). Other methods like nuclear magnetic resonance (NMR) and confocal microscopy will be only briefly introduced in Section 12.4.

The computed tomography scan—or CT-scan—is an apparatus that is used to generate three-dimensional images of objects using X-rays. The technique is based on the differences in X-ray interaction or attenuation of the material. The CT-scans are used in medical research as well as in other disciplines that need imaging without destruction of the sample, non-invasive imaging. For example, in geoscience the CT-scan is applied to obtain images of flow patterns, fractures, and fracture generation inside a porous medium, but also porosity, saturation, and permeability can be extracted from the images (Cnudde and Boone 2013; Mees et al. 2003; Wellington and Vinegar 1987; Ketcham and Carlson 2001).

The main components of a CT-scan are a **source**—to generate and focus a beam of X-rays—and **detectors** to register the intensity change of the X-ray after passing the object. The change of intensity is measured by sending the X-rays through the sample perpendicular to its rotational axis under different angles, creating a **projection**. This principle is comparable to recording shadows of an illuminated object on a screen where the object is semitransparent. The created set of 2D projections is reconstructed as a 3D image where the pixels contain local information on the **attenuation** of the X-rays, which is correlated to density variation throughout the sample.

12.3.1   X-Ray Absorption

When X-rays, consisting of photons, pass through the object, different interactions can occur dependent on the energy of these photons. These are photoelectric absorption, Compton scattering, and coherent scattering (Ketcham and Carlson 2001).

12.3.1.1   Photoelectric Absorption

With photoelectric absorption, a photon hits one of the *inner* electrons around an atom. There it transfers all its energy to this electron, which causes the electron to free itself from the atom; see Figure 12.13. The probability of photoelectric absorption per unit mass of a material is approximately proportional to $Z^{3.8}/E^{3.2}$, where Z is the atomic number of the material and E is the energy of the incoming photon. So, the photoelectric absorption increases with increasing atomic number or decreasing photon energy. Generally photoelectric absorption dominates till an E of 100 keV (Ketcham and Carlson 2001).

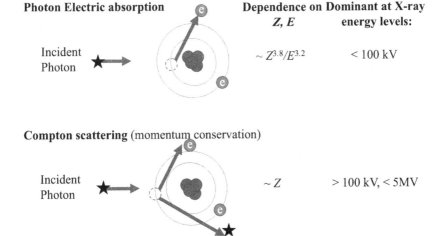

FIGURE 12.13   The illustrated interaction of a photon (star) with an atom for photoelectric absorption (top) and Compton scattering (bottom). *e* stands for electron, Z is the atomic number of the material, and E is the energy of the incoming photon.

12.3.1.2 Compton Scattering and Coherent Scattering

Above the $100–120keV$, the Compton scattering starts to dominate till $5–10MeV$ (Ketcham and Carlson 2001). The incoming photon causes an ejection of an *outer* electron together with a direction change of the photon due to a loss of energy; this phenomenon is called **Compton scattering** (see Figure 12.13). With Compton scattering, there is a conservation of energy and momentum. The amount of Compton scattering is proportional to the atomic number Z and natural logarithm to E as well. Compared with photoelectric absorption, the Compton scattering responds less to changes in atomic number. So, at lower energy levels where photoelectric absorption dominates, density changes are better detectable. Compton scattering is incoherent scattering.

The difference with **coherent scattering** lies in the fact that with incoherent scattering the frequencies or wavelengths of the incoming and emitted photons change while with coherent scattering the frequencies or wavelengths remain constant and it is therefore called an elastic interaction (Ketcham and Carlson 2001). It is comparable to light from a lightbulb, which has multiple frequencies and directions and a laser, having one frequency of light that moves parallel.

A standard medical or micro-CT scan works in the energy ranges of 40–225keV. There photoelectric absorption, Compton scattering, and coherent scattering occur when X-rays are sent through an object. This causes a decrease in intensity of the X-ray beam, meaning a reduced number of photons of a specific energy per time period. Derived from the change in intensity dI, dependent on the material attenuation μ, penetration length dh, and initial intensity I at position h:

$$-dI\left(h\right) = \mu I\left(h\right)dh \tag{12.6}$$

the total loss of photon intensity over an object with a penetration distance h is described in **Beer's law**.

$$\text{Beer's law: } I\left(h\right) = I_0 e^{-\int \mu(h)dh} \tag{12.7}$$

I_o is the initiate intensity of the X-rays, I the remaining intensity of the X-rays after passing an object, h the thickness of the object, and μ the material specific linear attenuation coefficient, depending on the photon energy loss due to Compton scattering and photoelectric absorption.

Beer's law assumes a narrow X-ray beam and a radiation of a monochromatic spectrum of photons (Akin and Kovscek 2003). Another assumption made to use Beer's law in the CT-scan is that the detectors must be insensitive to the scattered X-rays. In practice the generated X-ray photons are not mono-energetic and a spectrum of energy levels is being produced by the source, such that this equation is oversimplified.

For non-homogeneous, non-axial-symmetric materials, each material contributes differently based on density and pathlength. This makes it necessary to scan under different angles and obtain 2D projections.

An average composite attenuation coefficient μ_{av} can be derived from the sum of material attenuations times its atomic weight fraction f:

$$\mu_{av} = \sum_{1}^{n} f \cdot \mu_n \qquad (12.8)$$

Note that for a detectable density difference in materials with different components, the linear attenuation coefficients ideally differentiate sufficiently. Spiking fluids can be a solution to create more contrast for fluids like cesium chloride (CsCl) or sodium iodide (NaI). Tracer gasses like Xenon can fulfill the same function (Akin and Kovscek 2003). The presence of highly X-ray adsorbing metals shall be avoided. In Table 12.2 some attenuation coefficients of common materials are listed.

McCullough (McCullough 1975) describes for a linear attenuation coefficient of a pure material the share of coherent scattering, photoelectric absorption, and Compton scattering on the decrease of X-ray intensity, while it passes a material with a specific density ρ [kg/m³] and atomic number Z;

$$\mu = \rho \frac{N_A \cdot Z}{M_g}(0.597 \cdot 10^{-24} e^{-0.0028(E-30)} + b\frac{Z^{2.8}}{E^{3.2}} + 1.25 \cdot 10^{-28}\frac{Z^{2.0}}{E^{1.9}}) \qquad (12.9)$$

Compton scattering   Photoelectric absorption   Coherent scattering.

In this formula the term $b\frac{Z^{2.8}}{E^{3.2}}$ represents the photoelectric absorption, $1.25 \cdot 10^{-28}$ $\frac{Z^{2.0}}{E^{1.9}}$ represents the coherent scattering, and the term $0.597 \cdot 10^{-24} e^{-0.0028(E-30)}$ is a

---

TABLE 12.2

Attenuation Values for Common Materials (Adapted from Akin and Kovscek (2003))

Material	Density	Mass Attenuation		Attenuation		Attenuation in Hounsfield Units	
		150 keV	100 keV	150 keV	100 keV	150 keV	100 keV
	[g/cm³]	[cm²/g]		[1/cm]		[H]	
Air 22°C	0.001	0.133	0.148	0.00	0.00	−999	−999
N-decane 20°C	0.730	0.154	0.171	0.11	0.12	−241	−246
Water 22°C	0.998	0.148	0.165	0.15	0.17	0	0
3wt% NaCl solution 22°C	1.014	0.148	0.166	0.15	0.17	14	17
1wt% CsCl solution 22°C	1.005	0.153	0.180	0.15	0.18	37	96
0.2M%CsCl (16wt%) 22°C	1.134	0.218	0.399	0.25	0.45	673	1743
Quartz	2.320	0.135	0.158	0.31	0.37	1126	1221
Bentheimer sandstone	2.400	0.136	0.158	0.33	0.38	1201	1298
Kaolinite Al$_2$Si$_2$O$_5$(OH)$_4$	2.600	0.136	0.157	0.35	0.41	1393	1477
Iron (Fe)	7.874	0.179	0.334	1.41	2.63	8517	14941

representation of the influence of Compton scattering. Further, E is the energy of the X-ray beam in keV, N_A is the number of Avogadro: $6.022 \cdot 10^{23}$ [1/mole], and M_g is the molecular mass [kg/mol]. Equation 12.9 is often reduced to:

$$\mu = \rho \frac{N_A \cdot Z}{M_g} \left(a + b \frac{Z^{2.8}}{E^{3.2}} \right) \tag{12.10}$$

with the a as the Klein-Nishina coefficient and b as a constant in the order of $9.8 \cdot 10^{-28}$ (Vinegar and Wellington 1987). The fact that the Compton scattering depends on the energy level can sometimes be neglected. It is then a constant, which is called the Klein-Nishina coefficient, a. This assumption needs to be checked for every application. Also, the share of the coherent scattering is generally limited for SCAL applications. It is included in the constant in the term of the photoelectric absorption, the constant b.

So, the attenuation coefficients of materials are linearly dependent on its density and atomic number, and it is inversely proportional to the energy level of the X-ray beam. The higher the energy of the X-rays (E), the smaller the interaction and the lower the level of attenuation of the material. Density contrasts are then less distinguishable. The higher the density or atomic number, the higher the attenuation.

12.3.2 COMPONENTS OF THE CT-SCAN

In a CT-scanner a **source** is needed to generate X-rays and a **detector** to record the X-rays' intensity after passing through the object.

12.3.2.1 X-Ray Generation

The X-rays are generated in a so-called **tube** or **electron gun**. The electrons are released from a filament made from tungsten, through which a high current is flowing, slowly "burning up" the filament; see Figure 12.14.

The electrons are accelerated from the cathode to the anode, due to a voltage difference. When the electrons under high speed hit the anode surface made of e.g., tungsten, a spectrum of X-ray radiation is generated according to the Bremsstrahlung principle (braking radiation) and the characteristic radiation or K-series peak. In Figure 12.15 an example is given for 60kV and 100kV voltage difference over the cathode and anode.

Bremsstrahlung is electromagnetic radiation generated by deflecting an electron by a core of an atom. This causes a continuous spectrum of wavelength within the electromagnetic spectrum (Index 2024). The X-ray wavelengths are between 0.1 to 100 Å (angstroms; 1 nm = 10 Å). The maximum energy of the photons created represents the maximum voltage difference between the anode and cathode.

The **characteristic radiation** is generated when an electron collides with an inner electron of the target. This electron is consequently emitted from its position. Its position is taken by another, outer electron. When this electron "falls back", the electron loses energy, which produces photons of a characteristic frequency, specific for the source material.

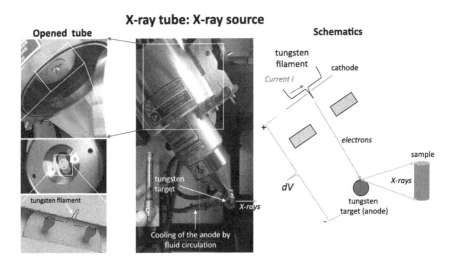

FIGURE 12.14 An example of an X-ray source of a (micro-)CT scan (Nikon XT 225).

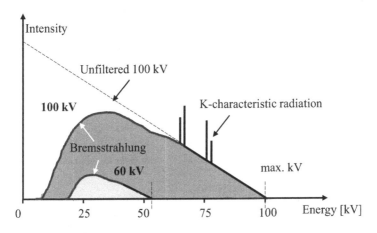

FIGURE 12.15 Examples of energy spectra of X-rays dependent on tube voltage. It shows the continuous spectrum caused by Bremsstrahlung and the characteristic radiation of the tungsten (intensity n.d.).

The maximum energy of the X-ray spectrum is determined by the voltage difference over the cathode and anode, here 100kV and 60kV. Note the unit for X-ray energies is in kilo electron volts [keV], describing the energy of an electron in field with a voltage difference V. So, the maximum energy an electron can obtain in a 60kV setting is 60keV.

When the voltage difference is higher or the number of X-rays generated or the intensity of the X-rays is higher, more electrons are emitted from the filament. Additionally, they hit the target with a higher speed, creating higher-energy X-rays.

The intensity can also be influenced by the tube current, being the current flowing through the filament. A higher current gives a larger stream of electrons that can be emitted from the filament, such that more X-rays can be generated at collision with the tungsten, but the acceleration itself is not affected.

The generation of X-rays happens in a closed tube to avoid leakage of radiation. The anode and cathode are placed under high vacuum, so that the electrons only collide at the anode and not with gas molecules from the air or other impurities. The X-rays are controllable, released through a window. Based on Beer's law, preferably X-rays of only one energy level are generated. Filters can be used to reduce the low energy level to the right energy levels, though this decreases the signal-to-noise ratio, so more noise is present (Akin and Kovscek 2003). A collimator is used in front of the window to only let X-rays through parallel to the subject.

The X-ray beam is formed on one location, a spot on the target. This spot size directly influences the resolution of the image in combination with where the sample is placed in the conical beam that is created. Resolution is a measurement of the amount of datapoints measured on a surface. The larger spot size will give less areal coverage of the detectors and therefore a lower resolution; see Figure 12.16A.

12.3.2.2   X-Ray Detection

The sample casts a shadow of reduced X-ray intensity on the detectors behind the sample. An area behind the sample is filled with detectors placed there to register the resulting X-ray intensity. The "shadow of the object" shall be divided over an area as large as possible to enable use of as many detectors as possible. Spot size affects this, but also the placement of the sample. Placing the sample as close as possible to the source magnifies the shadow and therefore coverage of the detectors;

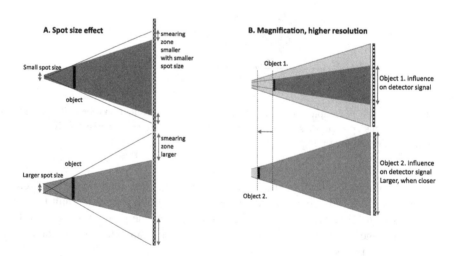

FIGURE 12.16   A. The effect of the spot size on resolution, represented by maximum coverage of the detectors. The principle of shadow magnification consequently giving a higher resolution is shown in B.

see Figure 12.16 B. Additionally, more X-rays are sent through the object when it is positioned closer to the source, enhancing the contrast.

The size of an individual detector determines over what surface area a single average intensity is recorded; the number of detectors determines how large the total area is that can be detected. These both will affect the resolution of the created images (Ketcham and Carlson 2001).

There are three different types of detectors or scintillators, characterized by the output signal. The **counting type** detects the total number of photons incident on it. The **scintillation type** gives the total measured photon energy, and the **ionization type** records the energy deposition per unit mass (Kak 1988). Mainly the detectors are of the scintillation type. Here the photons collide and are adsorbed by a crystal, mostly sodium iodide (Kak 1988). When colliding, light flashes are created, which are led through a photon-multiplier, which transfers the light photons into an electric current. The current is proportional to the detected intensity of the X-ray.

The ionization type detector consists of a chamber with a collecting electrode at the center and a high voltage strip on each side. The chamber is filled with xenon, which ionizes when the X-ray photons enter the detector. The ionization of xenon creates a current through the electrodes, representing the incidents X-ray intensity.

12.3.2.3 Resolution

Spatial Resolution and Contrast

The limiting size of detecting features, the **resolution**, is dependent on several settings.

Note that this includes number of data points – so coverage of the detectors but also achievement of sufficient contrast to distinguish boundaries.

For one projection: First, the kind and amount of X-ray exposure will affect the absorption of the different components in the object of interest, such that a good contrast is created. This is affected by the settings:

- Area covered with detectors and the number of detectors per unit area; the large area with a higher density of detectors gives a higher resolution.
- Distance: Closeness of sample to the source (target) at a fixed source-detector separation. The closer gives a higher beam density and exposure to X-rays (see Figure 12.16) and the cast "shadow" is spread over more detector area.
- Spot size: The focus of the accelerated electrons affects the casted shadow on the detectors.
- Tube current: Determining the number of electrons sent to the target. At a high current, more photons of an X-ray spectrum are created, leading to more contrast at a given exposure time.
- Tube potential: Creating X-rays with different beam energies and interaction with the sample. A higher potential results in higher X-ray energies created, having less photoelectric absorption. A lower potential gives a better contrast.

- Exposure time: Duration of X-ray beam sent through the sample, meaning the number of photons and duration of detectors collecting the signal. More signal collection generally gives a better contrast between materials of different densities.

More projections give more information to process and increase the resolution also.

Temporal Resolution

The temporal resolution depends on the speed of scanning; the exposure time per projection and number of projections can influence this. Fewer projections can be made if required but then the spatial resolution will decrease. A reduced exposure time will reduce the contrast. Depending on the dynamics of the experiment, the time resolution can be adapted. Current commercially available CT scanners are limited in time resolution, with common scanning times of 20 minutes to one day. Within the scanning time the changes in the experiment shall be limited.

So, to get an optimal image when preparing for a scan, the spatial resolution, including contrast, shall be considered and, second, how long the scan shall take depending on the available time and dynamics. The parameters that can be used to tune are the tube voltage, current, sample position, exposure time, and number of projections. For example, scanning at 60kV will give fewer electrons per time unit compared to e.g., 120kV, but it will give better contrast. The reduced intensity can be compensated for if the exposure time is increased, leading to longer scanning times.

12.3.3 Kinds of CT-Scan

Over the years different generations of CT-scanners have been developed, classified by the number and placement of sources and detectors. See Figure 12.17 for an overview.

First generation. The first CT-scanner was introduced by G.N. Hounsfield in 1970. This scanner had one source and one detector. With a pencil shaped X-ray beam parallel to the object, different attenuation measurements were done. This was repeated by rotating the source and detector to different angles and then repeating the parallel measurements. This scanner is also called the **translate/rotate scanner** (Amersham n.d.) (see Figure 12.17 I).

Second generation. The second generation only differs from the first-generation CT-scanners by using more detectors; up to 70. The emitted beam diverges from its source. So, more detectors are used to catch the radiated X-rays. Using more detectors decreased the scanning time and improved the image quality; see Figure 12.17 II.

Third generation. All further generations are meant to improve the scanning speed and image quality. In the third-generation scanners the fan beam was introduced. An arc of detectors is used to record the attenuation intensities of the incident X-rays. Both the source and arc of detectors are rotating around the object; see Figure 12.17 III. No parallel movements of source and detector along the object are necessary anymore. The scanning speed is therefore increased a lot. However, there were problems with energy supply; using cables to rotate the devices made the rotation discontinuous in one rotation direction. This problem was solved in 1987 by

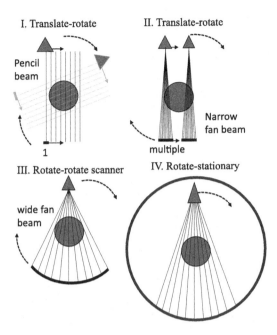

FIGURE 12.17 An overview of the first generations of the CT-scanners, based on Ketcham and Carlson (2001).

introducing a continuous rotating gantry (Nondestructive 2024). For this generation, ring artifacts and aliasing are introduced as problems. This scanner generation is also known as the rotate-rotate scanner (Amersham n.d.).

Fourth generation. Here 1,440 detectors are placed in a ring. The ring of detectors is stationary and only in the fourth generation has the number of detectors extended even more; now the X-ray generator is moving, see Figure 12.17 IV. This geometry makes the scanner more sensitive to scattered radiation, but the sensitivity to ring effects is reduced. This generation is also called the **rotate stationary generation** or rotation only (Amersham n.d.).

Fifth generation and more. Over time more CT-scanners have been developed that cannot easily be classified in a generation. In the nineties, experiments started with continuous data acquisition and, simultaneously, moving the object in the gantry. The data is then constructed as a spiral. To obtain slices of the object at one position, one extra processing step has to be taken: The spiral interpolation (Nondestructive 2024). This scanner is known as the **Spiral or Helical scanner** (Amersham n.d.).

A **gantry CT-scan** is a CT-scan where the source and detectors are rotating, and the sample remains in place. This is useful for a medial application, such that patients can lie on a static table. Also, for flow experiments it is easier to work with because of the flow lines attached to the flow device that is to be scanned.

CT-scans can also be classified by the resolution they can achieve. Besides a conventional CT-scan there are micro-CT-scans and nano-CT-scans (Table 12.3).

TABLE 12.3

Types of CT-Scanners and Data

CT scan	Range of Highest Resolution	Common Sample Size (Cylindrical Dimensions)	
		length	diameter
(medical) CT-scan	1–3 mm	3–30 cm	1–10 cm
Micro-CT scan	3–5 μm	10–30 mm	2–10 mm
Nano-CT-scan	< μm	μm	

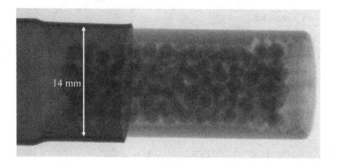

FIGURE 12.18 A single projection of a bead pack. Note that here the darker regions have higher absorption (courtesy of T. Vukovic 2023).

Imaging in synchrotron might be an alternative. That is a particle accelerator where radiation can be generated of different energy levels, as well as X-rays of single energy and high intensity. With this system the temporal resolution of 3D imaging can go down to seconds or minutes (Jahanbakhsh et al. 2020).

12.3.4 SIGNAL PROCESSING: 2D PROJECTIONS TO 3D IMAGE

The data obtained after a CT-scan is performed by the detectors is a 2D projection showing the X-ray intensities, also named **sinograms** (Ketcham and Carlson 2001), shot under different angles. Figure 12.18 shows an example of a single projection. It can be seen, dependent on the overlap of the beads, that the registered absorption differs.

As part of the processing, an axes system is chosen, generally with the center of rotation of the object as the origin. The projections are then processed with 3D images of the original object being a 3D matrix containing information of the average linear attenuation coefficient for each pixel.

In this paragraph, the principle behind the data processing will be described only superficially. For more information, a reference is made to Kak and Slaney (1988).

In the computed tomography, the main objective for the data processing is to convert the obtained intensity data, being projections in 2D, under different angles, back to an object or 3D image, which converts the initial X-ray intensity to the recorded X-ray intensity after passing the object.

As part of the reconstruction, Radon came up in 1917 with a transform where 2D image data could be transformed into a data collection dependent on angle and distance from a defined path through the center of the object (Radon 1917). Following the Radon transform, the Fourier slice theorem is used. **The Fourier Slice Transform** of the Radon transformed X-ray intensity profiles creates in the frequency domain a 2D line; see Figure 12.19 (left). Transforming all the projections, a web is obtained of 2D lines around the point of rotation; see Figure 12.19 (right).

A correction must be made for the fact that a lot of information is in low frequencies, near the center of the web. A ramp filter is used for this correction (Romberg 2004). The final step is the back projection where the inverse Fourier transform of the data in the frequency domain are placed in the image plane and added; see Figure 12.19 (Kak and Slaney 1988). The conversion can be adapted to the different scanning procedures and the use of a fan beam. So then the contributions of each spot in the object to the intensity decrease of the X-rays becomes visible. The more projections, the better the final result; see Figure 12.20.

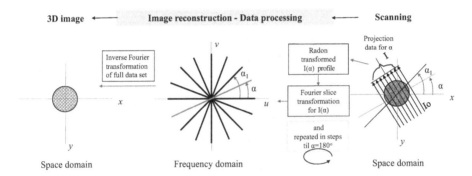

FIGURE 12.19   The Fourier transformation from the data from space domain (left) to frequency domain (middle), which gives, when repeated for several angles, a spread of the data over the frequency domain. When converted back into the space domain, this set forms the 3D image (modified from Kak and Slaney (1988), Figure 3.6, 3.7 (Pan et al. 1983)).

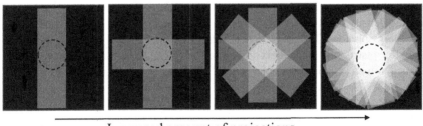

Increased amount of projections

FIGURE 12.20   The results of adding different amounts of back projections; the more back projections that are added, the better the final result (based on Kak and Slaney 1988) Figure 3.10).

Hounsfield applied the Fourier slice theorem so that he could calculate backwards how a location in a sample affects the projection. Hereby he created the X-ray tomographic computed scanner for which he obtained, together with A. Cormack, the Nobel Prize in 1972 (Kak and Slaney 1988). Today the tomographic conventional back projection algorithms, first developed by Ramachandran and Lakshminarayanan and improved by Shepp and Logan, are used for the data processing of the CT-scan (Kak and Slaney 1988).

The more the intensity of the X-ray beam is decreased, compared to its origin, the higher the density of the material. In general, the image scales illustrate the highest X-ray absorption as the brightest phase. White represents the highest attenuation and black the lowest.

12.3.5   ERRORS

Different errors in the final obtained image can be introduced during the process from X-ray generation to data processing. This is important to consider for the interpretation of the images.

Beam Hardening

The most common error in the measurements is called beam hardening. The X-ray generator creates a spectrum of X-ray energies (as explained earlier in "X-ray generator"). The lower energy protons are preferentially absorbed by the material, so during the path through the object, the number of low-energy protons decreases, leaving the beam with the high-energy protons. These are less sensitive for density differences, artificially reducing the measured intensity. So, it is possible that the outgoing beam has a higher average energy than the incident beam (Ketcham and Carlson 2001). This effect occurs along the longest pathlength, in the center of the object. Here the signal is showing an artificially lower attenuation than around the boundaries. In the circular images, beam hardening is seen as bright edges with an artificial darkened center; see Figure 12.21. In non-circular images the presence of beam hardening is not seen as clearly.

There are different ways to reduce beam hardening, either by avoiding the effect or correcting for it afterwards. To avoid the effect, first of all a shift can be made to higher energy levels (Akin and Kovscek 2003; Ketcham and Carlson 2001) so that the beam hardening is negligible, though higher energy levels of X-rays are less sensitive to density differences. Also, a beam filter can be used to reduce the number of low-energy photons. The problem here is that this can lead to a larger signal-to-noise ratio. Reconstruction algorithms are available and can best correct for beam hardening in symmetrical, uniform objects.

Ring Artifact

The ring artifacts are rings appearing in the images around the center of the object (Ketcham and Carlson 2001), which are caused by shifts in output value of a single detector or group of detectors. The shift is caused by a different response to changes in scanning conditions, like temperature or beam intensity or hardness. Temperature

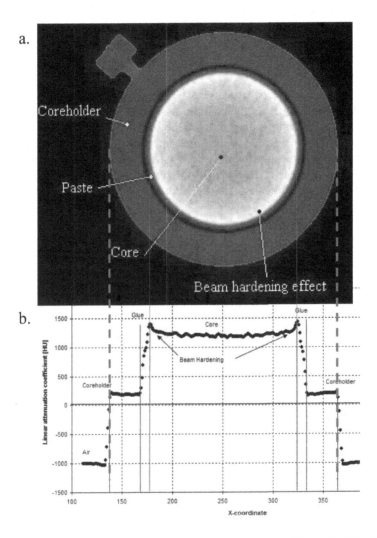

FIGURE 12.21 An example of beam hardening in a slice of a core filled with 15 bar Xenon, at 120keV. a. A CT-scan image of core sealed with epoxy paste, mounted in a core holder, b. a cross-section through the middle of the core showing a non-homogeneous the linear attenuation coefficient of the core, the effect of beam hardening.

and beam intensity can be better controlled during the experiment, though the influence of beam hardness has to be reduced by similar methods such as filtering the beam or moving to higher energy levels (Ketcham and Carlson 2001). Figure 12.21 shows an example of beam hardening in a slice of a core filled with Xenon, 15 bar at 120 keV. The different colors stand for differences in linear attenuation coefficients. The linear attenuation coefficient increases from blue to red, orange to white. The core holder has a blue color, the glue is red, and the core is orange and white at the edges due to the beam hardening (top image). A cross-section through the image is shown at the bottom. At the edges of the core an increase in attenuation can be noted, the beam hardening effect.

X-Shaped Artifact

By not using circular objects, artifacts can be introduced, as in the X-shaped artifact (Akin and Kovscek 2003). In the data processing the assumption is made that the object is circular and all the diagonals are the same size, so the traveling paths are assumed to be the same size. The measured intensity decrease is averaged over the path. If the path is longer, the intensity decrease is higher; then the average is higher if considering a smaller path than in reality. In the image this is seen as larger attenuation coefficients in the longest diagonals, forming the X shape. A change in data processing should be made.

Decentralization

Errors can be introduced when subtracting images for porosity or saturation calculations. If the images are not centralized equally, position errors can arise (Akin and Kovscek 2003).

Miscellaneous

Then there can be errors in the regulations of the machine. The set energy level generally is not the real beam intensity. The manufacturer of the CT-scan can provide the correction factor for the intensity.

12.3.6   DATA ANALYSIS AND ERRORS

Finally, when using images for further quantitative data analysis it must be taken into account that the voxels in the image can show an average attenuation coefficient over the voxel volume, depending on the CT scan resolution. This average can be an average over different materials with different attenuation coefficients, e.g. if the resolution is smaller than the pore size, an average attenuation will be given based on the grain volume and pore content. The surrounding materials also influence this average (Ketcham and Carlson 2001). This can be a problem when images are interpreted using quantitative attenuation levels.

Units of the image values are in Hounsfield units or in linear attenuation (m^{-1}). They are correlated as follows (Akin and Kovscek 2003):

$$H[HU] = 1000\left(\frac{\mu}{\mu_w} - 1\right) \tag{12.11}$$

μ_w is the linear attenuation coefficient of water, depending on the energy level, and μ is the linear attenuation coefficient to be transformed into Hounsfield units. The mass attenuation coefficient is the linear attenuation relative to the material's density [$\frac{m^2}{kg}$]

Depending on the reconstruction, the CT images are obtained in the form of matrices of greyscale values where the scale needs to be assigned by calibration.

Property Extraction

Similar image processing techniques described for 2D imaging can be used for 3D.

Properties that can be extracted for example are porosity, object shapes/orientations etc., contact angles, or the quantification of the different material concentrations. Image processing programs such as Dragonfly®, Avizo®, VG studio®, ImageJ, or one's own programmed codes can be applied.

12.4   OTHER 3D IMAGE TECHNIQUES

For the imaging of porous media, more 3D techniques are available. The two other widely used imaging techniques are nuclear magnetic resonance imaging (NMRI), and additionally confocal (fluorescent) microscopy or confocal laser scanning microscopy are discussed.

12.4.1   NUCLEAR MAGNETIC RESONANCE IMAGING

Nuclear magnetic resonance imaging (NMRI) is based on the principle of magnetic field relaxation of hydrogen molecule. Hydrogen can be found in water (H_2O) or hydrocarbon liquids like alkanes (C_nH_{2n+2}) and alkenes (C_nH_{2n}). In a normal state, before excitation, hydrogen molecules are spinning with axes oriented randomly. In the presence of an external imposed magnetic field, they will align according to the magnetic field given (B0). A second oscillating magnetic field or pulse (B1) will move the spin axis 90°. When the secondary magnetic field is turned off, the axis will re-align in time to the initial magnetic field, dependent on the location of the hydrogen.

The relaxation under influence of the second magnetic field will create an electric pulse in the receiving coils that can be detected, based on Faraday's law of induction. The amplitude change and decay time can be measured. By having a secondary magnetic field that is location dependent, the signal can be traced back to the location the hydrogen molecule had at time zero. The decay time of the secondary magnetic field B1 is called **the traverse relaxation time T2.** The decay of the NMR signal due to the primary magnetic field B0 is referred to as **the longitudinal relaxation time T1.**

The relaxation time depends on (McPhee et al. 2015):

1. Molecular motion in the fluid (bulk liquid relaxation).
2. Molecular diffusion, moving particles away from their starting point.
3. Surface relaxation, due to collision with the paramagnetic ions in the wall (e.g., iron, manganese).

For more information, refer to McPhee et al. (2015).

12.4.2   CONFOCAL MICROSCOPY

Transparent systems can be lighted and observed using a standard light microscope. Standard rock samples are not transparent, but model porous systems can be prepared for research purposes where this method can be applied. A variation to that is fluorescent material that is lit and excited such that it sends out fluorescent light, which can be visualized with a fluorescent light microscope (Fellers and Davidson 2023). The challenge for both visualization techniques is that light that is observed originates from all regions of the sample, resulting in a limitedly sharp image. In confocal microscopy or in the case of fluorescent light, fluorescent confocal microscopy is solved using a pinhole. Due to this pinhole only light from one point on a plane is in focus and can pass. This creates a sharper image of that region in comparison to a "wide field" microscope; see Figure 12.22. Additionally, selective illumination in the plane of focus by a laser helps to reduce the background scattering.

To get a sharp image of a 2D plane or a 3D image the sample is scanned respectively in the horizonal plane or additionally in different planes. The data from all points are afterwards reconstructed to 2D or 3D images (Krummel et al. 2013; Parsa et al. 2020). Transparent beads and fluids are needed to make observations in porous structures. So only model systems can be used for this application. To reduce the light scattering, matching of the refractive indices is advised.

Advice on additional reading on confocal microscopy can be found in Fellers and Davidson (2023).

Other imaging techniques not further discussed here are positron emission tomography (PET), synchrotron imaging, scanning electron microscopy (SEM), and transmission electron microscopy (TEM).

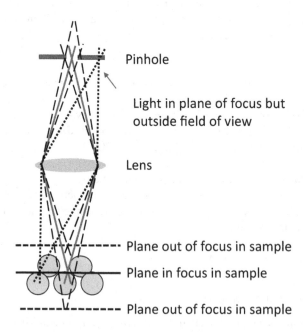

FIGURE 12.22 Schematic picture showing the principle of the pinhole to filter out background light (--- lines) or to filter out light in the same plane of focus but out the field of view (. . . lines).

EXERCISE 12.2

1. What does CT and NMR stand for?
2. Name three 3D imaging techniques.
3. Describe the principle of CT scanning. What is the difference to NMRI?
4. What is measured during a CT scan? How are 3D images created?
5. Describe how the spatial resolution of a CT scanning image is affected. Name three factors.
6. How can porosity be measured from a CT image?
7. Which particles are used to create X-rays for a CT scan?
8. What errors can be found in CT scan images? Explain two types of errors.
9. Determine from Figure 12.10B the porosity for both the actual circular pore shape and the imaged shape. Discuss the result. One pixel is 1 × 1 mm; the histogram of the image is shown in Figure 12.23 and data is shown in Table 12.4.

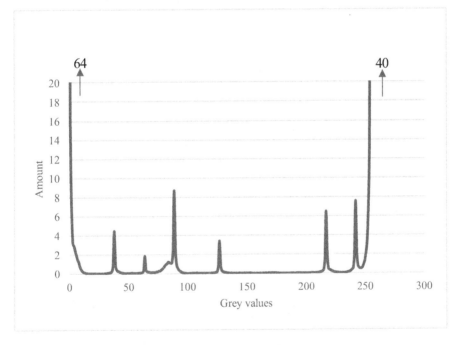

FIGURE 12.23 Histogram of the image in Figure 12.10B.

TABLE 12.4
Greyscale and Pixel Data

Greyscale	0	38	64	89	127	217	243	255	Total
Number of pixels	64	5	2	9	4	8	12	40	144

SOLUTION

1. CT = computed tomography, NMR = nuclear magnetic resonance.
2. CT scanning, NMRI, confocal microscopy.
3. CT scanning is based on differences in absorption of X-rays, due to material density differences. X-rays are sent through an object where the intensity change of the X-rays is measured. In NMRI 2 magnetic fields are used to align H atoms, after which one is switched off and the realignment is monitored and depends on the confinement of the H atom.
4. Intensity maps of X-rays, called projections, when X-rays are sent through an object under different angles.
5. Amount and size of detectors, how close the sample is positioned to the source, the spot size (the smaller, the better). Considering contrast enhancement as a resolution; number of X-rays sent through (high voltage difference over tube, high current in filament).
6. Identification of the pixels that belong to the pore space and total volume or identification of pixels representing the grains.

$$\varphi = \frac{\text{pixel volume*}\left(\text{number of pixels that is pore}\right)}{\text{pixel volume*}\left(\text{number of pixels that is pore or grain}\right)} = \frac{V_p}{V_p + V_g}$$

7. Electrons.
8. Ring artifacts caused by defect detectors or deviating scale of a detector becoming visible after image processing as rings. Beam hardening, due to filtering of the low energy X-rays in the sample leading to artificial lower X-ray attenuation in the object center of higher at the boundaries.
9. Actual porosity: Radius is ~5mm.

$$\frac{\text{Area circle}}{\text{total area}} = \frac{\pi r2}{\text{length} \cdot \text{height}} = \frac{\pi 5^2}{12 \cdot 12} = 0.54 \text{ or } 54\%$$

Imaged porosity:
The choice of binarization is a judgment call of the interpreter. If the cut-off value is set between 127 and 217, 84 pixels belong to black (0) and 60 to white (255)

$$\frac{\text{Area circle}}{\text{total area}} = \frac{\sum pixel}{\text{length} \cdot \text{height}} = \frac{84 \cdot 1 \cdot 1}{12 \cdot 12} = 0.58 \text{ or } 58\%$$

See Figure 12.24 and Table 12.5.
Due to the limitation in pixel size and choice of greyscale cut-off, the porosity can be under- or overestimated. In this case the porosity is overestimated. At e.g., cut-off between 89 and 127, porosity becomes 56%; between 64 and 89, it becomes 49%. The estimation would be better if the pixel size was reduced significantly.

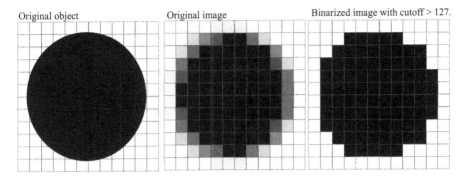

FIGURE 12.24　Original object and binarized image.

TABLE 12.5
Greyscale Values

Color								
Greyscale value	0	38	64	89	127	217	243	255
Amount	64	5	2	9	4	8	12	40
Binarized			84				60	

12.5　EXPERIMENT 12.1: CT-SCAN OF DRY CORE POROSITY COMPARISON

Description: In Experiment 5.1 the porosity of a core was determined by using helium porosimetry and the liquid porosity method. The experimented was asked to compare and address the differences. Hereby new data can be obtained from a dry scanned core, as extension of Experiment 5.1. The objective is to determine the porosity by image analysis of a scan of a dry air-filled core and to compare the results with data obtained in the earlier experiment 5.1.

CALCULATIONS AND REPORT

Procedure

- Before the core saturation with liquid, a dry core scan needs to be taken. Alternatively the liquid saturated core can be used, using liquid with a contrast agent like cesiumchloride or sodiumiodide (0.2M) to enhance the liquid-rock contrast.
- Based on the scanning equipment available, the settings of energy level, current and exposure, and number of projections are to be set such that the contrast is largest within an acceptable scanning time. A scan of rock generally gets a good contrast within using energy level 80–120kV.

- Image reconstruction: Consider the use of filters to reduce the impact of imaging and processing errors on the image analysis to performed.
- Image analysis: There are several ways of determining pore volume and bulk volumes from images. Here one option is presented:
  - Import the image using the correct voxel dimensions and bit type.
  - Crop the core to the size of the bulk volume.
  - Create a region of interest, segmenting either pore volume or grain volume, by setting a grey value threshold. The contrast between the grains and pores needs to be large enough to enable a good selection.
  - Having selected the pore space, the pores can be closed; respectively the voxels within the pores not selected can be added by a dilation-erosion (close) action.
  - Count the amount of voxels in the region of interest for the pore space and count the amount of voxels in the bulk volume. The ratio will give the total porosity.
- Compare the result with the helium and liquid porosities obtained. Consider a sensitivity study on how the segmentation and region of interest post processing steps affected the result and that the CT scan gives the total porosity and not effective porosity.

REFERENCES

Aadland, R.C.G., Akarri, S.S.F., Heggset, E.B., Syverud, K., Torsæter, O. 2020. "A Core Flood and Microfluidics Investigation of Nanocellulose as a Chemical Additive to Water Flooding for EOR." *Nanomaterials, 10.*

Akin, S., Kovscek, A.R. 2003. "Computed Tomography in Petroleum Engineering." In *Applications of X-ray Computed Tomography in Geosciences*, edited by F. Swennen, R. Van Geet, M. Jacobs, P. Mees, 23–38. London: Geological Society.

Anbari, A., Chien, H.-T., Datta, S.S., Deng, W., Weitz, D.A., Fan, J. 2018. "Microfluidic Model Porous Media: Fabrication and Applications." *Small, 14.*

Berg, S., Ott, H., Klapp, S.A., Schwing, A., Neiteler, R., Brussee, N., Makurat, A., Leu, L., Enzmann, F., Schwarz, J.-O., Kersten, M., Irvine, S., Stampanoni, M. 2013. "Real-time 3D Imaging of Haines Jumps in Porous Media Flow." *Proceedings of the National Academy of Sciences—PNAS, 110* (10), 3755–3759.

Blunt, M.J., Bijeljic, B., Dong, H., Gharbi, O., Iglauer, S., Mostaghim, P., Paluszny, A., Pentland, C. 2013. "Pore-Scale Imaging and Modelling." *Advances in Water Resources*, 51, 197–216.

Chaternever, A., Calhoun, Jr. J.C. 1952. "Visual Examination of Fluid Behaviour in Porous Media Part 1." *Transactions of the AIME, 195.*

Chen, S., Doolen, G.D. 1998. "Lattice Boltzmann Method for Fluid Flow." *Annual Review of Fluid Mechanics, 30*, 329–364.

Cnudde, V., Boone, M.N. 2013. "High-Resolution X-Ray Computed Tomography in Geosciences: A Review of the Current Technology and Applications." *Earth-Science Reviews, 123*, 1–17. http://doi.org/10.1016/j.earscirev.2013.04.003.

Eikehaug, K., Haugen, M., Folkvord, O., Benali, B., Bang Larsen, E., Tinkova, A., Rotevatn, A., Nordbotten, J.M., Fernø, M.A. 2024. "Engineering Meter scale Porous Media Flow Experiments for Quantitative Studies of Geological Carbon Sequestration." *Transport in Porous Media, 151*, 1143–1167. https://doi.org/10.1007/s11242-023-02025-013.

Fellers, T.J., Davidson, M.W. 2023. *Theory of Confocal Microscopy.* National High Magnetic Field Laboratory, Florida State University. Accessed 06 26, 2023. www.olympusconfocal.com/theory/confocalintro.html.

Gidaspow, D., Jiradilok, V. 2009. *Computational Techniques: The Multiphase CFD Approach to Fluidization and Green Energy Technologies*. Vols. Energy Science, Engineering and Technology Series. New York: Nova Science Publishers.

Haines, W.B. 1930. "Studies in the Physical Properties of Soil. V. The Hysteresis Effect in Capillary Properties, and the Modes of Moisture Distribution Associated Therewith." *The Journal of Agricultural Science*, 20 (97).

Hele-Shaw, H. S. 1898a. "Letters to Editor." *Nature*, 58 (1509).

Hele-Shaw, H.S. 1898b. "Flow of Water, Nature." *Nature*, 58 (1489), 34–36.

ImageJ, wiki. 2023. *ImageJ Wiki*. Accessed 07 01, 2023. www.imagej.net/.

Index, Fact. 2024. *Fact Index*. www.fact-index.com/b/br/bremsstrahlung.html.

Intensity, Energy. n.d. *Energy Intensity Picture*. https://i.stack.imgur.com/vbXFU.jpg.

Jahanbakhsh, A., Wlodarczyk, K.L., Duncan, P., Hand, R., Maier, R.J., Maroto-Valer, M. 2020. "Review of Microfluidic Devices and Imaging Techniques for Fluid Flow Study in Porous Geomaterials." *Sensors*, 20, 4030.

Kak, A.C., Slaney, M. 1988. *Principles of Computerized Tomographic Imaging*. New York: The Institute of Electrical and Electronics Engineers, Inc.

Karadimitriou, N.K., Hassannizadeh, S.M. 2012. "A Review of Micromodel and Their Use in Two-Phase Flow Studies." *Vadose Zone Journal, 11* (3), 1539–1663. http://doi.org/10.2136/vzj2011.0072.

Karadimitriou, N.K., Musterd, M., Kleingeld, P.J., Kreutzer, M.T., Hassanizadeh, S.M., Joekar-Niasar, V. 2013. "On the Fabrication of PDMS Micromodels by Rapid Prototyping, and Their Use in Two-Phase Flow Studies." *Water Resources Research*, 49 (4), 2056–2067. https://doi.org/10.1002/wrcr.20196.

Ketcham, R.A., Carlson, W.D. 2001. "Acquisition, Optimization and Interpretation of X-Ray Computed Tomographic Imagery: Applications to Geosciences." *Computer & Geosciences*, 27, 381–400.

Krummel, A.T., Datta, S.S., Münster, S., Weitz, D.A. 2013. "Visualizing Multiphase Flow and Trapped Fluid Configurations in a Model Three-Dimensional Porous Medium." *AIChE Journal*, 59, 1022–1029. https://doi.org/10.1002/aic.14005.

Mattax, C.C., Dalton, R.L. 1990. *Reservoir Simulation*. Vol. Henry L. Doherty Series 13. Henry L. Doherty Memorial Fund of AIME, Society of Petroleum Engineers. ISBN 9781555630287. Richardson, Texas.

McCullough, E.D. 1975. "Photon Attenuation in Computed Tomography." *Medical Physics, 2* (6). http://doi.org/10.1118/1.594199.

McPhee, C., Reed, J., Zubizaretta, I. 2015. *Core Analysis. A Best Practice Guide in "Developments in Petroleum Science, Volume 64"*. Amsterdam, Netherlands: Elsevier.

Mees, F., Swennen, R., van Geet, M., Jacobs, P. (eds). 2003. *Applications of X-ray Computed Tomography in the Geosciences*. Special Publications. London: Geological Society, 215.

Nadrljanski, M., Campos, A., Chieng, R., et al. "Computed Tomography." Reference article, Radiopaedia.org. Accessed 09 17, 2024. https://doi.org/10.53347/rID-9027

Pan, S., Pan, X., Kak, A.C. 1983. "A Computational Study of Reconstruction Algorithms for Diffraction Tomography: Interpolation vs. Filtered-Backpropagation." *IEEE Transactions on Acoustics, Speech, and Signal Processing, ASSP-31*, 1262–1275.

Parsa, S., Santanach-Carreras, E., Xiao, L., Weitz, D.A. 2020. "Origin of Anomalous Polymer-Induced Fluid Displacement in Porous Media." *Physical Review Fluids, 123*, 1–17.

Radon, J. 1917. "Über die Bestimmung von Funktionen durch ihre Integralwerte längs gewisser Mannigfaltigkeiten." *Berichte der Sachsischen Akademie der Wissenschaft, 69*, 262–277 (Session of April 30, 1917).

Romberg, J.K. 2004. *Image Projections and the Radon Transform.* Rice University. www.owlnet.rice.edu/~elec431/projects96/DSP/bpanalysis.html.

SAS, GAGGIONE. 2021. "Technical_documentation_material_PMMA_july_2021.pdf." *mise en oeuvre, normalisation', J.P. Trotignon, Nathan 2006 Information from 'Matières plastiques: structures-propriétés.* Accessed 04 29, 2024. www.optic-gaggione.com/wp-content/uploads/2021/07/Technical_documentation_material_PMMA_july_2021.pdf.

Solomon, C., Breckon, T. 2011. *Fundamentals of Digital Image Processing: A Practical Approach with Examples in Matlab.* Chichester, UK: John Wiley & Sons, Ltd.

Specialties, Industrial Specialties Mfg. & IS Med. 2024. *Acrylic Aka PMMA Chemical Compatiblity Chart—G & H.* Accessed 04 10, 2024. www.industrialspec.com/resources/acrylic-aka-pmma-chemical-compatiblity-chart/acrylic-aka-pmma-chemical-compatiblity-chart-g-h/.

Sun, Z., Santamarina, J.C. 2019. "Haines Jumps: Pore Scale Mechanisms." *Physical Review E, 100,* 023115.

Taheri, A., Lindeberg, E.G.B., Torsæter, O., Wessel-Berg, D. 2017. "Qualitative and Quantitative Experimental Study of Convective Mixing Process During Storage of CO2 in Homogeneous Saline Aquifers." *International Journal of Greenhouse Gas Control, 66* (2017), 159–176.

Taylor, R., Coulombe, S., Otanicar, T., Phelan, P., Gunawan, A., Lv, W., Rosengarten, G., Prasher, R., Tyagi, H. 2013. "Small Particles, Big Impacts: A Review of the Diverse Applications of Nanofluids." *Journal of Applied Physics, 113,* Article ID: 01. http://doi.org/10.1063/1.4754271.

Ulzheimer, S., Bongers, M., Flohr, T. (2018). Multi-slice CT: Current Technology and Future Developments. In: Nikolaou, K., Bamberg, F., Laghi, A., Rubin, G.D. (eds) Multislice CT. *Medical Radiology.* Springer, Cham, Switzerland. https://doi.org/10.1007/174_2018_187

van der Net, A., Blondel, L., Saugey, A., Drenckhan, W. 2007. "Simulating and Interpretating Images of Foams with Computational Ray-Tracing Techniques." *Colloids and Surfaces A: Physicochemical and Engineering Aspects, 309,* 159–176.

Vinegar, H.J., Wellington, S.L. 1987. "Tomographic Imaging of Three-Phase Flow Experiments." *Review of Scientific Instruments, 58* (1).

Vukovic, T., Røstad, J., Farooq, U., Torsæter, O., van der Net, A. 2023. "Systematic Study of Wettability Alteration of Glass Surfaces by." *ACS Omega, 8* (40), 36662–36676.

Wardlaw, N.C. 1982. "The Effects of Geometry, Wettability, Viscosity and Interfacial Tension on Trapping in Single Pore-throat Pairs." *Journal of Canadian Petroleum Technology, 21* (3).

Wellington, S.L., Vinegar, H.J. 1987. "X-Ray Computerized Tomography." *Journal of Pertoleum Technology, 39* (8), 885–898.

Xu, K., Liang, T., Zhu, P., Qi, P., Lu, J., Huh, C., Balhoff, M. 2017. "A 2.5-D Glass Micromodel for Investigation of Multi-Phase Flow in Porous Media." *Lab on a Chip, 17,* 640–646.

Index

Printed in the United States
by Baker & Taylor Publisher Services